T0419457

MATERIALS SCIENCE AND TECHNOLOGIES

CONSTRUCTION MATERIALS

MATERIALS SCIENCE AND TECHNOLOGIES

Additional books in this series can be found on Nova's website under the Series tab.

Additional E-books in this series can be found on Nova's website under the E-books tab.

CONSTRUCTION MATERIALS AND ENGINEERING

Additional books in this series can be found on Nova's website under the Series tab.

Additional E-books in this series can be found on Nova's website under the E-books tab.

MATERIALS SCIENCE AND TECHNOLOGIES

CONSTRUCTION MATERIALS

LEONID DVORKIN
SUNNY NWAUBANI
AND
OLEG DVORKIN

Nova Science Publishers, Inc.
New York

For permission to use material from this book please contact us:
Telephone 631-231-7269; Fax 631-231-8175
Web Site: http://www.novapublishers.com

Additional color graphics may be available in the e-book version of this book.

LIBRARY OF CONGRESS CATALOGING-IN-PUBLICATION DATA

Dvorkin, L. I. (Leonid Iosifovich)
Construction materials / authors, Leonid Dvorkin, Sunny Nwaubani, Oleg Dvorkin.
p. cm.
Includes bibliographical references and index.
ISBN 978-1-61728-693-3 (hardcover)
1. Building materials. I. Nwaubani, Sunny. II. Dvorkin, Oleg. III. Title.
TA403.D87 2010
624.1'8--dc22
2010026087

Published by Nova Science Publishers, Inc. ✣ New York

CONTENTS

There is considered the structure of crystalline, glassy and amorphous materials at different scale level of the influence on physical and mechanical properties of construction products. General ideas on construction materials as disperse systems are given. General properties of colloidal systems, powders, suspensions and emulsions are shown.

There are considered strength, rheological and strain properties of materials. General concepts on materials properties related to water, temperature and aggressive factors are given. There are shown general calculated dependences for prediction of materials properties and methods for materials testing.

There are given the data about most common technological processes of construction materials obtaining: grinding, mixing, burning, steam curing. The samples of typical equipment and technological circuits for implementation of basic processes are also shown.

There are considered basic schemes of construction materials classification and also the ways of the experimental estimation of their quality. The value of standardization for specifying technical requirements to materials and methods of their testing are uncovered. Basic systems of standardization are given.

Classification of rocks is given. Brief characteristic of composition and structure of rocks is shown. Characteristic of basic properties of construction materials based on natural stone and their production technology are covered. There are lighted up the ways of longevity improving of natural stone materials.

There is given general characteristic of metallic materials, applied in construction. There are covered the data on production technology of basic metallic alloys, construction elements based on them and methods of metals protection a gainst corrosion.

There are considered the types of raw materials for obtaining of ceramic materials, general schemes of manufacturing, characteristic of basic types of ceramic elements.

Peculiarities of glass structure and glass ceramic, characteristic of plate, profiled glass, blocks and pipes etc. are considered. Properties of glass and glass ceramic and technology of their obtaining are shown.

There are given the composition and properties of basic types of binding materials of air and hydraulic hardening. Basic factors which have influence on binders' properties are covered. Basic processes of binding materials hardening are under consideration.

The requirements to initial materials, properties of concrete mixtures and concrete are considered. The peculiarities of concrete structure are covered. There are given calculation – experimental method of concrete proportioning at providing required properties.

There are discovered the peculiarities of basic types of heavy-weight and light-weight concrete. Separately the concrete performing under severe service conditions and modern effective types of concrete and ways of there properties management are considered.

Characteristic of masonry, finishing and special mortars is given. There are shown classification of dry-pack mixtures, applied in construction and their characteristic. Basic types of admixtures – modifiers for dry-pack mixtures are discovered.

There are considered asbestos-cement products, lime-sand bricks and concrete, basic products made of gypsum and non-reinforced cement concrete

The essence and meaning of concrete reinforcement, peculiarities of pretension are considered. Basic ways of reinforced concrete elements manufacturing applied in civil and industrial construction are given. There is discovered the essence of dispersed reinforcement. Brief description of fiber concrete, glass cement and glass-fibre plastic are given.

There are covered the peculiarities of bitumen and tar binders and basic products based on them, Brief description of basic roofing and damp proofing construction materials is given. Characteristic of asphaltic concrete and peculiarities of its technology is given.

There is given concept on basic types of synthetic polymers and plastic masses based on them. The characteristic of basic types of plastic masses is considered. There are also shown composition, peculiarities of technology and properties of polymer concrete.

Characteristic of basic types of non-organic and organic heat-insulating materials is given. There are considered their distinctive properties and peculiarities. There are shown briefly sound-absorbing and sound-insulating materials applied in construction.

There is given the classification of varnishes and paints. Basic compositions of water, oil, emulsion polymer varnishes and paints and their distinctive properties are considered. Brief description of backing finishing materials is given.

There is considered structure and properties of wood. Basic construction-technical properties of basic types of wood are covered. There is given the characteristic of basic types of construction products based on the wood. There are shown separately materials with application of grind wood and wooden wastes.

There are considered possible ways of application of metallurgical, fuel and energy, chemical and other industrial wastes, and also municipal service in production of construction materials. There is given the characteristic of the most common construction products with application of construction wastes.

PREFACE

Training of the skilled civil engineers must include the learning of the main principles of construction materials technology.

Learning of the materials properties which are determined by regularities of influence on them of such parameters as structure, composition and different manufacturing features is the basis of construction materials technology. Such technology basis gives the possibility to design materials and control their properties and also create new materials ***with the necessary*** set of properties.

Progress in the field of construction materials *technology is the important condition for development of general construction technology. Main* principles of physics, chemistry and also wide manufacturing experience are actively used in construction materials *technology.*

The given manual is a summary of "Construction Materials *Technology*" course prelected by authors at National University of Water Management and Natural Resources Use (Ukraine). It includes two parts: the basic concepts about structure and properties of construction materials and also the main processes of their production are reviewed in the the first one and the classification and the brief characteristic of the most widespread materials applied in the construction are given in the second one.

At preparation of the manual authors realized that in one book it is impossible to fully review "ocean" of the available data, therefore they have concentrated on matters they consider to be the most important.

First of all the manual is assigned for students of civil engineering specialities. It can be used also by practical workers for deepening of knowledge in the area of construction materials technology.

Authors are grateful to colleagues for advices and the help in the book preparation, and also Nova Science Publishers, Inc., which took the trouble on its edition. We would like to express our thanks to Ph.D. N. Lushnikova for assistance in preparation of this book. Authors will be grateful also to all readers of the book for estimations and remarks which will be considered in its next editions.

INTRODUCTION

The construction materials science has started to develop when builders requied the generalization of knowledge, which was stored by manufacture and application of separate kinds of materials, used at erection of buildings.

Already in certain ancient Roman treatises the important place was given to building art including making of building materials. So, more than two thousand years ago (in 160 B.C.) the Roman consul Marcus Porcius Cato Major included in the book devoted to building art, such advice to lime burner: “Put stone in the luggage furnace good, the whitest one without any mixed character”. The Roman architect and engineer Marcus *Vitruvius* Pollio who lived in the first century B.C. in his famous “Ten Books on Architecture" wrote: "There is also a kind of powder which from natural causes produces astonishing results. It is found in the neighbourhood of Baiae and in the country belonging to the towns round about Mt. Vesuvius. This substance, when mixed with lime and rubble, not only lends strength to buildings of other kinds, but even when piers of it are constructed in the sea, they set hard under water”. Studying of the ancient Roman treatises after many centuries when the corresponding scientific base has already been created promoted development of some fundamental ideas of building materials science, for example, ideas pozzolanization of cements, mortars and concrete for the purpose of their increase water- and corrosion proofness, addition of organic admixtures in mortars and concrete for increase in their plasticity and durability, etc.

Romans had different names for the material, which is similar to concrete. So, they called cast masonry with the Greek word “emplekton”. There is also a word “rudus”. However, more often at a designation of the mortars used at erection of walls, the arches, the bases and other structures Romans used a word-combination “opus caementitum”, which was used to name the Roman concrete. The word "caementitum", transformed to "cement", meant a crushed stone or rubble stone.

The history of the concrete origin leaves far in depth of centuries. The earliest concrete, which was found out by archeologists, could be carried to 5600 BC It was found on the bank of Danube in Lapinsky Vir settlement (Serbia) in one of ancient settlement huts of the Stone Age, where the floor was made of concrete in the thickness 25 cm. The material for this floor was made of gravel and local lime.

It is a well-known fact that the Middle Ages were characterized by certain decrease in building technologies and qualities of applied materials. The beginning of intensive researches in the manufacture of new effective building materials, ordering of available empirical experience at the application of traditional materials was connected with the general development of building in second half of the 18th century. During this period hydraulic lime and Roman cement were invented and began to apply, for producing of waterproof mortars and concrete and a bit later in beginning of the 19th century – Portland cement.

Hydraulic lime was invented by John Smithon (1793) in connection with the decision of building of a beacon on the Eddiston rock which was accepted in 1756 As a result of extensive experimental researches Smithon determined the possibility of lime producing, slowly hardening in water, from limestones with raised content of clay impurities. For construction of beacon Smithon has selected the mortar based on the hydraulic lime and pozzolana additives. Smithon highlighted the results of these researches in his book. But for a long time hydraulic has not found wide application for and builders preferred the mixture of air hardening lime and pozzolana for waterproof mortars.

In 1796, developing the researches J.Smithon, James Parker patented Roman cement (the name of this cement is given from advertising reasons and does not answer an essence of the "Roman cement"), capable quickly to fasten and harden without preliminary slaking and to develop the raised durability. J. Parker's cement was widely applied in building to the middle of the 19th century when Portland cement started to supersede it.

In 1818 the French scientist Louis Vicat published the work “Experimental researches of building lime, concrete and usual mortars”. In this work classification of hydraulic limes depending on a mass ratio silica and alumina to calcium oxides is offered. This classification is applied till now.

The idea of hydraulic binders obtaining with kilning of an artificial mix of calcareous and clay component was developed independently from each other by Joseph Aspdin (England) and Egor Cheliev (Russia). J. Aspdin received (1824) the patent for "improvement of the way of an artificial stone obtaining" which is called by him Portland cement (because of similarity to limestones from stone quarries on the isle of Portland). J.Aspdin has recommended to make cement of the calcareous dust collected on roads, paved by limestone. The mass parity of components in the patent was not specified, and mix firing was recommended only to the full removal of CO_2. J.Aspdin's cement was used in 1828 in tunnel construction works near the Thames.

Isaac Johnson substantially improved Portland cement technology, having found a necessary parity of limestone and clay in a raw mix and having applied raised temperature for raw materials vitrification and clinker formation.

Portland cement era, which structures and technology came nearer to modern, has begun since the 60th years of the 19th century.

The first cement plants arose in England and then cement manufacture extended worldwide.

The basic unit for manufacturing cement clinker, the rotating furnace was invented in 1877 (the patent of Krempton and Rensom).

One of the first furnaces had length of 11 m and diameter of 1.5 m and in 1900 there had already been furnaces in diameter of 2 m and length of 35 m. Their daily productivity was 30 ton (modern rotating furnaces have diameter up to 7 m, length up to 230 m and productivity up to 3000 ton a day). A crusher (mill) for crushing clinker was invented in 1892 by the Danish engineer Davidsen. It was lined inside with quartz tiles, and the sea pebble was used as grinding bodies.

The scientific materials technology development became possible only after the natural sciences basic laws and structure and properties of substances, conditions of passage and feature of chemical interaction of substances, processes of dissolution, crystallization, sintering, fusion etc. have been formulated.

The considerable contribution to the science about construction materials was brought by the great Russian scientists M.V. Lomonosov and D.I. Mendeleyev.

So, M.V. Lomonosov in the “Course of True Physical Chemistry” gave a scientific substantiation of processes of sintering, glass formation, roasting and coloring of glasses. D.I. Mendeleyev in work “Glass Manufacture” for the first time stated the scientific glass formation conditions and theory.

In the 19th century there have been formulated classical bases of material science theories. Connection between temperature and structural changes of steel was established, the scientific description of steel crystallization process was given, and the microscope was applied for the first time for steel research.

Henry Le Chatelier, using optical methods and X-rays, showed in 1882 that Portland cement clinker contains 4 basic minerals named in 1897 by Tornebom alite, belite, celite and felite. In 1887 Le Chatelier published the crystallization theory of cement hardening. During the same period (1893) V. Mihaelis offered the colloidal theory of hardening.

The basic constructional material of our time reinforced concrete was offered in the 19th century. By the influence on the world civilization development reinforced concrete invention could be compared with the discovery of electricity or aircraft occurrence. In the European and North America countries practical application of these discoveries began approximately at the same time. Originally a boat (J. Lambo, 1848), garden tubs (J. Mone, 1849) was made of reinforced cement, and then in 1867 Mone received patents for building products manufacturing of reinforced concrete.

In 1872 J. Mone constructed the first tank for water in capacity of 130 m^3 from reinforced concrete. Reinforced concrete starts to be applied at erection of bridges, pipelines and other constructions.

While developing chemical science, creating physical and colloidal chemistry, physical chemistry of silicates, chemistry of high-molecular connections, the base for studying of binder materials hardening processes, manufacturing of ceramics and glass, asphalt materials and plastics was created.

By this time thanks to efforts of several generations of researchers such sections of chemical science as cement and concrete, ceramics, glass, wood, polymers chemistry was being successfully developed. Last decades the physical and chemical mechanic of concrete and other building materials develops, allowing actively to operate their properties, changing their dispersion, superficial activity, applying additives-modifiers, etc.

Studying of composition and structure of natural and artificial stone materials appreciably became possible due to the successes of geological sciences - crystallography, mineralogy and petrography.

Studying of construction materials structure and composition, phase transformations in materials while their obtaining and operating became possible due to wide application of chemical and physical mechanical methods of researches. More and more widely are one applies mathematical methods of researches of the materials, allowing receiving quantitative dependences and models for management of technological processes, the analysis of the basic laws forming properties of materials.

In the 20th century the construction materials technology was intensively developed in all basic directions. It was promoted by intensive development of various branches of building and creation of the powerful industry of building materials in countries with advanced economy.

The science about concrete was generated as one of the basic sections of building materials technology.

The 20th century became the "Golden Age" of concrete. Intensive development in all areas of building gave a powerful spur in development of the concrete theory and technology. Questions of durability of concrete, its various properties, interrelations of structure, composition and properties of concrete of various kinds were discovered. Concrete and reinforced concrete became the basic materials used for capital structures erection. With great success concrete is applied in shipbuilding, mechanical engineering and other branches of engineering. The technology of concrete and reinforced concrete products was intensively developed. In 1917 the French engineer E. Freyssinet applies external vibrating by means of pneumatic hammers for bridges and hangars construction. For construction of dams at first surface and then poker vibrators are started to apply in the USA. In 1903 in Germany production of ready-mixed concrete at factory by dispensing and mixing cement with other initial materials was started. Industrial concrete mixing machines were developed. In the twenties in the USA there were the first motormixers and already in 1930 about hundred concrete factories functioned there.

In 1963 in the USA about 78 % of total amount of concrete mixes was made with motormixers and among concrete factories 68 % were dry mixes factories. In the countries with the advanced economy manufacture of ready-mixed concrete made 500…1000 l per man in a year.

In 1880 V. Mihaelis has started to apply concrete hardening acceleration and obtaining lime-sand stones the direct steam. The steam curing becomes the basic way of concrete hardening acceleration in factory technology of reinforced concrete products. The technology of steam-cured silicate products was also developed.

To a number of revolutionizing ideas in manufacture of reinforced concrete designs one could refer the idea of prestressing, which belongs to E. Freyssinet, who received patents and published the essence of a method of prestressing in 1933 in article under the name "New Ideas and Sights". These ideas were specified further in the book "Statement of the General Ideas of Prestressing" (1949).

Methods of concrete structures design, basis of the theory of strength and technology of concrete (D.Abrams, F.McMillan, T.Powers, J.Bolomey, R. Lermit, B.Skramtaev, etc.) were developed.

Along with the development of heavy-weight concrete technology in the XX century the "know-how" of lightweight and cellular concrete was developed. Application of natural porous aggregates for lightweight concrete was known from an extreme antiquity. Still Pliny suggested mixing 1 part of lime, 2 parts of pozzolana and 1part of pounded sinter. This recipe was applied within many centuries at erection of hydraulic engineering constructions.

In 1917 an American Haid was granted the patent for expanded clay and slate, prototypes of porous aggregates like claydite manufacture.

Application of lightweight concrete with porous aggregates has reached the greatest development in USA and Russia. The area of their application covers the extensive range of structural and heat-insulating products and bearing designs. Lightweight concrete become one of the cores of wall materials. They are also applied in industrial building, in road bridges, in road and air field coverings, etc. In the USA there is an experience of lightweight concrete application in buildings with height up to 200 m. In 1954-1955 on Alaska through the river Kanai the four-flying claydite reinforced concrete bridge was built.

For concrete modifying polymeric additives was found the increasing application. Originally natural rubber latex (Kresson, 1923) was applied for this goal, and since 1932 synthetic latex, and then polyvinylacetate additives, various polymeric pitches, organic-silicon compounds, methylcellulose were offered for cement materials, etc.

Since 80th years of last century the possibilities for radical change of concrete mixes quality indicators and concrete in connection with manufacture superplasticizer and then microsilica additives were presented. Manufacture of superplasticizers was organized for at first in Japan and a bit later in Germany.

Superplasticizers allowed developing new technologies of cast concrete consistency with the moderate water content, high-strength concrete, and concrete with the high tightness, the improved quality of a surface. High-strength concrete of new generation have allowed to realize tunnel projects near La Manche, to build a 115-storeyed skyscraper in Chicago with height of 610 m, the bridge through passage Akasi in Japan with the central flight of 1991 m, etc.

In a complex with superplasticizers effective microsilica additives appeared. They started to be used since the 50th years at first in Norway, and then in other countries. With microsilica application a number of unique objects, for example, sea platforms from constructional lightweight concrete (Japan), a building of skyscraper Union Plaza (USA) is erected now, etc.

Last decades considerable successes of building materials technology are caused by the development of the theory connecting the materials basic properties with features of their structure both at micro- and macro-level.

The knowledge of materials structure formation laws opens possibilities of the directed regulation of their properties, designing of materials with the set properties.

Progress in technology of construction materials is appreciably caused by active influence on processes of structure formation and synthesis of properties of materials. One can refer technological parameters optimization, effective additives introduction and structure regulation mixture and compositions structures to such influences.

The researches directed on application of technogenic raw materials are important for increasing of construction materials production efficiency.

Extra-rapid-hardening and high-strength cements, multicomponent binders with low water requirement, new gypsum, slag, silica, phosphate and other binders impart new possibilities in technology of concrete and building composites.

In construction materials technology all the progressive processes providing high efficiency and quality of production, lower power consumption are widely applied (cast technologies, vibrocompression, extrusion, low-frequency and percussive ways of products formation, electric curing, calendering, etc.).

There are more required mechanical-chemical processes of mixes activation, high-speed turbulent activators, barothermal influence, electropulse and wave units, end item impregnation with hardeners.

The science about construction materials continues to develop intensively, bringing the powerful contribution to scientific and technological progress.

Part I. Structure, Properties and General Technological Processes of Construction Materials

Chapter 1

MATERIALS STRUCTURE

The *structure of materials* determines the mutual location, form and sizes of structural elements. Atoms, ions, molecules, solid particles, aggregates of particles, pores(i.e. voids between particles, filled with liquid or gaseous phase) can be structural elements. Particles are minute parts of a substance, which can be obtained by mechanical means (dispersion) or physical and chemical means. The aggregates of particles form as a result of their agglomeration, in particular, when dispersion grows and the surface energy increases, and also as a result of intergrowth, as in the process of crystallization.

The general signs and feature of structural materials, can be considered at different levels depending on the sizes of the structural elements (l). The four main structural levels which may be defined for construction materials: i.e., atom-molecule ($l<10^{-9}$ m); submicroscopic ($l=10^{-9}...10^{-7}$ m); microscopic ($l=10^{-7}...10^{-4}$ m); macroscopic ($l>10^{-4}$ m) are explained below.

1.1. ATOM-MOLECULE STRUCTURE

Considering materials structure at atom-molecule level, it is possible to distinguish: crystalline materials – characterised by unit cells structure and amorphous ones; and those characterised by molecular, atomic or ionic aggregates, which does not form well-organized lattice.

A large group of natural or artificial materials and various minerals among them are crystalline in nature. Materials like cements, polymers, and slags contains both crystalline and amorphous components. The most persistent state is the crystalline one, as the energy of material here is minimal. Energy of ionic crystalline lattice is equal to the sum of the energies of separate constituent ions. Energy performance or ionic constant (IC) is determined by a formula:

$$IC = \frac{W_i^2}{2l_c}, \tag{1.1}$$

where W_i is ion valence; l_c is distance between the ions centers.

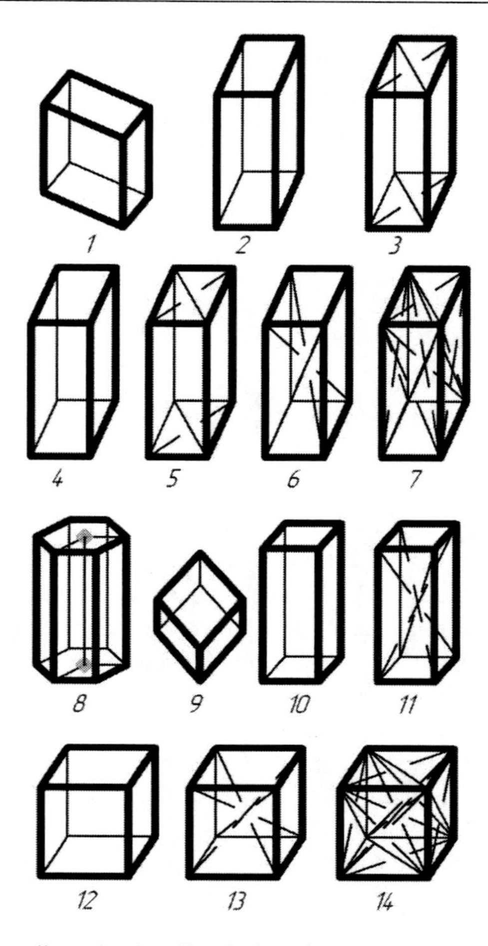

Figure 1.1. Simple and complicated unit cells of minerals.

The *unit cells* of crystals are divided into primitive and complicated (Figure 1.1). There are eight ions or atoms located on tops in primitive cells. In complicated elementary cells additional ions (atoms) are located on the middle of ribs or edges.

Table 1.1. Types of primitive elementary cells

Type of lattice	An angle between axes	Correlation of axes sizes	Designation Figure 1.1
Triclinic	$\alpha \neq \beta \neq \gamma \neq 90^{\circ}$	$a \neq b \neq c$	1
Monoclinic	$\alpha \neq \beta = 90^{\circ} \neq \gamma$	$a \neq b \neq c$	2,3
Rhombic	$\alpha = \beta = \gamma = 90^{\circ}$	$a \neq b \neq c$	4...7
Rhombohedral	$\alpha = \beta = \gamma \neq 90^{\circ}$	$a = b \neq c$	9
Hexagonal	$\alpha = \beta = 90^{\circ} \gamma = 120^{\circ}$	$a = b \neq c$	8
Tetragonal	$\alpha = \beta = \gamma = 90^{\circ}$	$a = b \neq c$	10, 11
Cubic	$\alpha = \beta = \gamma = 90^{\circ}$	$a = b = c$	12...14

Primitive elementary cells can be divided into seven types (Table 1.1), depending on length of crystallography axes and the size of angles between them.

Theory of crystal space lattice structuring was founded by E.S.Fedorov in 1890. He established the existence of 230 variants of space combinations of symmetrical elements with space lattices or space groups.

Minerals found in construction materials and in metals form different types of crystalline lattices. Thus establishment of their parameters constitute one of the main methods of their authentication. An example of the types of crystalline lattices of cement clinker minerals are shown in Table 1.2 below:

Most metals form highly-symmetric lattices with dense atomic package. These could be in form of a: cubic, body-centered cubic, face-centered cubic or hexagonal cell (Figure 1.2).

Distances a, b, c between the centers of neighboring atoms in an elementary cell are called the lattice periods and are measured in nanometers (1 nm = 10^{-9} m).

The density of cristallyne lattice is characterised by a coordination number (i.e the amount of atoms), nearest to the current atom. Half of the minimal distance between atoms or ions in crystal lattice is called the atomic(ionic) radius. When the coordination number increases, the atomic radius reduces because the distance between atoms increases.

The most stable state of crystalline structure of a substance is achieved, when cation come in contact with all the surrounding anions. In other cases the elements of the lattice reorder, forming an unstable structure with the diminished coordination number. Unstable structures with a low co-ordination number have greater chemical activity. For example, in metakaolin - a product of the process of dehydration of the clay mineral kaolinite, the coordination number for Al^{3+} ion is 4, whereas for the original kaolinite, the coordination number is 6. Consequently unlike kaolinite, metakaolin can chemically react at normal conditions, and also during autoclaving treatment with calcium and magnesium hydrosilicates.

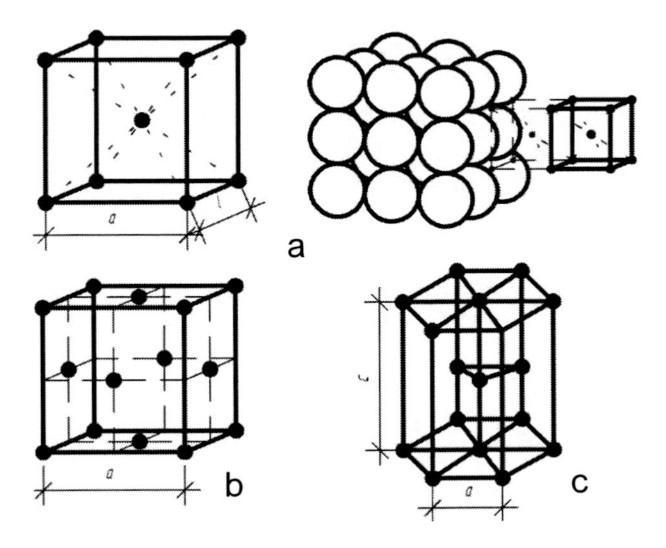

Figure 1.2. Crystal lattices of metals:
a – volume-centered cubic; b - face-centered cubic; c – hexagonal.

Table 1.2. Basic types of crystalline lattices of cement clinker minerals

Mineral	Grate
Tricalcium Silicate	Triclinic
β - Dicalcium Silicate	Rhombic
Tricalcium aluminate	Cubic
Tetracalcium aluminoferrite	Rhombic
Dicalcium ferrite	Monoclinic

The structure of crystals is divided into five classes by the parameters of their interatomic distance into: coordinative, isle, linked, layer and frame structures.

In *coordinating structures* which are representative of oxides, salts, etc, the distance between basic units (atoms, ions) is equal in order. In *isle structures,* there are groups of the separated atoms which form "islands" (calcite, pyrite, most of organic compositions). L*inked structures* are *characterized by* separated atomic groups which form continuous links (wolastonite $CaSiO_3$, diopside $CaMgSi_2O_6$, etc.). The *layer structures* have endless two-dimensional atomic layers (graphite, talc, etc.). *Framework structures* consist of volumetric three-dimensional anion frame of coordinating type and neutralizing "filling" types from cations or atomic groups. $KAlSi_3O_8$ exemplifies such orthoclase structures.

Because of the different density of ions or atoms in different planes and directions crystals properties depending on direction are different. This feature of crystalline solids is called anisotropy. It is not a characteristic of amorphous solids (glass, polymers and others like that). Anisotropy reflects mostly in linked and layer structures.

Mineral construction materials and metals are mostly in the polycrystalline state, including most of the chaotically oriented crystals which decreases anisotropy. Some treatment methods (e.g., cold hammering of metals) can be accompanied by the spatial orientation of crystals and can cause the anisotropy of such materials.

Depending on the chemical bond type between structural elements, it is possible to distinguish between ionic, atomic (covalent), metallic, and molecular bond (Figure 1.3) as well as crystals with hydrogen bond.

Ionic bond is typical for crystals which considerably vary in electro-negativity of their structural elements.

Most cations are smaller in size than anions, and the crystalline lattice of ionic compounds form as a result of placing cations in the voids between anions. The characteristic properties of ionic crystals are low conductivity, heat-conducting, brittleness and high temperature of melting.

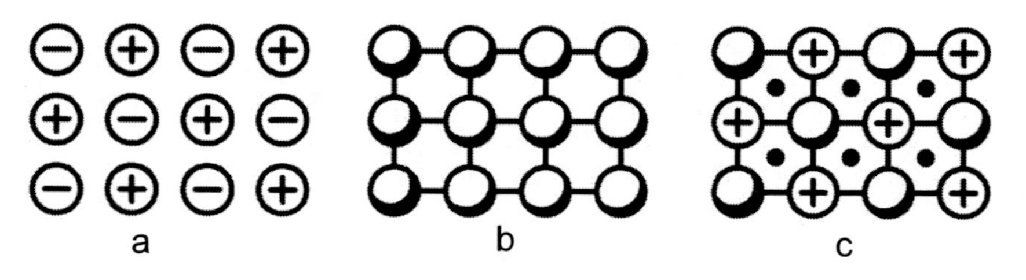

Figure 1.3. Schemes of chemical links of crystalline lattices: a- ionic; b- atomic; c- metallic.

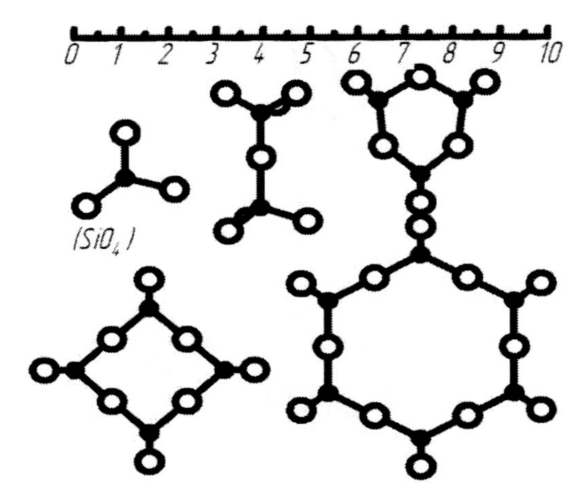

Figure 1.4. Forms of insulated silica groupings.

In the atomic crystalline lattice points there are neutral atoms, bounded by *covalent bond.* These bonds are very strong. Consequently, substances which have such lattices, are hard, refractory and practically insoluble (diamond, silicon, compounds of some elements, with a carbon and silicon - carbides and silicides). Covalent crystals are formed of atoms which have nearest values of electronegativity. If the difference in electronegativity of elements increases, the degree of transition of the covalent bond into ionic grows.

Typical bond of silicates Si - O is half ionic and half covalent. Oxide compounds of silicone, containing siloxane bond (Si - O), occupy an important place in modern developments in construction materials. Silica (SiO_2) is the most spread and stable silicone compound, it occurs naturally and exist mostly in crystalline state. According to modern conceptions silica consists of tetrahedron $[SiO_4]^{-4}$, bonded by points into separate complexes in such a way that every atom of oxygen is common to two neighboring tetrahedrons and is connected with two atoms of silicone (Figure 1.4).

Silica tetrahedron is a basic unit of all of natural and artificial silicates. Complexes of silica tetrahedrons form a closed annulus or endless links. Combinations of links form endless strips which can form layers by turn. Known examples of stratified silicates are talc, mica and kaolinite. Part of the Si^{4+} ions can be substituted by the Ai^{3+} ions.

In *metallic crystals,* electrons play an outstanding role. They move freely among the atoms. Positive metal ions oscillate in the points of such crystals, and valent electrons move through the lattice in different directions. The totality of free electrons is sometimes called electron gas. Such lattice structure cause high heat and electric conductance and plasticity of metals. Mechanical deformation of crystalline lattice within certain limits does not causes crystals destruction since the ions, composing them, as if float in the electronic gas cloud.

The groups of atoms or molecules, bonded between themselves by Van der Waals forces and dipole interactions, are placed in the points of *molecular crystalline lattices*. Van der Waals forces grow when the amount of atoms in a molecule and their polarity increase. Molecular forces are comparatively weak, therefore molecular crystals, characterized by row of organic matters, fusible, and volatiles, have low hardness. For example, the crystals of paraffin with a molecular lattice are very soft, although covalent bonds C - C between atoms of hydrocarbon molecules are strong enough.

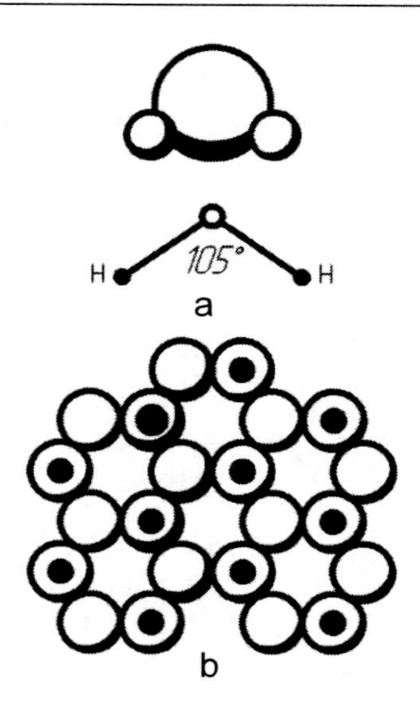

Figure 1.5. Molecule structure: a - water; b – ice.

The type of widespread connection in the inorganic crystals is due to hydrogen ion bond, between two anions tightly linked together. Formation of *hydrogen bond,* involving hydrogen and oxygen atoms, matters very much in the structures of water and many other compounds. Formation of hydrogen bond explains presence of associated molecules $(H_2O)_n$ in water. The doubled molecules $(H_2O)_2$ are the most strong, and its formation is accompanied by two hydrogen bonds appeal which are the most stable.

According to polymerization theory the molecules of water can exist in the forms of hydrol H_2O, dyhydrol $(H_2O)_2$, and also trihydrol $(H_2O)_3$. Ice (Figure 1.5) consists mostly of molecules of trihydrol, which are characterized by the highest volume of voids and therefore the least density for water, and water vapor forms from the molecules of hydrol.

Hydrogen bond explains anomalous properties of water to a great extent: the high dielectric constant, surface tension, capacity for moistening and dissolution of many substances.

Hydrogen bonds predetermine the polymerization of some organic acids and promote the formation of many inorganic polymers. The hydrogen bond formation partly causes hydration of polar groups, and also hydrophilic property at the proper surfaces of materials.

Energy of covalent, ionic and metallic bonds is 126...420 kJ/mol, molecular one - does not exceed 42 kJ/mol, and hydrogen bond is 8.4...42 kJ/mol.

A few types of bonds act in most crystals taking into account character of their estimated theoretical strength and other properties.

There are always different defects of crystalline lattice of real materials. By geometrical features they are divided into point, linear, surface and volume. *Point defects* can be caused by thermal vibrations in points, influence of radiation and electromagnetic waves (energetic imperfection), changes in electrons distribution on energy levels (electronic defects), displacement of atoms from a midposition, by the presence of admixture atoms, presence of empty points - vacancies (atomic defects).

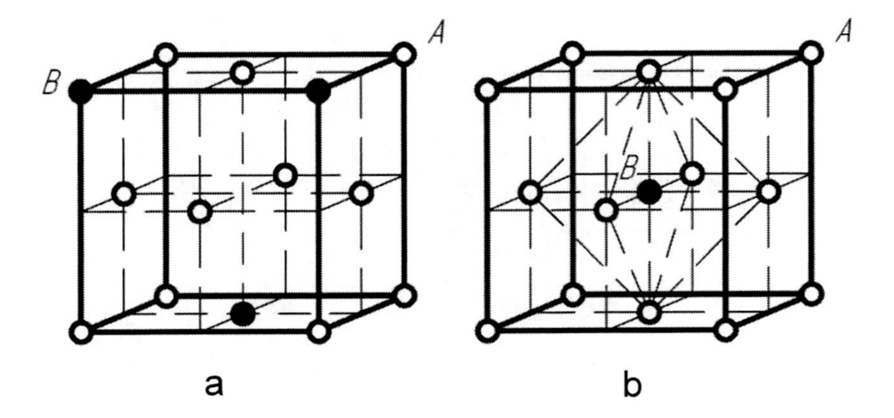

Figure 1.6. Crystalline lattice of solid solutions:
a- substitution b- penetration.

Among the point defects mentioned above, the atomic ones that increases atoms (ions) mobility in a crystalline lattice, resulting in increase of the diffusion permeability and ionic conductivity in crystals are of great significance.

If a substance is crystallizing from the solution or melt in the presence of foreign atoms, these atoms can enter into the lattice structure of the basic compound and form *solid solutions*. Foreign atoms penetrate into the lattice of basic crystal in two ways (Figure 1.6): 1) They occupy the key points of crystalline lattice, substituting the particles of basic component (solid solutions of substitution); 2) This takes place in the merithalluss of crystalline lattice (solid solutions of penetration).

Substitution formation of solid solutions (Figure 1.6, a) is a distinctive feature for obtaining most ceramic materials, cement clinker, etc.

Solutions of penetration (Figure 1.6, b) belong to the solid solutions which are characterized by changeable composition. As a rule, atoms and ions of small sizes adjusted to the voids of crystalline lattice are able to penetrate through merithallus. Mostly all solutions of penetration occur in metallic materials. Hydrogen, boron, carbon, nitrogen and oxygen form such solutions.

The basic types of *linear defects* of crystals are dislocations (Figure 1.7), along and near-by which order in the location of atomic planes disrupts.

Under the action of tangential stress dislocations can move, due to plastic deformations in crystals. Dislocations are the sources of internal stresses, areas of crystal close to them are in the plastic-elastic state. Even the negligible quantity of dislocations can reduce strength of materials by several digits.

Threadlike crystals have a small amount of dislocations and therefore extremely high strength close to theoretical one. These crystals can be effective microreinforcing fireproof materials and other elements used for special applications. Occasionally strengthening of materials is achieved by introduction of alloy elements to prevent dislocations movement.

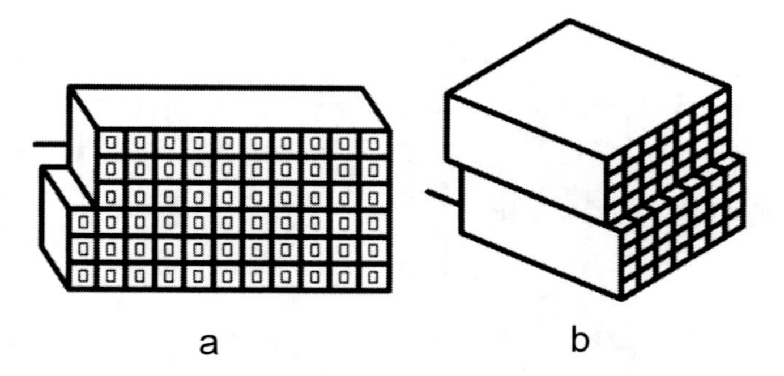

Figure 1.7. Dislocations, forming at plastic deformation of crystals: a-edge; b- screwi.

Structure of amorphous materials (Figure 1.8), as well as the structure of liquids is characterized by the so-called short-range order, when the well-organized state is observed only between the neighboring particles of a material. The main peculiarities of amorphous or glassy, structures are derived from the isotropic properties and absence of constant melting temperature.

The absence of crystalline lattice causes smooth variation of amorphous materials properties at the solid-liquid transition. The amorphous substances can be considered as supercooled liquids, but unlike liquids in them there is not a rapid exchange placed between nearby particles, what explains their high viscidity.

X-ray photography methods are the main methods used for the investigation of the structure of materials. In this method, X-rays, passing through crystalline lattices, are exposed to diffraction, as interatomic distances are compatible with length of X-ray waves. Every crystal has an X-rays diffraction pattern with characteristic lines which differ by location and intensity. By defining the distance between planes and relative intensity of lines, it is possible to define phase composition of the probed material by comparing with tabular information, preliminarily compiled for the known substances. Decoding of X-rays diffraction pattern also enables an investigator to define character of defects, type of elementary unit, position of atoms or ions and other peculiarities of atomic-molecular level of material .structure.

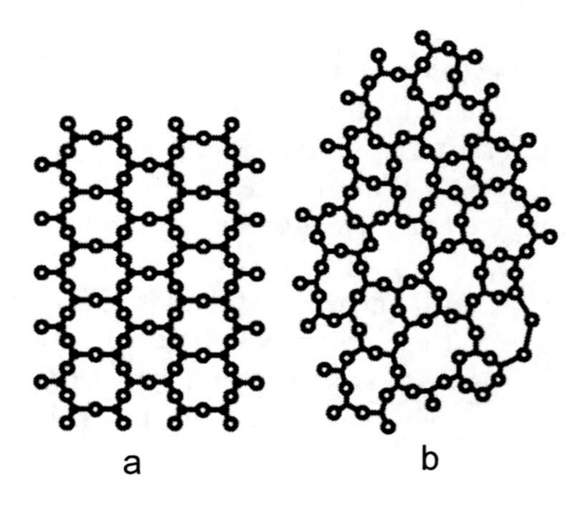

Figure 1.8. Scheme of space lattice of silica:
a- crystalline; b- amorphous.

Spectroscopic methods for assessing the structure of materials are based on quantum concepts. For a phase analysis infrared spectroscopy which is based on the ability of compounds to absorb the rays in the infrared area of spectrum preferentially is usually applied . The method enables investigation of the character of chemical connections in materials, their valency states and a series of other structural features. It is based on the phenomenon of electronic paramagnetic resonance, which consists in resonance absorption of energy of the radio frequency field in matters which contain paramagnetic particles at the imposition of magnetic-field. X-ray spectroscopy is applied in studying the energy features of atomic-molecular structure and in conducting chemical express-analysis.

Application of electron bunch with wave length in several times less than lengthes of visible light waves is the fundametnal statement of *electronic microscopy*, which permits to give a possibility to study objects with the sizes 6-100 μm at the enlarging up to 200 thousands times.

1.2. SUBMICROSTRUCTURE

The submicroscopic level concern the structures of materials, which are formed by colloid-size particles between 10^{-9} and. 10^{-7}m and can be distinguished in an ordinary light microscope. Most construction materials belong to the heterogeneous *disperse systems* which consist of two and more phases. In the disperse systems one or several substances (disperse phase) are fine particles or pores, distributed in a disperse medium. Dispersity of a system is characterized by the specific surface area(= ratio of general area F of disperse phase to its general volume) and the mass.

It is also possible to distinguish micro-heterogeneous systems, where disperse phase contains particles over 10^{-7} m in size , and colloid systems with the size of particles of disperse phase measuring between 10^{-9} and 10^{-7} m. Disperse systems are divided into three basic groups which differ by phase composition of disperse environment - solid, liquid and gaseous. Thus disperse phases in the systems of every group also can be in three aggregate states.

In construction materials which belong to the disperse systems, a disperse phase consists of solid particles more frequently. There are various powders, suspensions, pastes, yielding liquid-like mixtures, binders, plastics, paint-and-lacquer materials, ceramic masses, mortars and concrete mixtures, melts of glassy substances, etc. In some materials a disperse phase can be liquid (polymer emulsions) or gaseous (porous rocks, cellular concretes, foam glass, foam plastics, etc.).

For *the colloid systems* it is possible to apply a series of molecular kinetic theory positions. In particular, in colloid solutions (sols) the same as in true ones, disperse particles are able to take part in thermal motion. At the same time sizes of colloid particles, considerably larger than ordinary molecules sizes, predetermine the insignificant osmotic pressure of colloid substances, their slow diffusion. All the colloid systems are resistant to sedimentation, because the gravity in them becomes balanced diffusion.

Molecules of surface layer are under the action of molecular pressure, that is why the remains of energy forms at the phase interface, amount of energy (E), which is per 1 cm^2 of surface area (f) is called *surface tension* (σ):

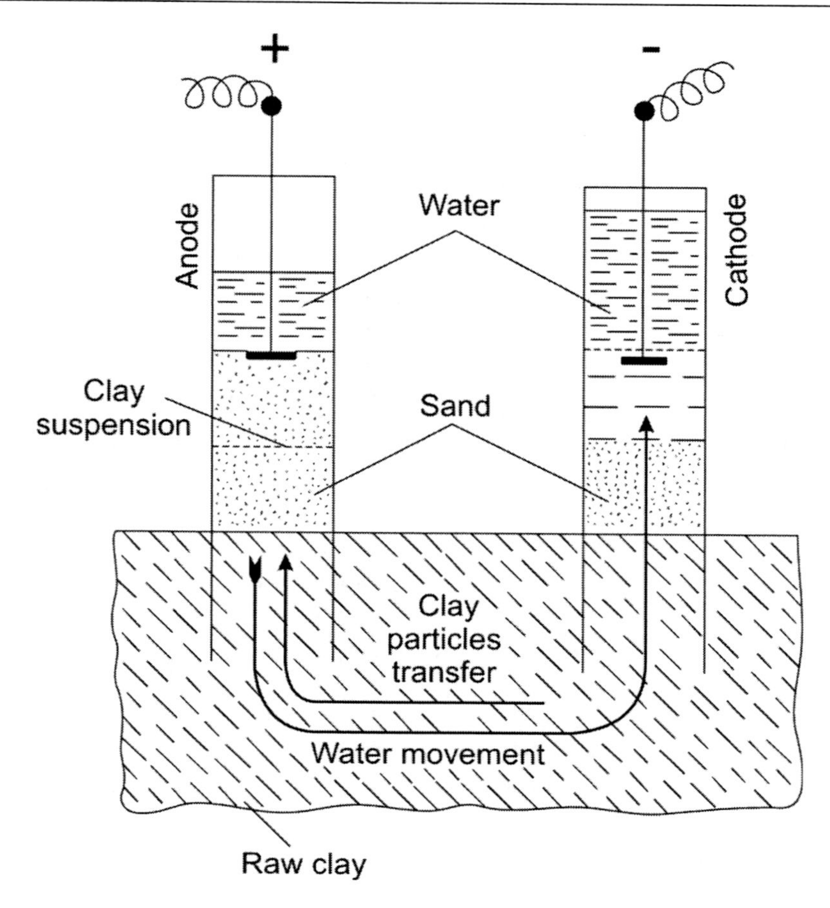

Figure 1.9. Scheme of electrophoresis and electroosmosis.

$$\sigma = \frac{E}{f} \cdot \tag{1.2}$$

Minimization of free energy and transition of the system in the thermodynamic stable state become possible due to diminishing of the phase interface, which is achieved by the arbitrary *coagulation* or agglomeration of particles in the colloid systems. Free energy can decrease due to the surface tension decline at active substances absorption on the phase interface – *adsorption.* This process can be described by Gibbs equation:

$$G = \frac{-d\sigma}{dc} \cdot \frac{c}{RT}, \tag{1.3}$$

where G - adsorption, mol/l; σ - surface tension, J/cm^2; C - solution concentration, mol/l; R - gas constant, J/mol·K; T - absolute temperature °K.

Most organic materials, containing both polar and non-polar groups in molecules are in the group of surface-active substances (SAS). This peculiarity of SAS molecules structure explains their ability to adsorb at the `phase interface and orient in such manner that polar groups (ON, COOH, NH_2 and others) are directed to the polar phases of the system (for example, to the water molecules), and non-polar (hydrocarbon chain) are directed to the non-polar phase (e.g, air). In technology of construction materials the phenomenon of SAS adsorption on solid surfaces is widely used, for the improvement of wettability of solid surface by the liquids and also reduction of their hardness (Rebinder effect), plasticity improvements and other property changes.

Adsorption on solid surfaces (adsorbents) of nondissociated or slightly dissociated substances (molecular adsorption) is inversed and decreases with temperature increase. The process of adsorption of strong electrolytes from water solutions (ionic adsorption) is irreversible and its intensity can increase with increasing temperature.

Colloid particles have certain charges and move in the electric field to the oppositely charged electrode (*electrophoresis*). Because of the external difference of potentials a liquid phase in the colloid system is able to move in relation to an immobile hard porous environment (*electroosmosis*) (Figure 1.9). The electrokinetic phenomena, which are characteristic of colloids, are used in the technology of construction materials. Thus, by electrophoresis it is possible to prepare ceramic masses for forming of porcelain and faiences, to separate fine particles, to get rubber products from latex. An electroosmosis is used for wood dehydration and in other technologies of industrial treatment of various porous materials.

As a result of redistribution of electric charge at the interphase boundary of two different chemical compositions there is a double electric layer which consists of two parts: denser internal and diffusive external. The difference of potentials between two parts of double electric layer is called *electrokinetic*, or *ζ -potential* (ζ -zeta). Potential is determined by speed of electro-osmosis or electrophoresis. It is a very important influence of the colloid systems, in particular, and represents their stability. The SAS and electrolytes additives make a significant influence on an index and sign of colloid solutions ζ-potential. If ζ- potential is equal to zero (isoelectric state), the system is unable to have electrokinetic characteristics. If ζ- potential is 25...30 mV, there is coagulation, i.e. aggregation of colloid particles. As a result of cohesion of disorderedly distributed solid particles of disperse phase in suspensions and colloid solutions, a coagulation structure is formed. Formation of such structures is typical for row of the materials, in particular materials based on mineral binders at the first period of their hardening. The distinguished feature of their hardening and coagulation structure formation is the presence of reverse contacts, i.e. freely renewable after destruction contacts (Figure 1.10). Strength of these contacts is caused by weak Van der Waals molecular forces of adhesion through thinnest interlayers of dispersion medium, thickness of which correlates to minimal surface energy.

Coagulation structures sometimes are called *gels.* Gelatinization is transition of colloid solution from the freely dispersed state (sol) to the bound dispersed (gel). Different factors make influence on gelatinization, that is coagulation process, in particular form of particles, concentration of disperse phase, temperature of mixture, types of mechanical actions (mixing, vibration). Coagulation is caused by electrolytes which contain the ions of opposite sign in relation to colloid solution. Coagulation force of ion-coagulator is related to its charge. For univalent cations it is approximately 350 times weaker, than for trivalent cations.

A process, which is the reverse coagulation, (i.e., the passing of aggregated particles to the initial colloid state), is called *peptization*. It can take place under the influence of substances - which promote disaggregation of sediments (for example, additions of electrolytes, SAS). Thus, clay slurry at the ceramics manufacturing are peptized, that is diluted using alkalis. The effect of peptization by SAS is applied for dilution of raw pulp at the cement clinker manufacturing. The peptization mechanism consists in extraction from the sediments of coagulating ions or formation of double electric layers in the colloid particles as a result of peptizer adsorption by them.

Coagulation structures are rarefied also under mechanical actions such as mixing, shaking or vibration. This isothermal process which flows as a type gel - sol is called *thixotropy*. The phenomenon of thixotropy proves that structure-forming in the coagulation colloid systems takes place due to Van der Waals forces. After mechanical actions stops, bonds broken in the process of coagulation structure formation, become renewed. A thixotropy is widely applied in technologies of construction materials, for example, for the vibratory compacting of concrete mixture.

It is possible to connect properties of colloidal solutions also with their micellar structure. *Micelle* is the minimum of colloid, which is a complex formation in which particles of disperse phase (nucleus) are in certain physical and chemical bound with solution through the ionic double electric layer (Figure 1.11).

Micellar structure impacts significantly the properties of construction materials. E.g., solid disperse phase of bitumen – asphaltenes form nuclei, covered by shell of liquid medium - from heavy resins to comparatively light oils. In the case of liquid medium excess micelle does not make contact with each other and move freely under the influence of Brownian movement. Such a structure is characteristic of liquid bitumen. On heating viscous bitumen, the gel type colloid solution destroys, but if the micelle concentration increases, the bitumen gains the gel structure.

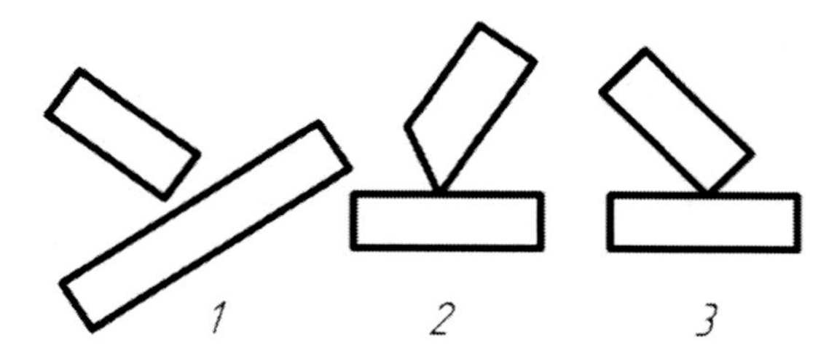

Figure 1.10. Types of contacts in the spatial structures: 1- coagulation; 2- point contact (after drying); 3- phase (after sintering or intergrowth).

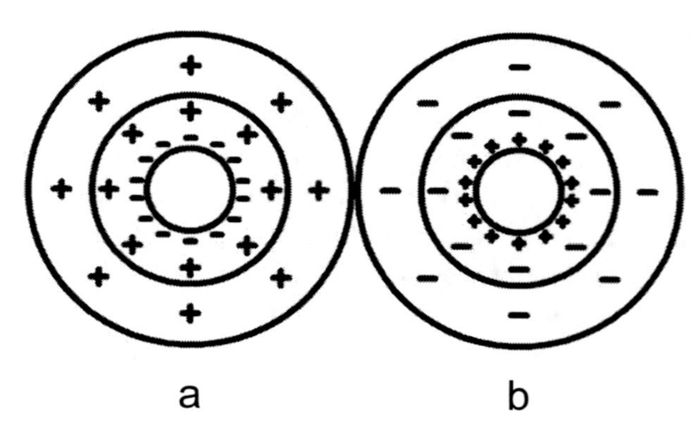

Figure 1.11. Micela structure: a- with negative charge of particle; b- with positive charge of particle.

Coagulation structures for many construction materials, in particular binders based, are primary. They transform into *condensation crystallization structure* in the course of time. The network of chemical bonds develop during formation of such structures (for example, during spatial polymerization, formation of gel-like silicic acid in water, burning-out of ceramic and other products). Condensation crystallization structures with characteristic irreversible contacts have high strength, low plasticity and do not renew after mechanical destruction.

Along with coagulation and condensation crystallization structures there can be structures of intermediate type. For example, if composition of solid phase and accordingly strength of coagulation structure exceeds some limit, its mechanical destruction becomes irreversible. Dried ceramic pastes, pressed by dry method moulding powders are among such objects.

1.3. MICROSTRUCTURE

The elements of materials structure which can be seen by use of an optical or electronic microscope are investigated at the microscopic level. Their size range is between 10^{-4} and 10^{-7} m, and they are characteristic of elements of the microheterogeneous systems. With reference to the concretes, they are characteristic of the structure of cement stone and contact layer, for ceramics they are the crystalline and glassy phases, and for metals they are linear, superficial and evident in volume defects, different phases, etc.

The typical microheterogeneous systems are powders, suspensions, emulsions and foams. For the microheterogeneous systems unlike colloid ones Brownian movement is not typical. Particles in such systems move under weight influence, therefore these systems are sedimentationally unstable.

Powders which are widely applied in construction materials can be considered as disperse systems, where air is a dispersion medium. Powders are obtained mainly, using the different methods of grinding. Dispersity of powders is controlled by the specific surface area (or just – by a specific surface) and grain distribution. Among the methods of specific surface determination of construction powders are methods based on measuring of layer resistance of the material under research to the air, passing through this layer is widely used. Blowing of air the heavier, the finer powder is. For porous powders an adsorption method is applied, which is based on dependence expressed by the following equation:

$$S = A_{ds} N S_0, \quad (1.4)$$

where A_{ds} is adsorption on the powder surface (for example, amount of nitrogen which is adsorbed on the cement particles surface; N– Avogadro constant; S_o - surface area which is covered by one molecule of adsorbable matter.

For determination of grain distribution of powder screen and sedimentation analyses are used. Sedimentation analysis is related to speed of particles setting in a liquid medium changes depending on their sizes.

Diminishing of grains sizes in powders below critical level causes their agglutination and granulation. Granulation of powders takes place due to diminishing of surface energy of the system at particles agglutination. The wettability of solid phase surface by a liquid favors the activation of this process. It provides formation of a layer with raised viscosity which increases adhesive interaction at the interface.

Grains forms in powder materials can be distinguished as: isometric (spherical, polyhedral) and anisomeric (fibrous or needle-like, lamellar, etc.) grains. There are a lot of transitional forms of grains. The Anisometry of grains influences their location in space and results in anisotropy of powders properties.

Suspensions and *emulsions* are micro-heterogeneous systems where solid or liquid dispersed phases are allocated in liquid disperse medium. Suspensions are widely used in construction materials production at obtaining raw sludges, slurries, and mortars. Emulsions are applied particularly as paint-and-lacquer materials. Concentrated suspensions are called *pastes*. Microstructure of bitumen paste is shown in Figure 1.12. For aggregate stability of suspensions and emulsions, that is for preventing coagulation (agglutination of emulsion drops, is called *coalescence*), it is required, that their particles should be covered by the shells made of disperse medium molecules (*solvate shells*). It is possible when disperse medium wets the disperse phase. Wettability of particles can be improved by SAS application. Formation of double electrical layer of ions around mineral particles facilitates system stabilization.

Emulsions can be direct and inverse (Figure 1.13). Disperse phase in direct or first type emulsions is nonpolar or weakly polar liquid (e.g., oil), and disperse medium is nonpolar. Water soluble emulsifiers promote formation of oil in water type emulsions (O/W), and nonsoluble – water in oil type (W/O). High-molecular compositions and soaps are representative emulsifiers. Powders which are well-wetted by disperse medium and have substantially smaller grains sizes than disperse phase particles can also be the emulgators. Practically in some cases there should be induced accelerated decay of emulsion. For this purpose substances, which have high surface activity and meanwhile do not form strong films in adsorbed layers (demulsification agents) are applied.

There are widely used emulsions based on the organic binders - bitumen, tar, polymers, etc. as construction materials

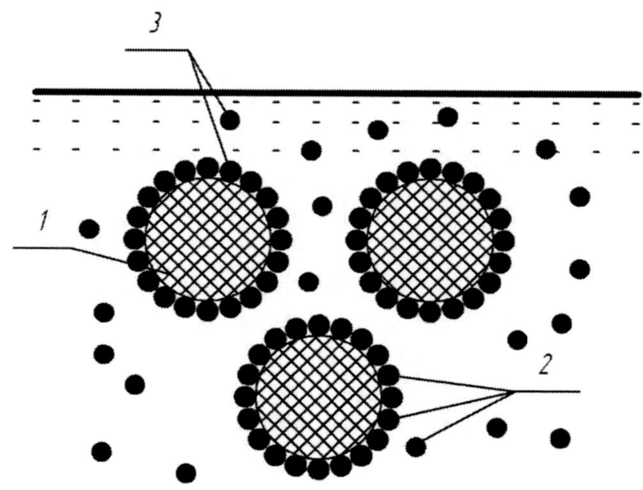

Figure 1.12. Structure bitumen paste: 1- bitumen; 2- emulsifier particles; 3- hydration water films.

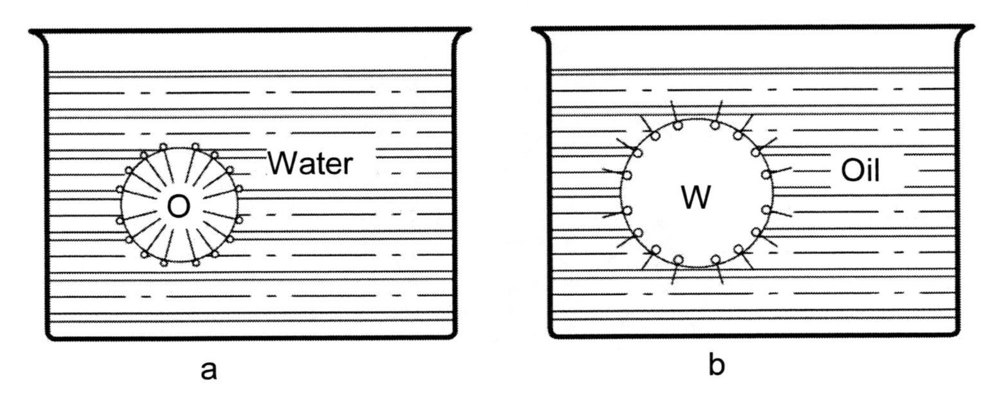

Figure 1.13. Types of emulsions: a- direct W/O; b- inverse W/O.

Bituminous emulsions are anionic and cationic. For preparation of the first type the anionic SAS are used (they are mainly high molecular-weight organic acids or their alkaline soaps). For the obtaining the second type emulsions emulsifiers of cationic type - salts of amines, amine-amide soaps, etc. can be used. For the acceleration or decay adjusting of emulsions 0.5... 1% potassium-aluminium alum, salts of chrome, magnesium, potassium are used.

It is considerably simpler to prepare *bitumen pastes*, which are the concentrated disperse systems consisting of bitumen, water and solid emulsifier, in the capacity of lime, cement or plastic clay.

Highly-concentrated systems in which gas is a disperse phase and liquid is disperse medium are *foams*. As construction materials, mainly heat-insulating ones, solid foams, where straps between gas bubbles are in solid phase (foam plastics, foam glass, gas- and foam concrete) are used.

To obtain stable foams foaming agents as highly molecular substances, soap and other compositions which have high activity of surface are used. The basic indexes of foams are multiplicity, dispersity and stableness. The multiplicity of foams refers to foam volume ratio to the volume of liquid or solid phase formed at sides of bubbles. Stableness of the foam is measured by the term of its existence and depends on the durability of films. Solid foams have unlimited stableness practically.

Foams, as well as other disperse systems, are obtained by two ways: by condensation - the association of ultrafine bubbles in larger ones; by dispersion - grinding of large bubbles and occluded gas.

A series of construction materials, in particular those based on mineral binders and aggregates, form conglomerate type of structure. Term *"conglomerate"* (latine conglomeratus) means mechanical aggregation of different components. The classic rocks which consist of rolled fragments of mountain origin, consolidated by clay, ferric oxide, silica, etc. is also called conglomerate (Figure 1.14).

At micro structural level there is a well-known binding part of conglomerates. It can be considered as the original microdispersed conglomerate where cementing substances and pore space are in. Cementing substances from the binders, which harden due to the chemical reaction with water or water solutions, are given hydrated new formations, and from synthetic organic based binders - by hardened polymers.

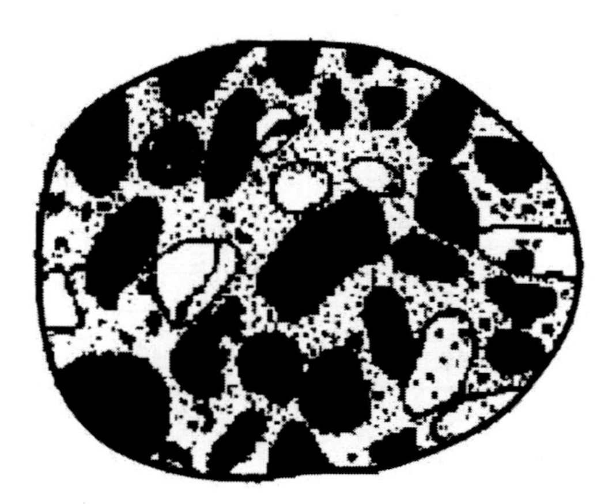

Figure 1.14. Structure of rock – conglomerate.

The peculiarities of materials microstructure substantially depend on the amount of fillers, their dispersity and physical and chemical activity of the surface. *Fillers* are fine-grained materials components, which do not form the hardening structure independently, but actively interfere in its forming jointly with cementitious substances. For mineral binders based construction materials fillers form primary adhesive contacts at the stage of coagulation structure forming, that transform with hydration in the irreversible contacts of intergrowth, durability and structure of which are determined by efficiency of filler. Fillers, diminishing energy at the, fasten crystallization of new formations in the same time. They also can intereact with the products of binder hydration and to increase, thus, volume of new formations.

Transition of binders in the filling systems from volume state to thin-film state gives the possibility to improve substantially their technical properties and decrease expenses.

The major elements of materials microstructure which determine their properties are pores. Finest pores (*ultramicropores)* appear as a result of anisotropy of properties of crystals and particles of condensation structures, and also their random orientation in space in growth process. The examples of such pores are pores in the particles of the hydrated cement (so-called gel *pores*), the size of which is (15...30) 10^{-8} m. Water in the pores is under the strong action of the field of forces of pores walls. Because of this most of water properties (density, viscosity, heat-conductance, etc.) have anomalous character. Larger pores of artificial materials are mostly technological by origin. They appear as a result of loose mixture placing, air entrapping, evaporation of excess water, destructive processes of leaching, dehydration or weathering, etc.

Pores can be divided into two groups: capillary and noncapillary. In *capillary pores* the surface of liquid acquires the form, caused by forces of surface tension and distorted a little due to weight. For capillaries with the radius (r) is representative the height of raising of liquid in the capillary (h), which is determined after the formula of Zhyuren:

$$h = \frac{2\sigma\cos\theta}{rg\rho_p} \tag{1.5}$$

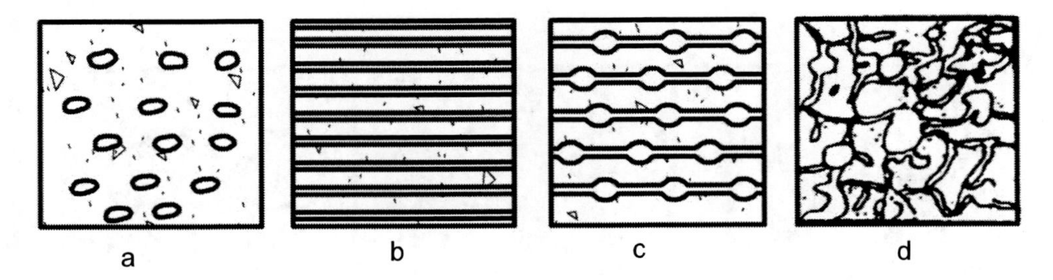

Figure 1.15. Types of cement stone structures: a- initial system with spherical pores; b- the same with cylindrical pores; c- the same with pores of variable cross-section pores; d- scheme of cement stone structure.

where σ is surface tension; θ is an angle of wettability; g is free fall acceleration; ρ_p – density of a liquid.

Microcapillaries (r < 0.1 mkm) as the result capillary condensation effect can be fully filled with liquid due to absorption of its vapors from environment.

Macrocapillaries (1.0 > r >0.1 mkm) can be filled with liquid only in the case of direct contact. Besides that, the peculiarity of macrocapillaries is that they not only adsorb the water from the air, but also vice-versa loss moisture into the atmosphere.

To estimate influence of structure on materials properties, the concept *of porosity(-* relation of pores volume to the general volume of material) is applies. The integral parameters of porous space are true (or total), open (apparent), conditionally closed porosity, etc. Thus it is important to distinguish pores by their sizes, form and character. The types of porous structures of cement stone are shown in Figure 1.15.

There are a series of methods for porosity analysis and pore-structure assessment. For determination of ultramicropores for example, the method of water and helium adsorption is applied. For micropores there are methods involving electron microscopy, adsorption of nitrogen and methanol, or mercury intrusion porosimetry.

1.4. Macrostructure

At the macroscopic level the materials structure is examined, if the sizes of particles are over 10^{-4} m. Macrostructure is studied with unaided eye or at insignificant magnification. Thus it is possible to define the peculiarities of structure and defects of materials, predefine the processes of their formation, production and performance (e.g., defects of casting origin in metals, defects of wood, bubble and impurity inclusions in glass, cracks and voids in concretes).

The study of conglomerate type materials macrostructure gives the opportunity to determine the relative amount of binding materials and aggregates, and sometimes their mineralogical composition, size and form of grains, character of surface, form and amount of macropores, etc.

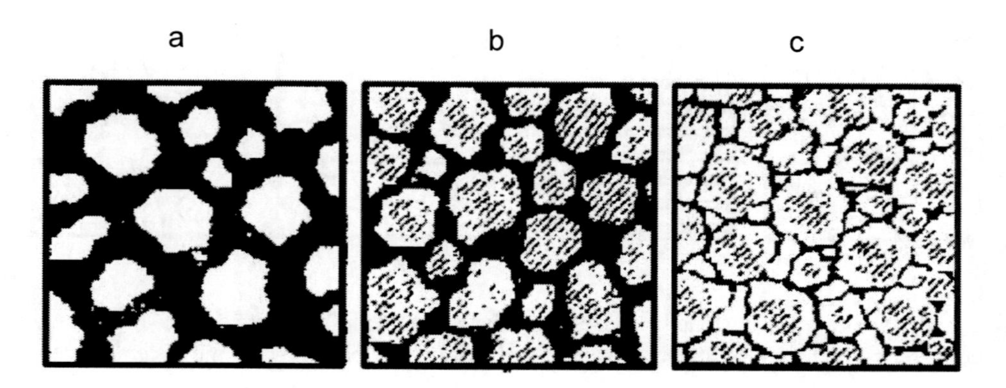

Figure 1.16. Schemes of macrostructures of conglomerate materials: a- with basal cementation; b- porous; c- contact.

In a number of cases complex multicomponent structures can be reduced to macrolevel to the two-component. For example, the macrostructure of concrete can be examined as a system "cement stone-aggregate" (sometimes the system cement-sand mortar – crushed stone is meant by the macrostructure of concrete), and macrostructure of pyrocerams - as a system "glassy adhesive substance - crystalline filler".

Two component structures can be divided into parallel comprised and serial comprised. Thus the most common are mixed parallel-serial structures.

Conglomerate two component structures (Figure 1.16) divide into three groups depending on the degree of separation of grains of aggregates. If material structure is *with basal cementation*, aggregates grains do not form contacts between themselves; they "float" in binding matter. Properties of material at such macrostructure are predefined mainly by properties of matrix part. Aggregates, acting as stress concentrator, can impair mechanical properties of conglomerate. With saturation of structure by aggregate grains forms dense framework, glued by a thin layer artificial or natural binder. Such structure is called *porous structure*. It is favorable both from the point of view of binder discharge and of giving the necessary technical properties to the materials.

Contact structure is characterized by maximum saturation of a material with an aggregate, when the binder amount is not enough for filling the voids between aggregate grains and in a number of cases for formation of continuous cover on their surface.

The index of macrostructure is a packing coefficient (K_p):

$$K_p = \frac{L_{pr} - D}{D} \qquad (1.6)$$

where L_{pr} is a projection of distance between the centers of neighboring grains; D is a diameter of grains.

At $K_p > 0$ aggregate grains shared by the interlayer of binder; at $K_p = 0$ – they contact; at $K_p < 0$ – they are anchoring, i.e. overstep each other.

Macrostructures may be differentiated also by their absolute and relative sizes of grains. Maximum sizes of grains for fine-, medium- and coarse-grained structures can be different depending on the material. For example, the structures of mountain rocks belong to *fine-*

grained, if the grain sizes is less than 2 mm, for concretes - 5 mm; to *medium-grained* — 2...5 mm and 5...20 mm; to the *coarse-grained* - over 5 and 20 mm correspondingly.

By their relative sizes they are differentiated as even-grained and non even-grained structures. P*orphyritic* structures are the typical kinds of uneven-grained structures, which are characterized by a presence in material of glassy or fine-grained great bulk where separate large crystals - inclusions are dissipated. Such structures are mainly effusive rocks, a series of conglomerate materials which are artificial by origin.

The structure of construction materials changes with time under the influences of processes, predefined both by their internal nature and environment. The development of new formations during hydration process increases concrete strength, improves a series of other properties, but at the same time under the influence of aggressive environmental conditions destructive processes like corrosion may occur with destructive character. The desired durability of material is achieved by formation of such structure which minimizes destructive processes. The typical example of the directed formation of such structure is formation in concrete air voids by SAS admixtures application, uniformly distributed porosity in all the volume. Such pores prevent growth of water pressure which increases as its freezing, and also diminish capillary water suction as a result of hydrophobization of capillaries surface.

Pores, cracks and others defects of materials structure influence on development of destructive processes. The most dangerous pores are capillaries, filled by water.

The criteria of structure efficiency are the parameters of the main properties of materials, such as strength, thermal conductivity, frost resistance, impermeability, etc.

Thus, the optimization of concrete structure from the strength point of view requires minimization of the volume of opened and closed pores, and from the point of view of frost-resistance, it is important to provide appropriate ratios between volume of closed and the volume of opened pores to be saturated with water.

Properties which are connected with each other definitely and formed under the influence of the same structural parameters acquire minimum or maximum value practically at certain structure which can be considered as an optimum one. At the ambiguous connections of different properties an optimum structure is in a compromise area.

The important parameters are the:- disperse phase content and disperse medium, phase relationships, degree of homogeneity of disperse phase particles distribution in the medium, etc. Optimization of materials structure requires the polystructural approach (-taking into consideration of interrelations of property – structure type, formed at all the levels considered).

1.5. Basic Processes of Structure Forming

Formation of natural and artificial construction materials occurs as a result of complex physical and chemical processes. The main ones are the processes of dissolution, hydration, coagulation, polymerization, crystallization and sintering. Either of these processes are decisive depending on the type of materials. For example, artificial stone materials are obtained by hardening of mineral binders, mainly by hydration and crystallization processes; but organic binders are obtained by polymerization and polycondensation processes.

Depending on the character of binder hardening, the processes which take place may be divided (Table 1.3) into three groups: hydration, coagulation and polycondensation (polymerization).

Ceramic materials formation is determined by the reactions at sintering processes, for metals by the processes of melting and crystallization.

Dissolution of solid in a liquid is destruction under the action of dissolvent for formation of solution - homogeneous system, which consists of dissolvent and molecules or ions which have passed into solution. If there is no chemical interaction, there can be reverse crystallization of dissolved substance. Substance ability to dissolve in contact with a solvent depends on the change of Gibbs energy ΔZ. The only substances that are soluble are those in which $\Delta Z<0$. The measure of solubility is a concentration of the saturated solution. Solubility of most substances grows with temperature increase but it can diminish also (Figure 1.17). There are also supersaturated solutions which are unstable, as they are not in thermodynamic equilibrium with a solid phase.

True or colloid solutions can form at dissolution of solids. True solution is characterized by the homogeneous distribution of molecular or ionic size particles. The systems with the size of disperse phase particles between 1 and.100 μm are considered to be colloid solutions. The requirements for colloid solutions formation is low enough solubility of disperse phase and presence of substances stabilizing colloid particles. Colloid particles can also appear due to polymerization or polycondensation on the basis of particles of ionic or molecular size.

Hydrationbinding materials processes in accordance with modern ideas take place mainly as a result of the chemical interaction of dissolved substance with water (hydration through solution).

Hydration is thermodynamically possible also in a solid phase without previous dissolution of initial substance (topochemical process).

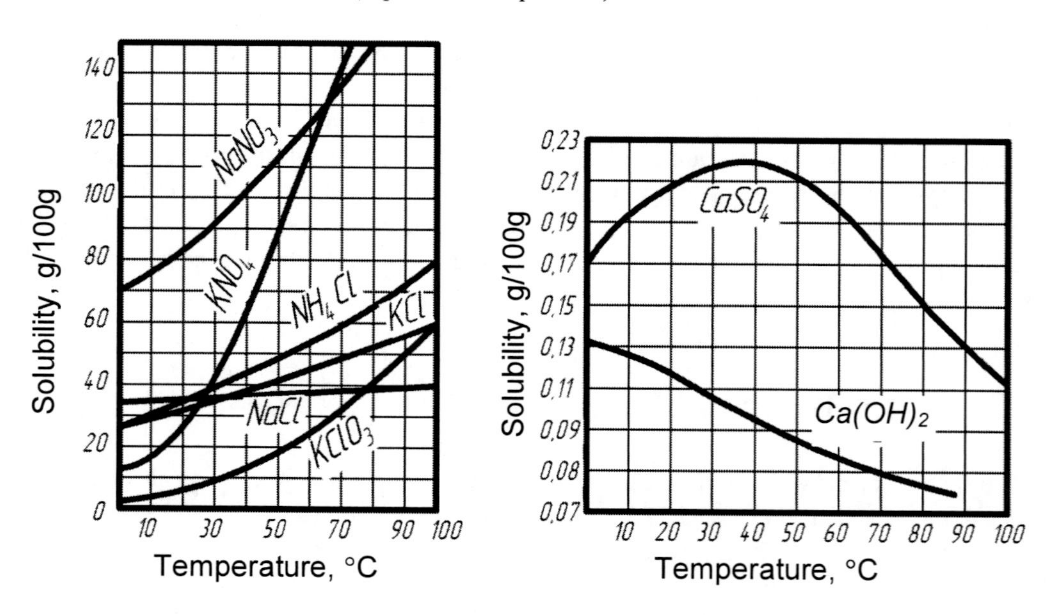

Figure 1.17. Relationship between temperature and solubility of some substances.

Table 1.3. Classification of binding materials

Hydration		Coagulation		Polycondensation (polymerization)		
Air	Hydraulic	Inorganic	Organic	Inorganic	Organic	Elemental-organic
Gypsum Air lime Magnesia	Hydraulic lime Portland cement Roman cement Pozzolanic cements Slag cements Alumina cement Expanding cement Autoclaved binding materials	Clay	Bitumen Tar	Waterglass Sulphuric cement Phosphoric acid cements	Phenol formaldehyde resins Furane resins Polyester resins Epoxy resins	Organic-silicon Hydrolyzate Ethyl siliconate

Dissolution and hydration are the major processes which predetermine hardening of mineral binding materials. The mechanism of hydration processes depends primarily on water-solid ratio. The particles size of hydration products varies from colloids to the crystals, which can be observed with microscope. Formation of mainly colloid hydration products is a characteristic for Portland cement, autoclaved materials, crystalline products - for gypsum binders, magnesia cements.

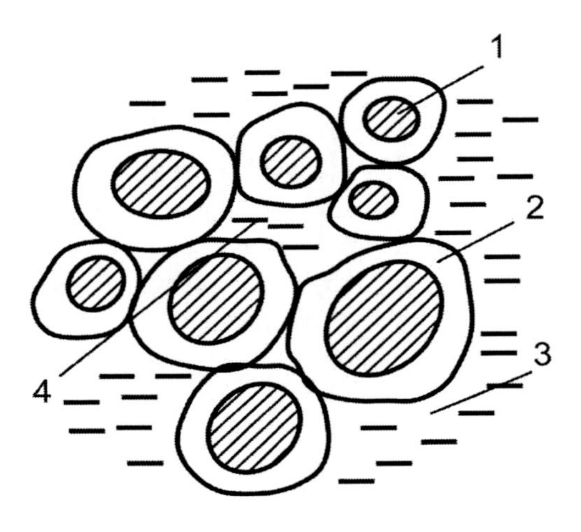

Figure 1.18. Scheme of coagulation structure of cement paste: 1- cement particle; 2- solvation sphere; 3- free water; 4- "entrapped" water.

Spatial *coagulation structures* form in highly-concentrated disperse systems with developed interface (Figure 1.18). Such structures are intermediate for hydration and polycondensation or polymerization binders, for binders of coagulation type they are basic. Thus, in clay suspensions dispersible clay particles – micelle - are negatively charged which prevents their cohesion. After the loss of charge (drying out) clay particle cohere and coagulate. The gels formed during coagulation lose water with time and are subjected to recrystallization.

Organic binders – bitumens - also form micellar colloidal structures. The solid part of (asphaltenes), is surrounded by protecting bodies (resins), forms micelle, suspended in oils. Stability of the system depends on surface interaction between micelle and oil medium.

Coagulation structure - gel, predefined by raised content of asphaltenes is characteristic of solid bitumens at the temperature range 20...25°C. For the bitumens of liquid consistency with high resins and oil content, the structure of colloid solution – sol is characteristic. The structure of bitumens under the action of temperature, changes inversely. Under the action of air oxygen oxidization (polymerization and polycondensation) occur in bitumen which results in its ductile-to-brittle transition (ageing).

Polymerization and polycondensation are underlying processes of forming polymer materials. The process of bonding of low-molecular products (monomers) without formation of by-products is called *polymerization.* Monomers which enter into polymerization reaction, are more frequently the compounds with multiple bonds (double, triple) or cyclic compounds. In the first case polymerization takes place as a result of multiple bonds opening under the influence of any power action, in the second one - as a result of cycles opening.

Polycondensation is the formation of polymers, which is accompanied by release of low-molecular substances (water, ammonia, hydrogen chloride, etc.). Thus mass of polymer obtained unlike the polymerization process is smaller than the mass of original substances and its elemental composition does not coincide with the elemental composition of compounds entering into reaction.

An example of polycondensation reaction for obtaining polyethers is as follows:

$$nH_2O - R - COOH \Leftrightarrow (n-1)H_2O + H - [O - R - CO]_n - OH \quad (1.7)$$

Polycondensation takes place stepwise. It can be conducted in melt, solution, emulsion, suspension, solid phase both with catalysts and without them.

Processes *of crystallization* occur in the supersaturated solutions of mineral binding materials. They can also pass as structure forming polymer compounds. The process of crystallization in Polymers is determined by their chemical composition. The degree of crystallization of most polymers varies from 10 to 90 %. Crystallization processes may be divided into elementary process of formation of crystal nucleus and a secondary process of their growth.

The increase of mechanical strength of binders and concrete during hardening process is predefined formation of crystalline aggregates. The hardening structure develops in two stages:

1) Formation of crystalline structure framework with formation of intergrowth contacts.
2) Framework creation without new contact formation.

Crystalline contacts between particles form, if they approach each other at thermal motion and diffusion at a distance not greater than the thickness of the doubled hydrate molecules adsorption layer. The value and kinetics of supersaturated liquid phase deeply influences the retained strength of hardening materials.

Crystallization processes take place not only in the supersaturated solutions of hardening binders but also during formation of different materials from the fusions - rocks, glass, slags, and metals. Crystallization from fusions flows at a temperature lower than equilibrium one.

Fusion hardens in the *glassy state* if between temperature of intensive formation of crystals nuclei and temperature of crystals growth is wide difference. Usually, the glassy state becomes at rapid cooling of silicate melts. Glass can be considered as the supercooled liquid, it is more similar to melts, than to the solids by its structure. Glass is the unstable form of a state and at certain regimes of heat treatment it can be crystallized. The process of glass crystallization is exothermic. Crystallization ability of glass depends on its chemical composition. For example, the presence of small amounts of MgO and Al_2O_3 simultaneously interferes with the crystallization of silicate glass, but the introduction of fluorine fastens the process.

Obtaining the glass-ceramic materials (pyrocerams) which have particularly high mechanical properties and resistance is based on the crystallization ability of the glass of certain composition.

For ceramic and some other fired material structure forms at *sintering(* -strengthening and compacting of a material, taking place at high temperatures and accompanied by shrinkage). The essence of sintering consists in slow filling with a substance caused by increasing of elements mobility of its lattice at the high temperatures in free space inside the grains and between them. The amount of unbalanced grains is also increased in the sintering process, the defects of their lattice diminish and the stresses occurring at the bonding areas of a material are also relaxed. It is normal to distinguish sintering in a solid phase and sintering in liquid phase. Sintering can be also mixed (so-called solid-liquid).

The basic factors which influence sintering are the radius of grains and temperature. The surface area of grains in mutual contact also influences the sintering rate. Surface area is higher for polydisperse powders than for monodisperse ones.

The most common sintering of materials often takes place in the presence of liquid phase. The rate of liquid-phase sintering is proportional to the radius of grains and viscosity of the liquid phase. As the grain size diminishes from 50 to 0.5 mkm, sintering speed increases a hundredfold.

Self-Assessment Questions

1. What are the basic types of materials structure?
2. What are the peculiarities of crystalline and amorphous structures?
3. How would you define the microstructure of materials?
4. What are the types of bonds between structural elements in materials?
5. What can be the defects of crystalline lattice in real materials?
6. Outline the basic research methods used in the research of materials structures.
7. What are the peculiarities of surface-active substances (SAS)?
8. What are the features of colloidal systems?

9. Explain the difference between the terms "gel" and "sole".
   What do you undersatand by the term “thixotropy phenomenon”?
10. Define the terms "suspension" and "emulsion".
11. Provide examples of construction materials relating to the different types of the disperse systems. What are their peculiarities?
12. What are the main types of pores, characteristic to construction materials?
13. What are the main types of macrostructures, characteristic to construction materials?
14. Explain the term "optimum structure" of materials. What are the technological facilities to achieve it?
15. What is the essence of the processes of dissolution, hydration, coagulation and crystallization?

Chapter 2

TECHNICAL PROPERTIES OF MATERIALS

2.1. MECHANICAL PROPERTIES

Mechanical properties are the most important properties for construction materials, which characterize their attitude to external power influences. Strength and deformation parameters determining the ability of materials to resist to destruction and deformation mainly under the action of external forces belong to mechanical properties. Mechanical properties are directly related to the structure of materials, cohesion forces between particles, and also the peculiarities of their thermal motion. Mechanical properties of the structured disperse systems, which most of the materials belong to, are called *rheological* properties.

In Table 2.1 basic mechanical properties of some construction materials are shown.

Table 2.1. Mechanical properties of materials

Material	Ultimate strength, MPa		Hardness, MPa	Abrasiveness g/cm^2	Impact strength, kJ/m^3	Module of elasticity, 10^3·МПа
	compression	tension				
Granite, syenite, diorite	100...250	5.0...6.5	-	0.05...0.1	-	30...60
Limestone	3,5…200	6.7...7.5	-	2.5	-	10...60
Heavy-weight concrete	5...80	1.1…3.5	220...1800	0.6…1.5	2.0…4.5	19...40
Grey cast-iron	800...1000	-	2000	-	10...20	80…160
Clinker brick	40…100	-	-	0.2...0.4	-	-
Pine-tree (along fibers, by humidity 12%)	36.2…48	70…130	21…26	-	42	12
Window glass	600…700	30…35	4000...6000	-	2.0	48...83
Slag glass-ceramics	500...650	-	8100...8400	0.015	4.5…6	-
Fiberglass	-	110…300	200…220	-	50...116	11...21

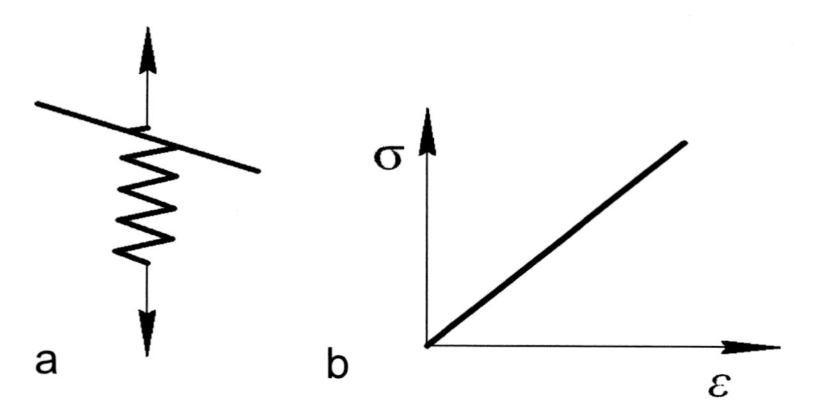

Figure 2.1. Model of perfectly rigid body (Hookean substance): a – model of elastic body; b – stress (σ) - strain (ε) diagram.

Deformation Properties

Force influence on material causes approach, exception or displacement of atoms. The ability of material to return lost shape and volume (solid materials) or only volume (liquid and gaseous materials) after removing of applied external forces is called *elasticity*. For crystalline materials elasticity is caused by attractive force between particles forming a space lattice. The elements of the lattice distended under the influence of mechanical forces after unloading try to return to initial position (Figure 2.1.)

Property of a material to acquire significant elastic deformations under the action of comparatively low loads and to return its sizes and shape after unloading is called elasticity.

High-elasticity materials (rubber, foam rubber, etc.) after unloading in practice instantly renew in shape and sizes. Elastic deformations have pronounced anisotropic character. At exceeding of certain limit stress value which is called the limit of elasticity there is observed irreversible (plastic) deformation.

Material strain in elastic range is in direct proportion to the stress acting (Hooke's law). According to this law:

$$\sigma = E\varepsilon, \tag{2.1}$$

where σ is normal stress; E– modulus of elasticity in tension ; ε elongation.

Modulus of elasticity determines atomic bonding strength; it correlates with other mechanical and physical properties (Table 2.1). It should be mentioned that brittle materials destroy under limit of elasticity (Figure 2.2.).

Materials which are under the action of external forces are able to the arbitrary decreasing of internal stresses without linear dimensions change. This phenomenon is caused by *relaxation* - gradual dispersion of elastic energy of the material strained under stress (Figure 2.3) and its transmission into heat. Relaxation is explained by the principle difference in mechanical properties and behavior under loads of solid and liquid bodies.

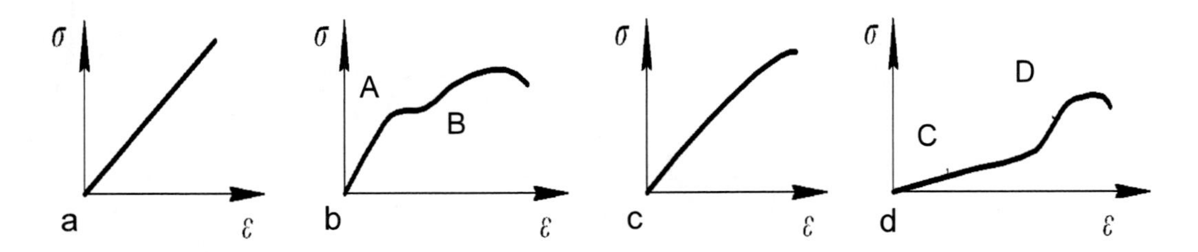

Figure 2.2. Schematic diagrams of deformations (ε) depending on stress (σ) : a – glass; b – steel ; c – concrete; d – elastomer.

Period of relaxation at which elastic stress decays until defined value is extremely high for solids and for liquids it is inverse.

Plasticity is a property of materials, opposite to elasticity. It determines the ability of a material under loads influences change shape and sizes without rupture and to keep them after removal of loading. Plasticity is a major property which determines processability of materials forming. The characteristic examples of plastic materials are high-concentration suspensions of lime, cement, gypsum, clay and other mineral substances in the water. These materials are widely applied for construction elements production. Plasticity of suspensions of the most mineral compounds is closely connected with properties which thin coats of water show on the surface of disperse phase particles.

Plastic deformations of crystalline materials are caused by shear into crystals, when one part of a crystal moves relatively to other one. The form of crystals changes at that, lengthening in certain direction. The plastic shears of crystals are predefined moving of dislocations.

With temperature increasing, materials plasticity is increased. It also increases with loss of deformation speed, with a transition from covalent to metallic bond. As far as loading increasing a period comes for plastic materials, when deformations continue developing, in spite of constant stress. The least stress at which material deforms without noticeable load growth is called yield stress (Figure 2.4). *Yield* is an important feature of the structured disperse systems (cement paste, concrete mixture, bitumens, etc.).

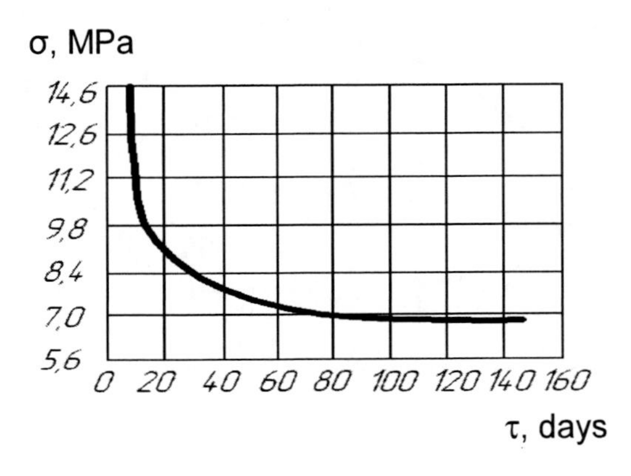

Figure 2.3. Relaxation stress concrete at constant strain.

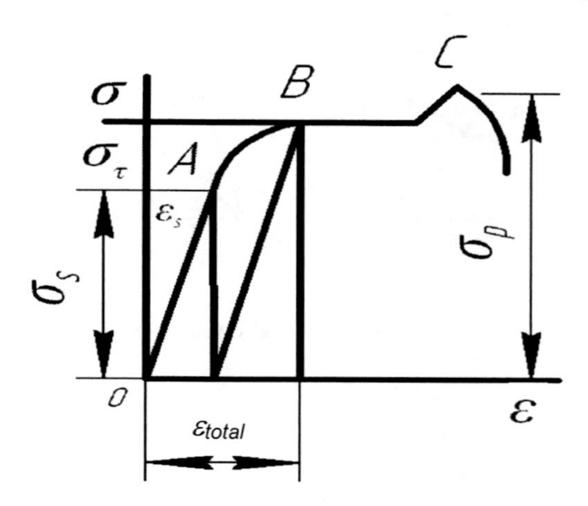

Figure 2.4. Diagram σ –ε at tension: σ_s – limit of elasticity;
σ_p – ultimate strength.

For solids *creep* is an important mechanical property; it is slow plastic deformations of materials increasing with time at the constant power loads (Figure 2.5).

Creep speed reduces at temperature and stress decreasing. Creep deformation is undesirable for the most of materials as if it leads e.g. to increased deflection forming. It should be taking into consideration at structure design.

Shrinkage is decreasing in linear sizes and volume of materials, caused diminishing, mainly, their humidity and porosity. At the comparatively low moisture content gradient (W) inside the material the change of linear sizes of material (l) is described by linear dependence:

$$l = l_0(1 + \beta W) \tag{2.2}$$

where l_0 is a size of absolutely dry material; β is a coefficient which characterizes intensity of shrinkage, %.

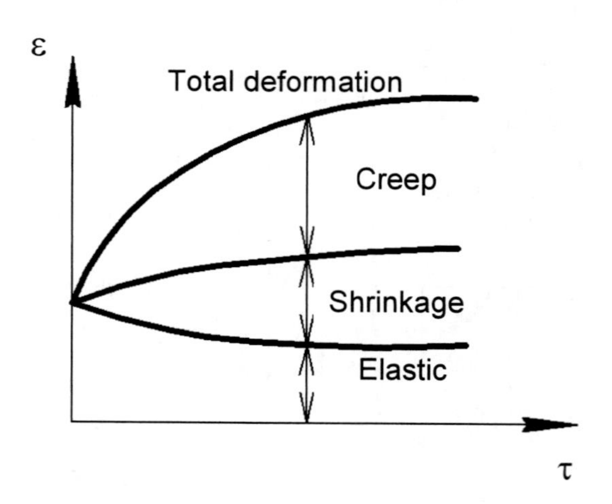

Figure 2.5. Deformations types (ε) of material, depending on time (τ) at constant load.

At the nonuniform distribution of moisture content and temperatures volume stress state develops in material that is why cracks can appear and even complete destruction of materials can take place.

Shrinkage is often accompanied by warping of a material, when there are shape deformations along with volume deformations. Dry surfaces are subjected to warping, that is why, that to decrease it, the proper rate of moisture emission from all the surfaces should be provided.

To decrease shrinkage and to prevent cracks formation, materials composition is proportioned (e.g., sand and other admixtures are added to the clay), the moist curing is applied, and also special coatings are used to retard the rapid drying.

Important parameters of deformation properties of materials (limit of elasticity and yield stress, module of elasticity, unit elongation and contraction after rupture, specific work of deformation before rupture, etc.) are determined by tensile test (in some cases compressive and bending tests), with further stress-strain diagram plotting. Loads are created on testing machines with mechanical or hydraulic drives.

Strength

Destruction is the final stage of power influence on material. An ability of materials to resist stress without destruction is called strength.

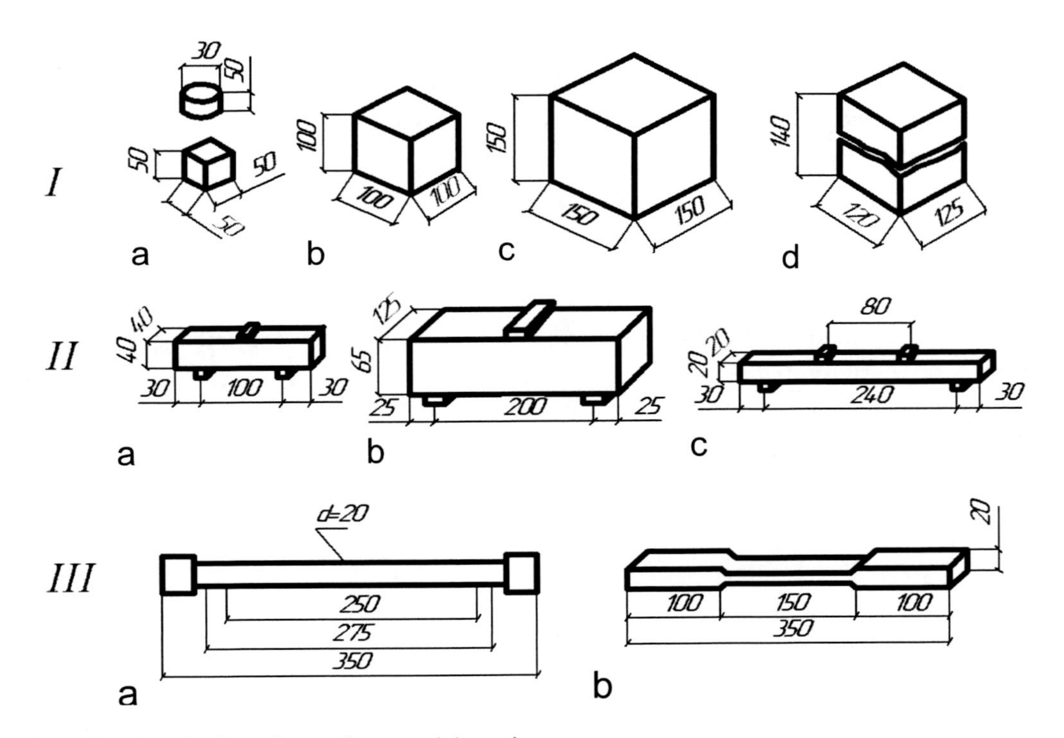

Figure 2.6. Standard specimens for materials testing:
I – compression: a – dense natural stone; b – porous natural stone; c– concrete; d – brick (cube glued of two half-bats. II – bending: a – cement mortar; b– brick; c– wood. III – tensile: a – steel; b – wood.

Strength designs of structural elements which are in complex stress state are based on the certain criteria. Basic classical mechanical strength theories are following: theory of maximum normal stresses (Galilean) according to which dangerous state is determined by maximum normal stresses in a structure, theory of maximum tangential stresses (Coulomb); theory of Huber - Mises - Henke, according to which dangerous state of a structure becomes in a moment when certain level of elastic energy is achieved in a material which leads for shape change. First two theories are applied for strength design of structures made of brittle materials, third and fourth – for plastic ones.

Strength is characterized by critical stress, when the integrity of material is destroyed. This stress is called *ultimate strength.* It is determined, usually, by acting on a material by static load growing with certain rate during a few minutes. The change of loading growth rate and character of its application (e.g. repeated interchangeable or dynamic loading) causes strength change. It can substantially vary depending on the type of the stress state (tension, compression, bending, torsion, etc.).

Ultimate strength of materials is determined at the standard specimens -cylindrical, cubic and other in shape (Figure 2.6).

Among all of methods of mechanical tests compression and tension are the most widespread; they are conducted by special lacerating machines and presses (Figure 2.7). Ultimate strength under tension (R_t) and compression (R_c) is determined by formula, MPa:

$$R_{t(c)} = \frac{kP}{F}, \tag{2.3}$$

where k is a coefficient which takes into account the sizes of the specimen, its humidity, etc.; P is a breaking force; F is an initial area of specimen cross-section.

Ultimate bending strength is:

$$R_b = \frac{M_b}{M_r} \tag{2.4}$$

where M_b is a maximum bending moment; M_r is a moment of specimen cross section resistance.

At bending of a specimen, for example, of rectangular section under the action of the concentrated force:

$$R_b = \frac{3Pl^2}{bh^2}, \tag{2.5}$$

where l - distance between supports, m; b, h - accordingly width and height to the section area.

Theoretical strength of homogeneous material (R_t) is characterized by maximum stress, required for separation of two layers of atoms. It is proportional to the module of elasticity (E) and surface energy of solid (σ_S) and inverse to interatomic distance (l_a):

$$R_t = \sqrt{\frac{2E\sigma_s}{l_a}} \tag{2.6}$$

Strength of the real solids is in hundred and thousand times less than theoretical one. Thus, for NaCl crystals theoretical value of rupture strength is $2\cdot10^3$ MPa, and for metals - $10^4...10^5$ MPa. However experimentally defined strength for NaCl does not exceed 5 MPa, and for metals – $10^2...10^3$ MPa. Such disagreements of theoretical and real materials strength are predefined by different defects in structure of solids, first of all microcracking.

The processes of materials destruction are mainly added up to gradual deformations increase, cracks formation and accumulation of local defects.

There are distinguished brittle and plastic ruptures of materials. The peculiarity of brittle rupture, representative for concrete, ceramics, glass, natural stone and other construction materials is absence of noticeable plastic deformation. Mechanical stresses appeared there, do not have time to relax therefore cracks are formed in the area, perpendicular to the action of fast developing stresses. Strength is divided into short-time, fatigue and long-term ones. The cyclic loads (vibrations, impact loads, etc.) cause brittle rupture, at which material fatigue, related to the damages accumulation, origin formation of micro- and macrocracks is developed. Temperature decreasing, increasing in rate strain, presence of surface-active medium are also promoting the brittleness of a material.

Figure 2.7. Hydraulic press scheme:

1 – frame; 2 – screw arrangement; 3 – top base plate; 4 – specimen; 5 – bottom base plate; 6 – plunger.

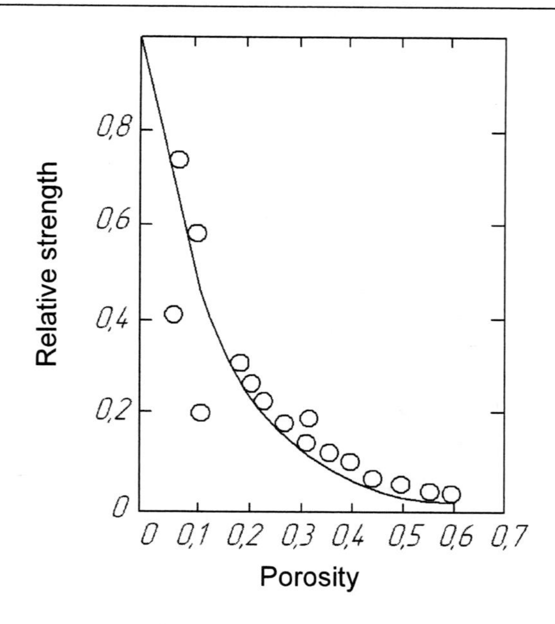

Figure 2.8. Relationship between materials strength and their porosity.

Rate of mechanical destruction of the loaded solid depends on the stresses, caused by load and temperature.

For the most of the materials (metals, polymers, glass, etc.) the following strength – time dependence is representative:

$$\tau = A_0 \exp(-\alpha\sigma), \qquad (2.7)$$

where τ is a time from a moment constant mechanical load application to destruction of a specimen; A_0, α are permanent coefficients which are determined by the properties of a material and its temperature; σ is a stress.

Along with total porosity (Figure 2.8) pore size also influences on the strength. Larger pores more deflate the strength, than fine ones.

Numerous experimental data confirm that strength of fine-grained materials with equal porosity is higher than coarse-grained ones.

Materials strength can be measured not only by direct method, that is specimen destruction but also by nondestructive methods which are based on correlation between strength and certain property of solids, in particular, by spreading rate of ultrasonic waves in solids, by hardness of material surface.

The separate type of strength is *hardness* (strength at indentation) - resistance of material to destruction by rigid force. As well as other types of strength, hardness is a structurally sensitive property; it depends also on the type of surface treatment, temperature and other factors.

Hardness is measured by pressing on the surface of the tested material or moving on it loading tips which have a spherical, conical or other form (Figure 2.9).

Hardness measure or number, at that is a relation of load to the surface area of indent. Hardness after the Brinell test is equal:

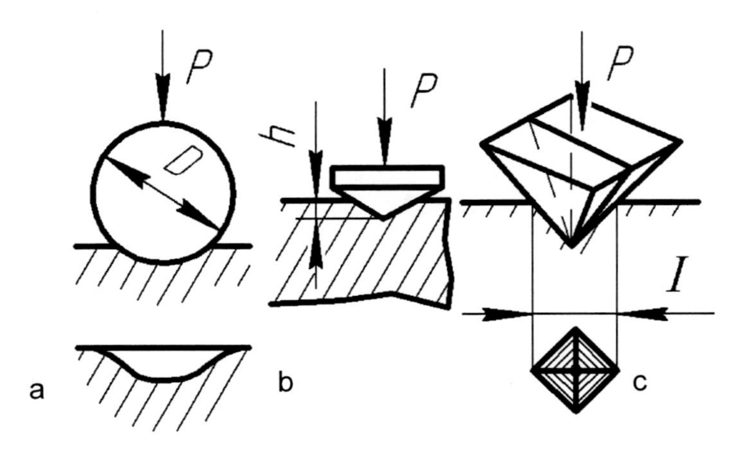

Figure 2.9. Scheme of hardness determination by:
a – Brinell; b – Rockwell; c – Vickers.

$$HB = \frac{2P}{\left(\pi D\left(d - \sqrt{D^2 - d^2}\right)\right)}, \quad (2.8)$$

where P is load; D is a diameter of spherical indenter (balls 10; 5 or 2.5 mm in diameter); d is a diameter of indent.

For approximate determination of rocks and other brittle materials hardness Mohs scale method is used which consists in scratching of testing material surface by etalon mineral. At that hardness is measured in arbitrary units, which correspond to material number ten-point scale: talc - 1; gypsum - 2; calcite - 3; fluorite - 4; apatite - 5; orthoclase - 6; quartz - 7; topaz - 8; corundum - 9; diamond – 10.

Hardness of some homogeneous materials is proportional to compressive and tensile strength which enables to use this property for the nondestructive methods of mechanical tests.

For the most construction materials (except plastics) their hardness is correlated to abrasive resistance - *abrasion*. It is measured by a ratio of mass reduction of a material Δm after test conductance to the area of abrasion F:

$$R_a = \frac{\Delta m}{F}. \quad (2.9)$$

Some rocks, stone casting, ceramic materials and plastics have high abrasion resistance. For example, abrasion of quartzite is 0.06...0.12 g/cm^2, floor ceramic tiles - 0.08, one-layer polyvinylchloride linoleum – 0.06 g/cm^2. For some road construction materials abrasion grades are defined, connected with the maximal possible mass losses during abrasion testing. Abrasion is determined on the special machines, where material destroys as a result of intensive friction.

Materials resistance to destruction when impact load applied is called *impact strength* and during the simultaneous action of abrasion and impact is called *wearability*. Impact strength is estimated by the value of fracture energy of specimens by rigs, and wearability - after the mass losses in drums, filled steel balls.

2.2. Physical Properties

Physical properties are characterized by the parameters of the materials state or their relation to the action of physical factors: water, temperature, electric current, magnetic-field, etc.

Parameters of the State

The most important physical parameters of material state are density and porosity.

A density is determined as the mass of material per unit volume. For construction materials there are distinguished *absolute* (ρ) and *average (ρ_0) density. Absolute density* (or simply density) characterizes mass (m) of a material per unit volume without pores and voids (V), and average one - with voids and pores (V_1):

$$\rho = \frac{m}{V}, \tag{2.10}$$

$$\rho_0 = \frac{m}{V_1}. \tag{2.11}$$

For the bulk materials along with average density of grains *bulk density* is determined, which takes into account intergranular voids.

Absolute density of the most inorganic materials is 2200...3300 kg/m^3, steel - 7600...7900, aluminum alloys - 2600…2900 kg/m^3, polyethylene - 910…970 kg/m^3. Average density of materials changes within the wide range (Table 2.2), for example, for the lightest porous plastics it is 10...20 kg/m^3 and for dense ones - 2000 kg/m^3 and more.

The density of materials increases, if pressure increases, and diminishes with the increasing of temperature. Water is an exception as it has a maximal density at 4°C. At phase transformations density of materials changes saltatory, increases in the process of transition from the liquid state to the solid. Water, and also cast-iron and a series of other materials become less dense at hardening. The density of materials are determined a pycnometer method, hydrostatical weighing and others.

There is on a number of occasions when *relative density* (ρ_r) is determined for construction materials as a ratio volume of material in a specimen to its total volume:

$$\rho_r = \left(\rho_0 / \rho\right) 100\% . \tag{2.12}$$

Average density can be regulated by porosity change – pore volume per material volume unit. Total porosity can be defined by the formula:

$$P = \left(1 - \rho_0 / \rho\right) 100\% . \tag{2.13}$$

Table 2.2. Materials characteristic

Material	Absolute density, kg/m^3	Average density, kg/m^3	Total porosity, %
Granite	2650.2700	2600.2700	0.2
Limestone	2700	1800...2700	11...13
Heavy concrete	2600	1800...2400	8...31
Foam concrete	2600	300...600	77...85
Wood:			
pine-tree	1500	400...500	67...73
oak	1500	610...750	50...60
Window glass	2400...2650	2450...2650	-
Steel	7800...7850	7800...7850	-

Porosity of materials substantially influences on the other properties, in particular thermal- and electrical conductivity, strength, permeability.

There is distinguished open and closed porosity. The open (apparent) porosity can be measured by volume water absorption of material. For bulk materials the total value of porosity is characterized by the degree of filling both separate grains and intergranular voids by pores. Porosity of construction materials varies in a wide range. Porosity adjustment is an effective technological mean of purposeful change of materials properties.

For disperse materials *specific surface* is the important parameter of the state, attributed to volume or mass unit of a material. Specific surface (S_s) is inversely proportional to the particles size. For the particles of spherical form:

$$S_s = {}^{3}\!/_{r}, \quad (2.14)$$

where r is a particle radius.

The internal energy of materials and reactivity grows with the increase of their specific surface. The specific surface of disperse materials is measured by determination of the resistance to air passing through the layer of powder, by adsorption and by other methods.

Hydrophysical Properties

This group of materials properties represents their behaviour to water. Interaction of water (as well as other liquids) with the surface of solid appears in moistening, caused by forces of molecular interaction of solids with solutions. It causes liquid spreading at the surfaces and impregnation of porous solids and powders. The only liquids moisten solid surfaces which diminish air-liquid interfacial tension. Water moistens materials with highpolar chemical bond: concrete, ceramics, natural stone, etc. There are not moistened materials with weak intermolecular interaction in a superficial layer: there is a series of polymers, bitumens, etc.

Property of material to be moistened by water is called *hydrophilicity*, not to moisten - *hydrophobicity*. The measure of wettability is a wetting angle (θ) formed by the drop of a liquid on solid surface. For hydrophilic materials an angle θ is acute, for hydrophobic it is obtuse. The degree of hydrophobicity of materials can be substantially changed under the influence of SAS that adsorb on their surface. The representative example of the hydrophobization is an obtaining of hydrophobic cement by the grinding at the presence of fatty acids additions or their salts.

Porous hygrophilous materials are hygroscopic, that are able adsorb the water from air. *Hygroscopicity* is predefined by water absorption at the surface and in microcapillaries of a material. It is measured by the ratio of absorbed moisture amount to mass of material and it increases with the increasing of humidity and diminishing of ambient temperature. Hygroscopic moisture varies depending on the peculiarities of materials structure: for sand it is 4...9%; for ceramic masonry materials 5...7; for wood - 12...18; cellular concrete - 20% and more. The hygroscopic moistening causes the unfavorable change of row of construction materials properties, in particular to the decline of cement activity, swelling of wood and decline of its strength, increasing of thermal conductivity of heat-insulating materials.

At the direct materials contact with water they are moistened due to capillary suction, diffusion and hydrostatical filling of open pores by water.

Capillary suction (ability of materials to absorb liquids as a result of raising of them on capillaries) is caused by surface tension forces, which appear at the solid-liquid interface and is characterized by height of arising (h), which is determined by formula:

$$h = \frac{2\sigma \cos \theta}{rg\rho_l}, \tag{2.15}$$

where σ is superficial tension;θ it is a wetting angle of moistening; g it is a free fall acceleration;ρ_l it is density of a liquid.

At the materials hydrophobization cos θ reverses sign and h<0, i.e. capillary pressure appears which counteracts to water rising. For example hydrophobizated material with capillaries about 10 mkm in diameter resists to hydrostatical pressure about 0.03 MPa.

Possibility of materials self-moistening due to capillary suction should be taken into account at buildings and structures performance. Thus, to prevent moistening by groundwater of buildings downstairs, dampproof layer which separates basement is arranged. The effect of capillary suction can be used in impregnation of porous materials by protective mixtures.

The degree of pores filling of a material by water is characterized by *water absorption.* There are distinguishing water absorption by mass (W_m) and by volume (W_v):

$$W_m = \left[\frac{(m_l - m)}{m}\right] \cdot 100\%, \tag{2.16}$$

$$W_v = \left[\frac{(m_l - m)}{V}\right] \cdot 100\%, \tag{2.17}$$

$$W_v = W_m \cdot \rho_0, \tag{2.18}$$

where m_l is a mass of a specimen, saturated by water; m is a mass of dry specimen; V is a volume of a specimen in the natural state; ρ_0 is an average density of materials.

Volumetric water absorption characterizes open or so-called imaginary porosity. Unlike mass water absorption porosity is always less than 100%. For metals, glass, porcelain water absorption is approximately equals zero, for granite is 0.5...0.7%, for dense concrete - 2...7, for ceramic brick - 8...20, to the row of high-porous heat-insulating materials - over 100% (by mass). As water absorption is correlated to the row of other materials properties (strength, frost-resistance, permeability, etc.), it is specified in the case of necessity.

For materials, which are used for construction of dams, reservoirs, collectors and other pressure structures, *water permeability* an ability to leak water under the pressure, is an important property. Water permeability is characterized by *filtration coefficient (k_f)* which shows the amount of water (V_w) that during time (τ) leaked out through unit of area (F) of the tested material thick (δ) at the difference of hydrostatical pressure P_1-P_2= 1 m of water column:

$$k_f = \frac{V_w \delta}{F(P_1 - P_2)\tau}. \tag{2.19}$$

Water permeability or watertightness is measured also by terminal pressure at which water does not leak through a specimen. Depending on the value of ultimate pressure (MPa×10) in particular for concretes grades by watertightness are set (W2, W4, W6, W8, W12, etc.).

During materials moistening their mechanical properties can substantially change due to formation of adsorption-active environment and disjoining action of water, can be also dissolution on the contacts of crystals intergrowth, swelling of the stratified structure of some minerals, etc.

The ability of a material to keep its mechanical properties in the saturated water state is called *water resistance* and characterized the softening coefficient:

$$k_s = \frac{R_s}{R_d}, \tag{2.20}$$

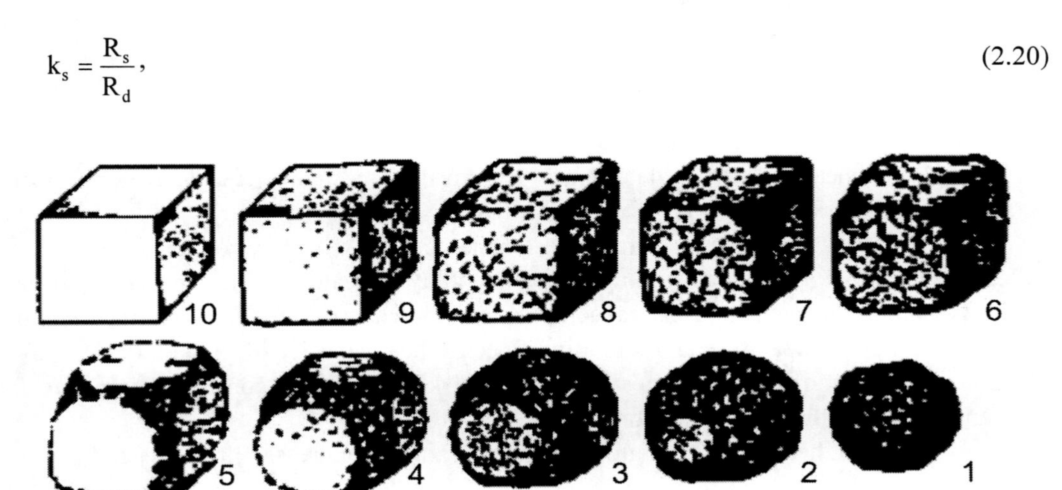

Figure 2.10. Estimation of specimen's state at testing of frost resistance by ten-point scale.

where R_s is a strength of a specimen, saturated with water; R_d is a strength of dry specimen.

Softening coefficient can change practically from a zero (lime, some types of clay, etc.) to one (steel, glass, porcelain, granite and others like that). Materials which have a softening coefficient not less than 0.8 belong to water resistant. Water resistance increase is reached by reduction of materials solubility and porosity, their hydrophobization or impregnation and coverage by water resistant mixes.

Destructive influence of porous water is particularly significant at the influence of alternate freezing and thawing. It is explained by the development of substantial internal stresses caused by crystallization pressure of water, freezing in pores of materials. As alternation freezing-thawing cycles residual strains accumulate in a materials inducing fatigue type destruction (Figure 2.10). Destruction intensity is connected with increasing of water saturation of a material by open pores, and also with temperature decreasing, i.e. increasing of formed ice volume in a material.

Ability of water-saturated material not to fail under the action of alternating temperature is called *frost resistance*. Quantatively frost-resistance is estimated by grade, that is the amount of cycles of alternate freezing and thawing, which is maintained by the material specimens without the strength decline 5...25% (depending on the type of a material) or the mass over 5%. Grades by frost resistance can vary in wide range depending on the type of a material, its composition, porosity nature, etc. E.g, the grades from F15 to F50 are set for a brick and from F50 to F500 for the hydraulic concrete.

In the capillary pores of a material as the most representative ones water starts crystallization at a temperature $-10...-20\,^{\circ}\mathrm{C}$, therefore water-saturated specimens are tested for frost resistance in cooling chambers by sequential freezing and thawing usually at a temperature $-20...+20\,^{\circ}\mathrm{C}$.

The increase in frost resistance of materials is achieved first of all by diminishing of open capillary porosity, and also increasing of amount of closed pores. Closed pores, filled with air, act as shock absorbers weakening the pressure of the ice formed. Frost resistance grows, if materials water resistance and tensile strength increase.

Thermal Properties

Attitude to heat influence determines the row of technological and performance parameters of materials.

The measure of thermal energy required for the material temperature increase for 1°C is called *a heat capacity*, it is determined experimentally by calorimeters, applying the heat balance equation the systems. Heat capacity depends on chemical composition and structure of materials, their temperature and humidity. Specific heat capacity of glass is 0.035...1.047 kJ/(kg·K), natural and artificial stones – 0.754..0.921 kJ/(kg·K). It is considerably higher, for inorganic materials than for organic ones.

The increase of heat capacity at growth of temperature is expressed linear dependence:

$$C = C_0(1 + \alpha T), \tag{2.21}$$

where C_0- specific heat capacity at 0°C; α - constant; T- absolute temperature.

At the difference of temperatures in material heat is transferred in a direction of the less heated surfaces. *Heat conductance (thermal conductivity)* is predefined by oscillating movement (ceramics, natural stone, and glass) or motion of free electrons (metals). In most of the materials heat-conductance grows at temperature rising, but in some (rocks, metals) - diminishes.

The measure of heat conductance λ is an amount of energy which is transferred per time unit through surface unit of a material at the difference of temperature of 1°C. It is determined experimentally, equation based:

$$Q = \frac{\lambda F \tau \Delta T}{\delta}, \tag{2.22}$$

where Q is a quantity of heat, J; F is an area of section, perpendicular to direction of heat flow, m^2; τ is a duration of heat flow transfer, h; ΔT is a difference of temperatures °K; δ is a thickness of material, m.

A value, reverse to heat conductance, is called *heat resistance*.

Heat conductance diminishes as chemical composition of materials and structure of their spatial lattice complicate and as crystalline structure transform to amorphous one. However it is most sensible to porosity change (Figure 2.11).

For comparison there are presented heat conductances of some materials λ, W/(m·K), which have an average density ρ_0, kg/m^3:

	ρ_0	λ
Granite	2600...2800	2.8...3.4
Pine-tree	530	0.17
Steel	7860	47
Mineral cotton wool	200...400	0.06...0.08
Brick	1900	0.80
	1200	0.44

With incresing of material porosity open pores fill with air, which has minimal heat conductance among the known substances – λ=0.023 W/(m K). For materials with low heat conductance finely porous structure is desirable as at that heat transmission due to convection, i.e. transmission of heated air relative to cold one is insignificant.

Heat-conductance is one of the key parameters of quality of heat-insulating materials used in the structures of walls and roofs of buildings. If humidity grows, heat-conductance of materials increases. In the materials pores 0.027...0.1mm in diameter a heat-conductance for air at 0°C is 0.024...0.031 W/(m·K), 0.58 – for water, and 2.326 W/(m·K) – for ice.

The increase of oscillation amplitude at heating causes the increase of average distances between atoms and, as a result *thermal expansion* of solids. For thermal expansion description thermal expansion coefficient is used, which is taken into account at expansion joints arranging, protective coatings applying, composite materials proportioning.

Heat stability is an ability of a material to withstand temperature variations without the decline of mechanical and deformation properties; it grows at diminishing of coefficient of linear expansion.

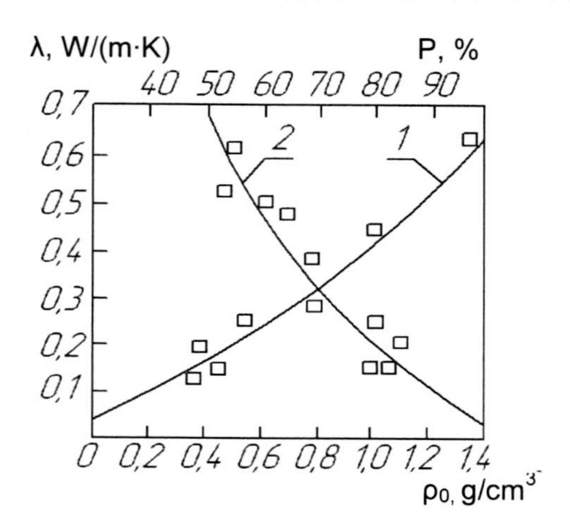

Figure 2.11. Heat conductance of different light-weight refractory materials depending on: 1 – average density (ρ_0); 2 – porosity (P).

Heat stability is determined by temperature, heating to which and rapid cooling deflates strength of material. Heat-resistance is represented also in form criterion dependence, for example as ratio material strength to thermal stresses.

Materials, which have a thermal coefficient of linear expansion (TCLE) less, than a $45 \cdot 10^{-7}$ K^{-1}, are high heat stable materials. Materials with TCLE exceeds $80 \cdot 10^{-7}$ K^{-1}, are considered as low heat-stable. For example, elements made of fused silica (TCLE$<7 \cdot 10^{-7}$ K^{-1}) do not collapse at the thermal shock at any intensity.

Property of material to resist the action of high temperatures, without melting, is called *refractoriness (fireproofness)*. It is characterized by a temperature which the specimen of pyramidal form deforms at, reaching the base by vertex. High fireproofness and melting temperature are characteristic of one-component systems (for example, pure oxides). For multicomponent materials melt formation and accordingly softening are observed at comparatively low temperatures.

Fireproof materials are materials with refractoriness 1580...1770°C (dinas, quartz, chamottere refractory products, etc.), high fireproof - over 1770°C (high-aluminous, chromite, carborundum refractory products, etc.).

Ability of materials not to change the physical-mechanical properties under the action of the open flame is called *fire resistance*. Fire-resistance rating of building structure is measured in hours of open flame action until appearance of through cracks or voids, through which the combustion products or flame penetrate freely. Fire-resistance rating is achieved also when the temperature increase on the unburned surface of structure exceeds 220°C and the structure loses load-bearing capacity. For unscreened metal structure fire-resistance rating is 0.5 hours, for reinforced concrete one - 1...2, for concrete one -2...5 by hours.

The main parameter of fire safety of materials is their *flammability*. According to their flammability construction materials are divided into three groups:

- incombustible are the materials, unable to burn in the air normal composition;
- hard combustible are materials, able to inflame under the action of source of ignition, but go out after the source is removed;

- combustible are materials, able to burn independently in the air of normal composition.

Natural and artificial inorganic materials are incombustible, nonflammable materials are such which consist both of incombustible and combustible components (asphalt concrete, gypsum and concrete elements on an organic aggregate, fibrolite, etc.), combustible are organic materials, not protected by fire-retardants (by ammonium compounds, borax– and phosphate salts, etc.).

Optical, Acoustic and Electro-physical Properties

For some of construction materials, especially finishing, optical properties - color, glitter, transparency have an important value.

The *color* of materials physically can be explained by selective absorptions of visible area of spectrum. There are differenced two groups of the coloring: achromatic (black-and-white) which has transitional tints, and chromatic with certain color saturation that is with the degree of approaching to the pure spectral color.

The color of natural mineral materials is predefined by the character of electronic interaction of component elements, and also structural defects and the amount of mechanical admixtures.

For obtaining various colored mixtures there are applied the paints of mineral origin (pigments), mainly oxides and salts of different metals or their mixtures.

Glitter is a property of materials to reflect the light that falls on them. It is expressed quantitatively by reflection coefficient, which is the function of refraction index, and also by coefficient of absorption for opaque materials.

Transparency is a property of a material to transmit light without dispersion. The measure of transparency is a coefficient of transparency:

$$k_t = \frac{L'}{L_0}, \tag{2.23}$$

where L', L_0 are the intensities of light, which accordingly transmits through the surface of output and falls on an entrance surface.

Material transparency decreasing can be possible due to application of different admixtures, microcracking, in relation to large dislocations.

Glass loss of transparency called opacification, it is provided by formation of the highly dispersed two-phase system. There is formed a phase from the finest crystals in molten glass, for example, oxides of tin, titanium, zirconium

Sound absorption of materials is characterized by degree or sound absorption coefficient. The ability of materials to absorb sound is caused by their porosity, it grows with increasing of open pores amount, the maximal diameter of which does not exceed 2 mm.

Sound-absorbing materials provide also required duration of reverberation of the gradual sound fading in the closed volumes (premises). Duration of reverberation depending on the interior of a premise and frequency of vibrations is 0.2...2.0 sec.

Mineral wool and glass-fibre boards are well-known sound-absorbing materials which have the open porosity not less than 75%.

High sound absorption in the broad band is provided also by combination of the perforated covering with porous material.

One of basic electro-physical properties of materials is *conductivity* - ability to conduct electric current.

The index for electroconductive materials is $10^4...10^6$ Om^{-1}·cm^{-1} semiconductors is $10^{-10}...10^4$ Om^{-1}·cm^{-1}, for electrically insulating material it is less than 10^{-10} Om^{-1}·cm^{-1}. For conductors and semiconductors electronic conductivity is characteristic, for the different types of electrical insulators (dielectrics) it is ionic prevalently. Ionic conduction grows with temperature increase. Admixtures, promoting coefficient of diffusion, also increase the conductivity of materials with ionic conduction. Influence of porosity on conductivity is similar to its influence on a heat-conductance. The negligible quantity of isolated equispaced pores diminishes conductivity almost proportionally to porosity growth. With the increase of porosity its influence on conductivity also increases.

Metals and their alloys are representative solid electrical conducting materials; ceramics, glass, mica, asbestos are electrical insulating materials.

Any dielectric can be used for voltages, which do not exceed maximum values, characteristic for it in certain conditions. At higher voltage the phenomenon of dielectric breakdown appears and total loss of their insulating properties comes. The ability of materials to stand the voltage *(electric strength)* is characterized by the value breakdown voltage stress of electric field.

Radiophysical Properties

Different materials absorb ionization radiation in a different measure. Thermal neutrons are effectively absorbed by the materials, containing the atoms of boron, cadmium, samarium, hafnium and others. For materials which absorb the radiation, there are often used the elemental boron, boron nitride or carbide, and also boron steel. γ- Radiation absorption increases in dense materials. Therefore for absorption of this type of radiation cast-iron, steel, lead, leaden glasses are used and also frequently ordinary concrete is used. If it is required to decrease the thickness of protective layer, usually extra-heavy concrete is applied, which density is 6000 кг/м^3 and more. Increasing of light elements content (hydrogen, boron) in concrete enables to utilize it simultaneously for absorption both γ- - and neutron radiation.

Among all of the known performance factors which act on materials, radiation is the strongest one. A degree and depth of quality changes of materials depends on the radiation dose.

Due to transmission to material large quantity of energy it can be heated and pass from the crystalline state to amorphous. The radiation defects of materials structure which form are resulting in internal stresses development, deformations, cracks and finally in complete destruction. Elastic characteristics, temperature deformability, heat conductance, density and other properties of material change also.

Table 2.3. Degree of influence of aggressive environment

Index of corrosion	Degree of aggressiveness of environment			
	N	L	M	H
Strength decreasing %	non-typical	< 5	5...20	> 2
External signs		Weak surface damage of a material	Angles damage or coating cracks	Striking damage of a material

Radiation resistance of materials depends on their composition and structure. For example, radiation resistance of rocks and other stone materials grows, if an amorphous phase in their composition increase and the size of crystals diminishes. For concretes and mortars a positive value has an increase of cement stone volume, coarseness of aggregate diminishes.

Stainless steel, dispersion-hardened alloys of chrome, zirconium, vanadium and niobium have raised radiation resistance.

Among non-metal materials concretes and other inorganic materials have high radiation resistance.

2.3. Corrosion Resistance

Corrosion (from lat. corrosio is eating away) is a destruction of materials under act of aggressive environment. By the degree of influence on building constructions environments are divided into non-aggressive (N), low-aggressive (L), medium aggressive (M) and heavy aggressive (H). The classification is based on the relative strength decreasing of a material in corrosion area and external display of corrosion signs as a result of performance of structure during one year (Table 2.3).

There are differenced gaseous, liquid and solid aggressive environments. Depending on the type and concentration, gaseous environments are divided into four groups: A, B, C, D (Table 2.4). As solubility of gases in water temperature and humidity growth, the aggressiveness of gas-air environments increase. E.g., if at relative humidity $W<60\%$ an environment of group C in relation to a concrete and asbestos cement is low-aggressive, at $W=61...75\%$ it becomes medium -, and at $W>75\%$ - heavy aggressive.

Oxidizing destruction of polymers develops under the act of oxygen or ozone. Light, moisture and heat action also promote destruction. Changes which are observed at that are called *ageing*.

The degree of aggressive influence of liquid environments is determined by the concentration of aggressive substances, temperature, pressure or fluid velocity of near-by the surface of material. Acids, caustic alkalis, salts, dissolved in water, can be aggressive agents. In particular in relation to cement concretes, aggressive influence of water depends mainly on the concentration of hydrions, free carbonic acid content, magnesium salts and sulfates.

Table 2.4. Classification of gaseous aggressive environments

Gas	Normative concentration of gas, mg/m^3, in the environment of group			
	A	B	C	D
Carbon dioxide	to 2000	> 2000	-	-
Sulfurous anhydride	to 0.5	0.5...10	10…200	> 200
Anhydrous hydrogen fluoride	to 0.05	0.05...5	5…10	> 10
Hydrogen sulphide	to 0.01	0.01…5	5...100	> 100
Nitrogen oxide	to 0.10	0.10...5	5…25	> 25
Chlorine	to 0.10	0.10…1	1…5	> 5
Hydrogen chloride	to 0.05	0.05...10	5…10	> 10

Aggressive properties of water are determined by the degree of its mineralization, hardness, and also acidity and alkalinity. Water of the rivers and lakes has alkalescent reaction. General content of salts in river waters does not exceed 0.3...0.5 g/l. Ground and underground natural waters contain the raised amount of mineral salts and other admixtures. Salt (ocean) water can contain up to 35 mg/l of salts, from them up to 78% of sodium chloride, about 11% of magnesium, calcium and potassium sulfates.

There are three basic types of chemical corrosion in cement stone: dissolution of cement stone in water; exchange reaction between the components of cement stone and soluble substances in water; destruction of cement stone as a result of crystallization in its pores of poorly soluble salts – reaction products.

Corrosion resistance of mineral materials is determined by the weakest components which are included in their composition (usually, cementing substances).

Resistance of mineral materials to alkaline and acid solutions depends on their composition and structure. Non-metal acid-resistant materials according to the chemical composition mainly consist of acidic oxides, alkali-resistant consist of basic oxides. For example, silicate materials (glass, quartz, asbestos, etc.) which contain mainly silicone oxide are proof to the action of acids, but they are able to react with alkalies. Cement stone, limestone, marble, which contain mainly the oxide of calcium are alkali-resistant, but easily deteriorate under the action of acids.

Polymeric materials are the most chemically resistant. However some of them at certain terms also attackable by chemical destruction. Heterochain polymers, which contain in the main chain atoms of oxygen, nitrogen, sulphur, etc. (polyamides, thiokol, siloxanes, polyethers, etc.) are comparatively easily disintegrate into hot water, acids and alkalis.

Polymers which have double bonds in a structure react easily with acids, for example, some types of artificial rubber and polyvinylacetate.

The serious damages of different materials can happen due to biological factors. For example, intensive corrosion of concrete is caused by bacteria. Denitrifying bacteria oxidize sulfureous compounds containing in waste waters, primarily to hydrogen sulphide, and then to sulphuric acid. Urolitical bacteria impact mainly on urea, hydrolyzing it. Ammonia and carbonic acid exude at that.

Bacteria can also actively influence on metals. For example, bacteria which form acids have the most influence on steel.

Basic biological factors which influence on wood are fungi. Cellulose is nutritional medium for them. Fungi secret the special enzyme, which transforms the insoluble in water cellulose $(S_6H_{10}O_5)_n$ into soluble glucose $(S_6H_{12}O_6)_n$.

For metals corrosion under act of liquid medium is electrochemical by character. The type of electrochemical corrosion is the so-called electro-corrosion under the action of both direct and alternating current. It acts not only on metals but also on cement stone, concrete. More frequently electro-corrosion is caused by stray currents, the sources of which can be tram-lines, railways, subway etc.

Salts, aerosols, dust and others like that are solid aggressive environments influencing on materials. The degree of their aggressive influence is determined dispersity, solubility in water, hygroscopicity.

Self-Assessment Questions

1. What are the peculiarities of elastic properties of materials and concepts of elasticity and relaxation?
2. What are the types of destruction and what is characteristic of materials strength?
3. Give definitions and formulas for determination of ultimate strength of materials.
4. What are the peculiarities of hardness, abrasiveness and impact strength of materials?
5. What is the difference between the concepts of absolute and average density of materials?
6. What porosity is, how does it is calculated and how it influences on basic properties of materials?
7. Give the definition of concepts of wettability of materials with water, to the hydrophilicity and hydrophobicity, hygroscopicity.
8. What are the possible methods of materials moistening and what are the peculiarities of concepts "capillary suction" and "water absorption"? Why is it important to take into account these properties of materials during performance in elements and structures?
9. Give the definitions of such properties of materials, as watertightness and water resistance, and tell about ways of their control.
10. 10. Give the determination of a concept of materials frost-resistance and tell about ways of its increase.
11. Describe basic properties of materials in relation to a temperature (heat capacity, heat conductance, heat resistance, heat expansion) and ways of control of these properties.
12. Expose the features of properties which characterize materials stability in relation to a temperature (heat stability, fire resistance, refractoriness).
13. Give the definition of basic electro-physical properties of materials.
14. What is understood under radiation resistance of materials, what factors determine this property?
15. What is understood under chemical resistance of materials and what is the essence of corrosion processes? How is it possible to improve chemical and corrosion resistance of materials?

Chapter 3

GENERAL TECHNOLOGICAL PROCESSES

Properties of materials and elements, technical and economic efficiency of their manufacturing are substantially determined by the peculiarities of the technology applied, which is the group of different processes, directed on production of the ready-made products from the initial raw.

The general process chart of construction materials production includes output and transportation of raw material at the enterprise, preparation of working mixture and its processing, products forming, their processing to produce the required qualitative indexes. The finished elements are transported on storage of the end products, where they are stored and then are delivered to the customer. Depending on the type of materials the certain stages of technological process can be eliminated (for example, output of raw material, if it is imported, making of working mixture - in the case of elements production made of natural stone, single-component raw, etc.).

The *technological processes* can be classified into:

- *mechanical* – crushing, grinding, screening, fractionating, mixing, forming, consolidation, etc.;
- *chemical* – antiseptic treatment, fire-retarding agents treatment, extracting, hydrophobization, oxidization, etc.;
- *thermal* - drying, burning, evaporation, steaming, autoclaving, etc.

There are also distinguished hydromechanical (transportation of gaseous, liquid, powdery-like materials, precipitation of the suspended particles, etc.); mass-transfer (separation of raw materials according to their density, reduction of humidity) and other processes. In many cases technological processes are complex. For example, thermal processes are often accompanied by the mass-transfer and chemical reactions.

3.1. GRINDING AND CLASSIFICATION OF MATERIALS

Grinding is one of the most widespread and power-consuming method of raw material treatment for construction materials manufacturing. For example, about 70...75% of the general consumption of electric power at the production of 1t cement is expended on grinding of raw material materials and clinker (Table 3.1).

Grinding of solid materials is possible for the acceleration of their chemical interaction, production of more homogeneous mixtures and facilitation of its subsequent processing.

The relation of average values of diameters of pieces of initial and end products is understood by the degree of grinding U. Depending on U value there are distinguished coarse (U=2...6), secondary (U=6...10), fine crushing (U=10...50) and milling (U=50 and more). Depending on physical and mechanical properties, initial fineness of grains and required degree of grinding of a material methods of crushing, impaction, abrasion, splitting and its combination are used (Figure 3.1, 3.2).

It is reasonable to grind hard and brittle materials by crushing, and hard and viscous – by crushing with abrasion. Coarse grinding of soft and brittle materials is executed by splitting and secondary and fine - by impaction. Crushing and abrasion are very often used at fine grinding – milling. Wood, peat and other fibrous materials, are ground down by impact and cutting.

The basic types of machines for grinding are jaw, cone, hammer (impact), roll crushers and grinding mills. The tumbling, hammer, vibrating, jet and other types of mills are applied for the material milling.

Resistance of materials to grinding can be estimated by the coefficient of grindability. If grindability of cement clinker is taken as one, the grindability of other materials can be described by the following coefficients: limestone – 1.2…1.8; granulated blast-furnace slag – 0.8…1.1; silica clay – 1.3…1.4; trass – 0.5…0.6; quartz sand – 0.6…0.7.

In accordance to the law of grinding, set by Rittinger, work of grinding (A) is proportional to the area of neogenic surface (ΔS):

$$A = \sigma \Delta S, \tag{3.1}$$

Table 3.1. Consumptions of electric power at Portland cement production

Operation	Consumptions of electric power, kW · hour/tone	Consumptions of electric power, %
Raw material mining	4	3,5
Grinding of raw materials	12	10,5
Milling of raw materials	18	15,5
Grinding of hard fuel	14	12,0
Burning of clinker	9	7,5
Milling of clinker	38	34,0
Drying of admixtures	6	5,0
Consumptions in auxiliary workshops	8	7,0
Other consumptions	6	5,0
In total	115	100

where σ is specific surface energy of solid.

Also, work of grinding is described by the formula:

$$A = \sigma \Delta S + K \Delta V, \quad (3.2)$$

where K is a coefficient, equal to specific work of deformation (per volume unit); ΔV is a part of volume, subjected to deformations.

The first member of equation (3.2) is energy which is expended for the formation of new surfaces, and second one - by energy of deformation. It is possible to neglect the second member in equation (3.2) at the high degree of grinding.

As a result of deformations analysis, appearing at grinding of materials, dependence was formulated, according to which, work of grinding is proportional to the volume or mass of the ground piece. This law is more frequently applied for the analysis of the coarse and secondary grinding, and it is expedient to use the Rittinger law for the analysis of fine grinding, when the surface of material increases in hundreds times.

The processes of deformation and destruction of solids substantially change under the influence of the physical and chemical interaction with the environment. If the surface energy of materials diminishes under influence of surface-active substances (SAS), its grinding is accelerated. According to the experimental data, grinding in active mediums is in 5...7 times higher, than grinding in an air environment at the identical energy consumptions. These data correlate with theoretical conceptions, in accordance with which adsorption of SAS is able to diminish considerably the elasticity limit, strength and hardness of materials. The wedge-shaped cracks, developed during deformation of solids, do not close up with the adsorption layers of SAS that assists to its development and destruction eventually.

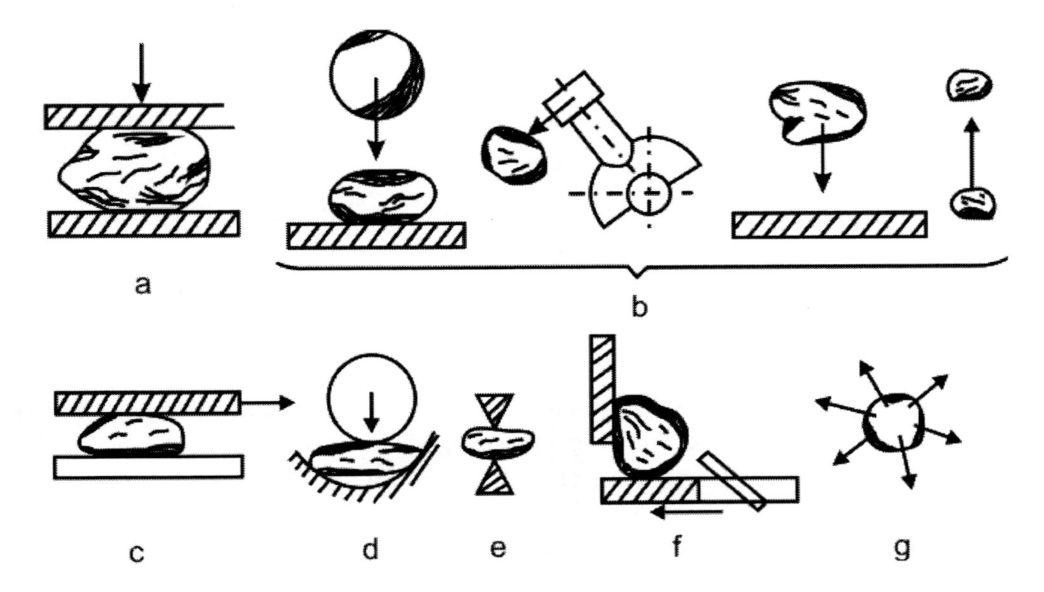

Figure 3.1. Methods of materials grinding:
a – crushing; b - impaction; c - abrasion; d - tension; e - splitting; f - cutting; g – explosion.

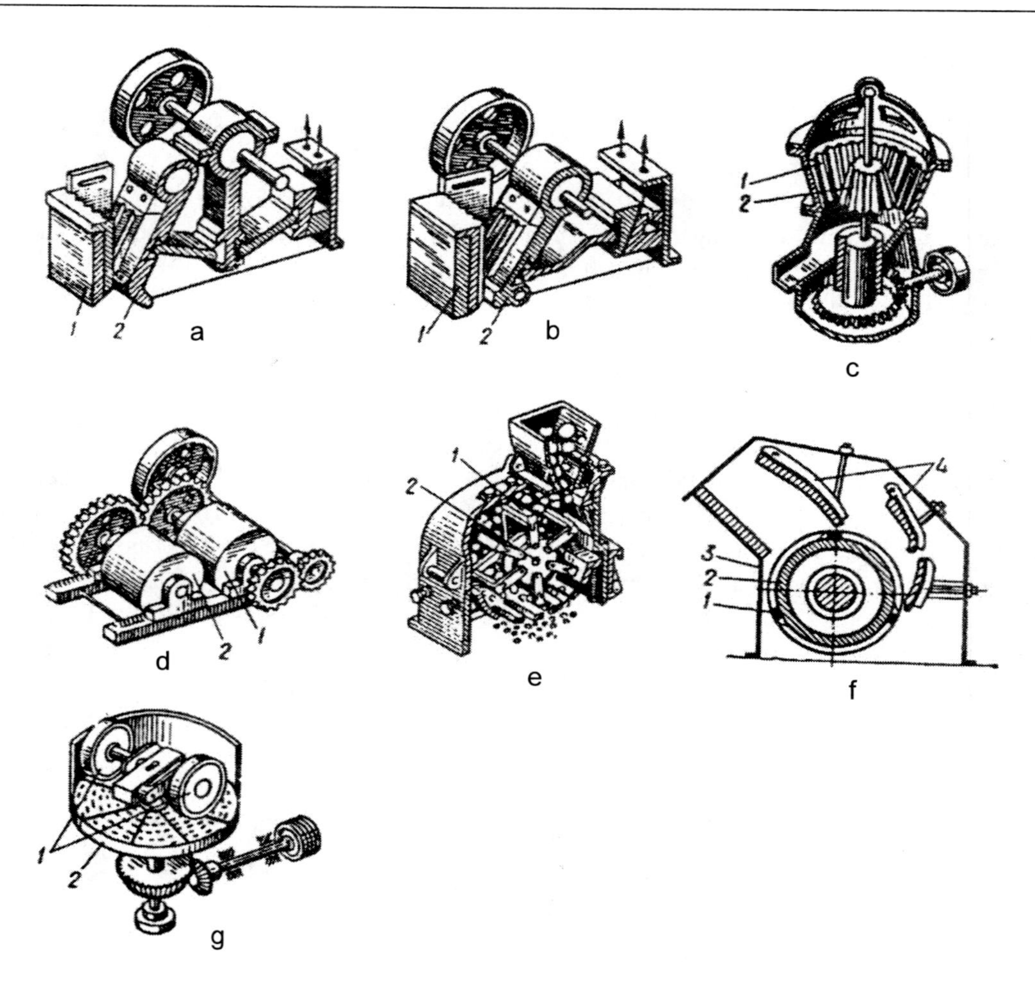

Figure 3.2. Crushers:
a - jaw with the simple oscillation of crushing jaw; b - jaw with the complicated oscillation of crushing jaw (1 - fixed jaw, 2 - movable jaw); c - cone (1 - immobile external cone, 2 - mobile internal cone); d - roll (1,2 - rollers, revolved to each other); e - hammer; f - rotor (1 – hammers, 2 - rotor, 3 - enclosure, 4- beds); g - grinding mills (1 - rolls, 2 - bowl).

In many technological processes, grinding is used in a complex with *classation* of discrete materials that means, distributing or sorting them depending on fineness. The classification, depending on the method of realization, can be mechanical (sieving), air (separation), hydraulic and electromagnetic. During the mechanical elutriation, the bulk materials are divided by machines or equipment (sieving machines) with plane or barrel-type fire bars, grates or sieves. The simplest sieving machine are movable ones, they can be also revolving, vibrating, etc.

Efficiency of sieving is estimated by the ratio of particles mass of undersize (M_u), that means the particles size of which is less than the size of voids on a sieve, to the actual amount of particles of this class in the initial product (M, %):

$$K_s = \frac{M_u}{M} \cdot 100. \qquad (3.3)$$

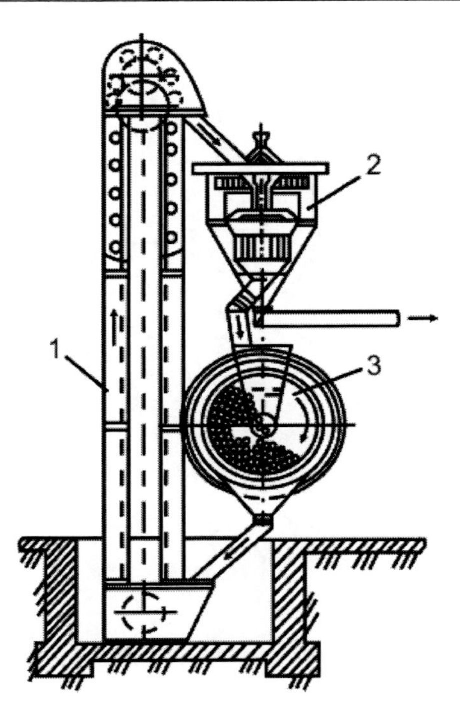

Figure 3.3. Scheme of mill device work with centrifugal separator in the closed cycle:
1- elevator; 2 - centrifugal separator; 3 - ball mill.

This index, called the coefficient of sieving quality, depends on the parameters of process, humidity, grain distribution of material and other factors. For the barrel-type sieves it is 50...80%, oscillating - 70...80%, vibrating - 90...95%.

It is possible to select particles of material during the grinding from the air environment under the action of gravity from vertical or horizontal circulation or under the influence of centrifugal force from a curvilinear stream. Distribution very often takes place at the combined action of these forces in the existent constructions of air separators.

Air- separation method is widely used at the work of mill aggregates by the closed cycle, if it is needed to separate the prepared product from the general stream of material, and to send more large particles for re-crushing (Figure 3.3).

The hydraulic classification is carried out in a water steam. A stream rate is selected to provide the carrying out from the classifier for wash the particles less than the certain fineness in size, and sedimentation of the particles with coarser fineness in a classifier (underflow). Hydraulic classification is widely applied at enrichment of raw meals that means separation of admixtures and fractionating. Washing of admixtures is usually carried out in spiral classifiers and hydraulic cyclones, and fractionating of sands - in vertical hydraulic classifiers. All of these methods and processes are based on the hydrodynamics laws.

At grinding, sieving and other types of material processing, dust emission into the atmosphere takes place. It results in worsening of the hygiene and sanitary conditions both at factory and near-by territories, work of mechanisms and also is a reason of financial expenses and decline of technical and economic parameters of manufacturing.

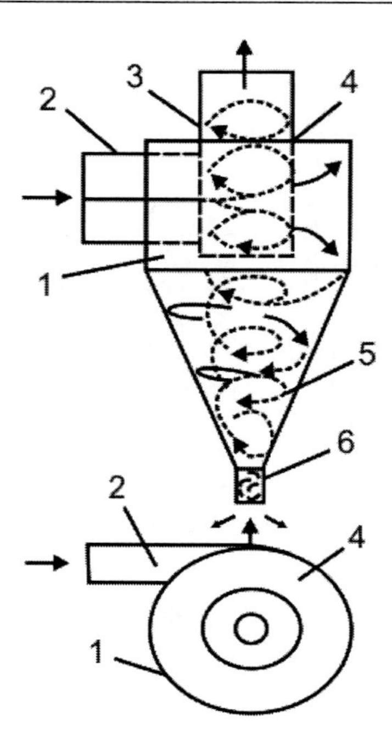

Figure 3.4. Scheme of cyclone action:
1 -cylindrical part; 2 - gas flue; 3 - branch for gas make; 4 - deck; 5 - conical part; 6 - branch for unloading.

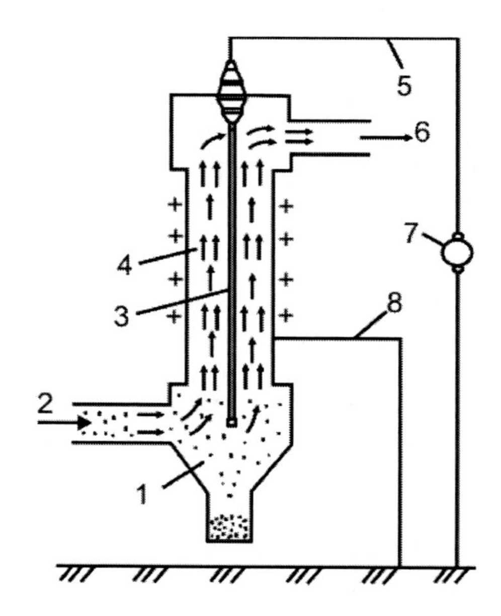

Figure 3.5. Scheme of electrical cleaner work:
1 - hopper; 2 - entering of unrefined gas; 3 - line; 4 - cylinder; 5 - wire with negative charge; 6 - output of the cleared gas; 7 - source of current of high voltage; 8 – grounding.

Table 3.2. Gas purification efficiency by different dust collectors

Aggregate	Particle sizes, which are settled, micrometers	Purification efficiency, %
The dust precipitation chamber	5...2000	40...70
Centrifugal air filter (cyclone)	3...100	45...85
Electrostatic cleaner	0,005...10	85...90
Wet-type collector	0,01...10	85...95

It is possible to divide technical methods of cleaning from dust into such basic groups:

1. *Mechanical* (dry) *gas cleaning*, at which particles are settled under the action of mechanical forces - gravity, centrifugal and inertial. The dust precipitation chambers, cyclones (Figure 3.4), centrifugal air filter of rotary action and inertia-type dedusters are applied for the mechanical cleaning.
2. *Electric gas cleaning*, at which particles are settled in the electric field of high voltage. The electric cleaning is carried out by plate and pipe electrostatic cleaners
3. *Gas filtration* - through the woven (bag) filters, bulked or prepacked and ceramic filters, which detain the particles, weighted in a gas stream.
4. *The wet cleaning* - spraying of gas stream or its washing at passing through the layer of liquids. The gas cleaners and hydraulic cyclones belong to the wet-type collectors.
5. Settling of particles with the use of the *ultrasound* for its aggregation.

The work efficiency of dust precipitation device is estimated by coefficient of efficiency or *by purification efficiency:*

$$K_p = (M_2/M_1)\cdot 100\% = [(C_1 - C_2)/C_1]\cdot 100\%, \quad (3.4)$$

where M_1 – the dust amount, which enters the dust collector; M_2 – the amount of the recovered dust in dust collector; C_1, C_2 – the concentration of dust before and after settling, g/m^3.

Purification efficiency, achieved by the dust collectors, depends on many factors, but the main one is the size of dust particles, entering settling chamber, type or construction of dust collector. The limits of application of various types of dust collectors depending on the size of particles are shown in Table 3.2.

3.2. Mixing of Materials and Molding of Products

The raw meals - batch mixtures are composed of the prepared components. The batching and mixing are the basic processes at the mix production. The mass *batching* is the most accurate. Water and other liquids can be measured out both by mass and by volume with identical accuracy. The considerable errors of volumetric batching occur during the work with bulk materials because of the large variations of their bulk density and humidity in particular.

Devices for batching - batchers - can be continuous and cyclic, with automatic and hand control in equipment for construction materials production.

Primary purpose of mixing is providing of the uniform distribution of the mixture components and achievement of required homogeneity. Its properties, and also quality of materials and elements based on them, are closely related to homogeneity of mixtures. Two basic methods of mixing are widespread: mechanical one - by machines-mixers and pneumatic - by the compressed air (barbotage). Mixing can be carried out both periodically and continuously. Homogeneity of mixture depends on the amount of movements of particles and its trajectory. Thus, concrete mixtures are mixed up the best in the center of concrete mixer. The important role trajectory plays at the presence of coarse grains, as if here mixture can segregate under the gravity forces.

For overcoming of inertia rotation is considered to be the most advantageous. Powder-like materials are mixed up by barrel, screw, band, circulation, centrifugal, vibro- and pneumatic mixers. Barrel mixers are the simplest; however mixing in them takes place comparatively slowly. The units of the intensive mixing are tandem- and two-shaft screw mixers. Circulative, centrifugal and vibromixers belong to the effective types of mixers. In some of them materials are mixed up by intensive circulation in the sprayed (pseudodiluted) state, in other one - due to centrifugal force, developed at the rotation of rotor, in the third one - due to creation of mobile layer on-the-spot, which oscillates with high-frequency.

At manufacturing mortars and concrete mixes, emulsions, paste, mixing in liquid environments takes place by paddle, propeller, turbine and other mechanical mixers (Figure 3.6), jets, pumps, and also at barbotage.

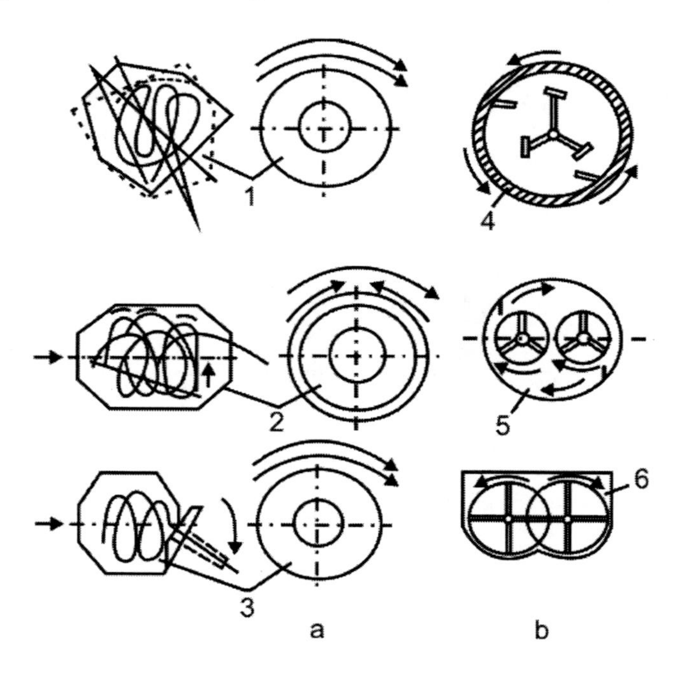

Figure 3.6. Schemes of concrete mixers:
a - gravitation action; b –forced action: 1 - reversible; 2 - unreversible bidirectional; 3 - unreversible with an unloading tray; 4 - counterflow action with revolved bowl; 5 - with two mixer devices; 6 - twin-shaft.

Intensity of mixing depends on the rheological characteristics of materials, initial degree of heterogeneity and quantitative correlation of the mixed components.

Efficiency of mixing (E) can be characterized by the degree of homogeneity of mixture:

$$E = (x_1 + x_2 + \ldots + x_m)/m, \quad (3.5)$$

where $x_1 \ldots x_m$ - particles of the mixed components; m - amount of tests.

The construction elements which have certain geometrical form and sizes are shaped from the well mixed homogeneous mixes of the specified composition and also the required properties. The most representative for technology of construction materials is the application of such methods of *shaping,* as pressing, plastic forming, vibrocompaction, casting and extrusion (Figure 3.7). Another ways of forming also used are following: ramming, centrifugation, rolled stock, shortcreting, etc.

Shaping by *pressing* is applied mainly at the use of low-plastic, bulk raw meals. The solid components of mixture forcedly move and are mutually drawn together during pressing, which leads to the most compact placing them in a certain volume, to extrusion of uncombined water and air.

Pressing can be single-action and double-action compacting, continuous (single-stage) and irregular (multi-stage). Three stages in the process of pressing are distinguished:

1. Comparatively easy particle approach under the action of loading.
2. Discontinuous change of settling-out of mass because of elastic deformation of particles.
3. Achievement of critical pressure with stopping of compression of mass and achievement of maximal contact of surface area between particles.

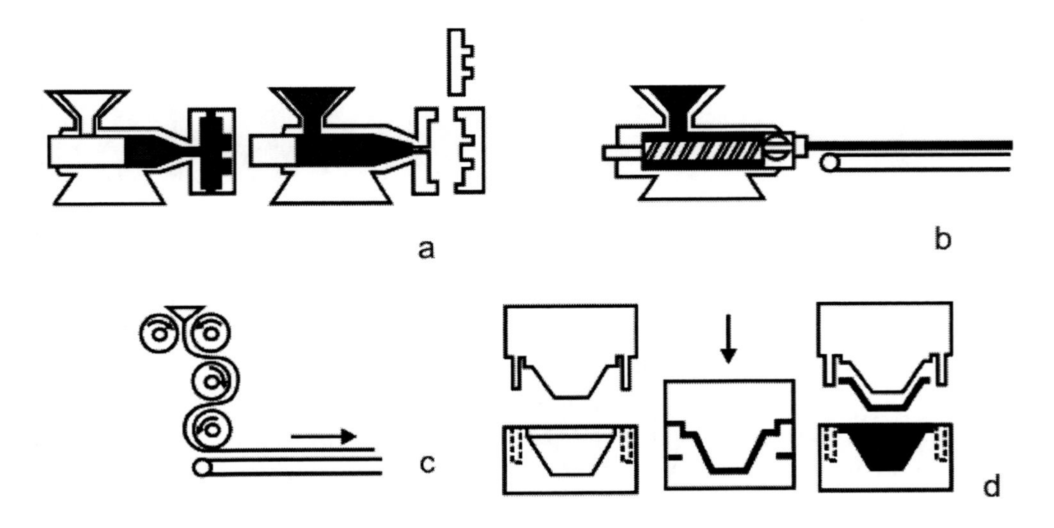

Figure 3.7. Scheme of shaping the elements from the plastic mixtures:
a - injection casting (injection); b - continuous profile extrusion (extrusion); c – rolling; d – pressing.

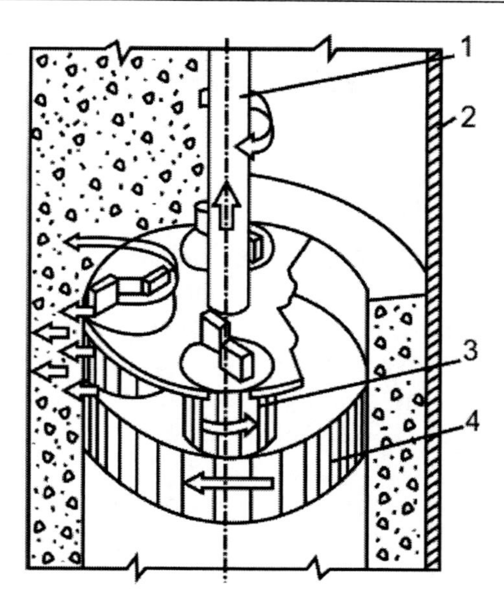

Figure 3.8. Chart of shaping of barrel-type element by a member of roller type: 1 - input drive; 2 - form; 3 - rear roller; 4 –smoothing bowl.

Optimum compaction pressure is selected in experimental way. Thus, pressure, required for taking of mass to the specified density and overcoming of friction of mass at the walls of molds, and also some excessive pressure because differences between humidity, grain distribution and others are taken into account.

At the dry pressing of the powder-like masses under the pressure 10...200 MPa, dependence between porosity of raw product (P_r) and compacting pressure (P) is described by the formula:

$$P_r = a - b\lg P, \quad (3.6)$$

where a and b - empirical constants.

Values a and b depend on composition and rheological properties of the masses. For example, for the refractory masses it is suggested to assign a=50, b=15. Pressure at which the volume of mass is equal to the sum of volumes of solid and liquid particles is called critical. If pressure exceeds critical, the overpress comes, elastic counteraction develops, which results in destruction of appeared contacts and cracks formation.

The optimal compacting pressure the most frequently is 5...15 MPa, it depends on the humidity and grain distribution of the mix. The internal friction of particles diminishes with the increase of humidity; however it interferes to air release and formation of dense structure. Usually, the humidity of powder mass for the dry pressing is 6...12%. The best conditions of pressing are provided at a continuous size of distribution, and maximal sizes of grains up to 3 mm. Shaping by pressing can be executed with *stamping* with transfer the pressing loading through a stamp, which recovers all of element's area and forms facial and back surfaces; *pressing by rollers*, moved on the placed mix in the form (Figure 3.8); *vibrocompaction*, combining the static pressure and vibration itself.

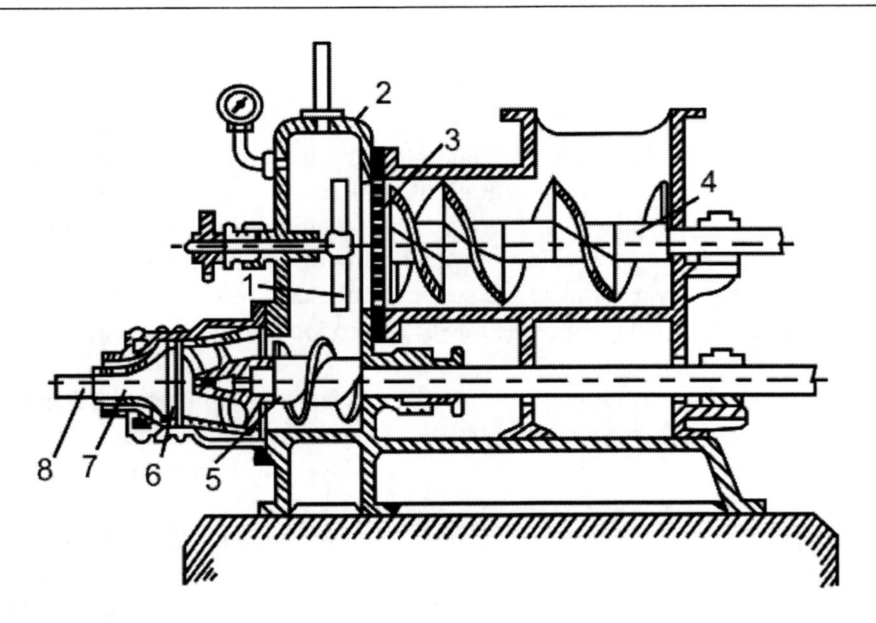

Figure 3.9. Band vacuum-press:
1 - knife; 2 - chamber; 3 - grate: 4 – clay mill; 5 - screw roller; 6 – squeezing knob; 7 - jet; 8 - clay squared beam.

Ramming is a type of pressing, which consists in repeated application of momentary compacting pressure to the particles of the material under compaction.

Elements from the plastic (for example, clay) masses are made by the plastic shaping on vacuum augers (Figure 3.9).

The plastic shaping allows giving complicated shape to elements at comparatively low compacting pressure. The required condition of the plastic shaping of elements is the use of comparatively viscous masses for which sum of forces of the internal adherence (cohesion) is higher, than the cohesive forces with the working surface of molding element (adhesion), and coefficient of internal friction is higher, than external one. Adobe elements, formed by such method, have the lower strength, than pressed, such defects of structure, as knottiness and S-liken cracks are characteristic for them. The plastic shaping is very often used for production ceramics and other elements. In the vacuum augers with the help of an auger mechanism, the mass is transported, compacted and punched shear through die carrier, intended for giving the elements the exact sizes, form and additional compression of the squared beam.

Extrusion is similar to plastic shaping (see Figure 3.7, b); this is a production of materials and elements from polymeric and some other raw meals by squeezing through the die carrier. Various plastic profiles, pipes, sheets and films are shaped by this method. The screw extrusion machines are mainly applied for squeezing out the plastic masses.

Shaping by *casting* consists in filling of the appropriate volume with the raw meal and its compression under the gravity. This method is applied for raw meals, which have sufficient fluidity, depending on properties of solid phase, water-solid ratio, presence of additions of electrolytes and surface-active substances. Along with required flowability (workability), the mixes for casting should have sufficient structural strength and viscosity for counteraction to segregation caused by sedimentation (settling) of particles of solid phase. Casting is applied for production of gypsum elements, and also for the production of cellular concretes, ceramic sanitary - construction elements from ceramic slurries.

The method of slurry production is based on ability of clays to give stable water suspensions, and also on absorption of liquid phase by the capillaries of gypsum form with formation of solid layer on its surface. Rate of wall set of element depends on the rate of slurry absorption of liquid phase by the form, grain distribution of solid phase and ratio of solid and liquid phases. At the considerable advancing of water absorption rate by form, the excessively compacted layer forms at the surface of the cast resulting in the exfoliation and cracks appearance. At the delaying of the rate of absorption of water by the form, its diffusion and element adherence can occur.

At recent time the application area of method of element shaping by casting broadens. It is applied, for example, for production concrete and concrete products from the cast concrete mixes, diluted by superplasticizers, which are introduced in concrete mix on the stage of mixing.

The method of *vibration* belongs to the most widespread methods of shaping. More than 90% of elements from concrete and concrete products are made with the application of vibration. It is explained by the conditions, which are created for thixotropic dilution and more compact placing of particles in the process of the oscillation influence in the concrete mixes. The vibration is an oscillating process, which is characterized by the values of amplitude, frequency and intensity of vibrations. Sinusoidal and periodic nonsinusoidal (poly-frequency) oscillations are effective during the shaping of concrete mixes. The different types of vibrations can also be combined.

Intensity of vibration is comparatively reliable index of vibrating influence (I_{nt}) which can be determined for sinusoidal oscillations by a formula, cm^2/sec:

$$I_{nt} = A^2 f^3, \tag{3.7}$$

where A - amplitude of oscillations; f - frequency.

Index of vibration intensity for majority of mixes, which are used for reinforced concrete structures shaping, is 80...300 cm^2/sec. On Figure 3.10 the hatched area corresponds to the most widespread in practice amplitudes of vibrations.

Amplitude of vibrations depends on the sizes of particles of concrete mix aggregates. For coarse-grained heavy-weight concretes it is equal 0.3...0.7 mm, it is increased with the increasing of rigidity of mixes, it is expedient to increase frequency of vibration. Too large amplitudes of vibrations without the use of loading can cause the aeration of concrete mixture and worsen concrete properties.

With diminishing of aggregate coarseness frequency of vibrations increases accordingly. For example, at the coarseness of the aggregates 40 mm optimum frequency is 38 Hz; 20 mm - 50 Hz; 10 mm - 100 Hz (1 Hz=60 oscillations per minute).

The use of the modes with different frequency of vibration gives the possibility to improve the compaction of different aggregate fractions.

The duration of vibration, which is selected by the experimental way is optimal for every concrete mixture at the accepted parameters of vibrations. It is possible to shorten the time of vibration due to the increase vibration intensity (to a certain limit) and creation of pressure on the surface of the compacted mixture with the cantledge.

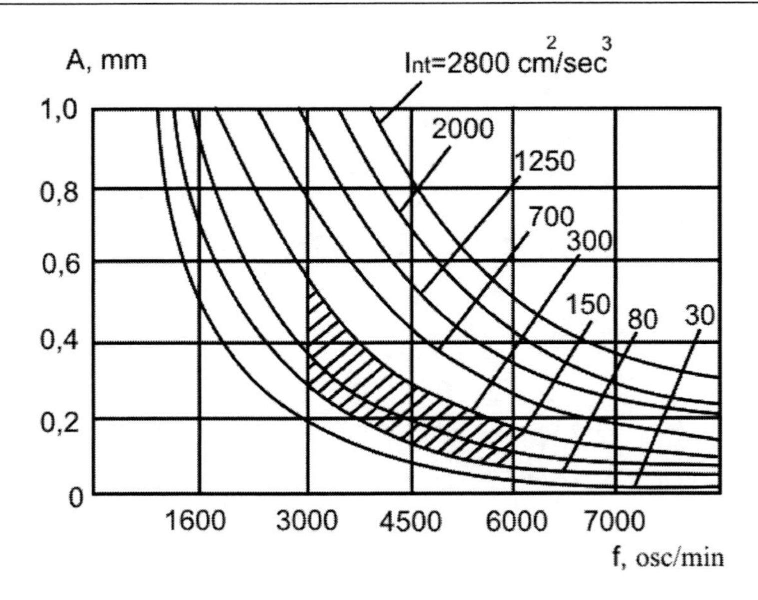

Figure 3.10. Relationship between amplitudes (A) and frequencies (f) of vibrations at different vibration intensity (I_{nt}).

Vibratory compacting of mix is provided by different machines among which there are internal vibrators, surface vibrator, volume vibrators and vibroplatforms. They have inertia vibroexciters of unbalance or self-balancing type at all the variety of arrangement of vibrating machines. Vibroplatforms are characteristic equipment for the enterprises of concrete and reinforced concrete products (Figure 3.11). More than 80% of concrete and reinforced concrete products are compacted on them. Many of industrial vibratory compactors have frequency of vibration of 3000 oscillations per minute (amplitude 0.25...0.70 mm). It is expedient to apply the mode of low frequency (10...15 Hz) and large amplitude of oscillations (3...10 mm) for the large elements of considerable thickness.

Vibratory machines are divided into three groups: shockless, shock-and-vibratory and resonance. The majority of applied vibration machines and devices: vibroplatforms, vibrators, vibroinsertions, vibroshields, vibrostamps and others belong to the *shockless*. In the *shock-and-vibration* machines, the oscillations of actuator are accompanied with impacts at other elements of machine or at medium, which is processed. The example of such machine is a shock vibroplatform, the platform of which together with a form continuously rises and falls, strucking at metallic beams. In *resonance oscillation vibration machines* the frequency of the forced vibrations is near to the natural frequency of the oscillating system.

Vibration is often applied together with other methods of compactions (vibrostamping, vibration pressing, vibrovacuumizing, vibrorolling).

Shaping by *centrifugation* is applied at production structures of pipe section (pipes, chimneys, piers, etc.). Essence of process consists in that the mixture, which is placed in the revolved form, under the action of centrifugal forces, is thrown aside to the walls of form, distributed on uniform layer and compacts.

Centrifuges differ one from other by the construction of drive of form (roller, axial, belt and other).

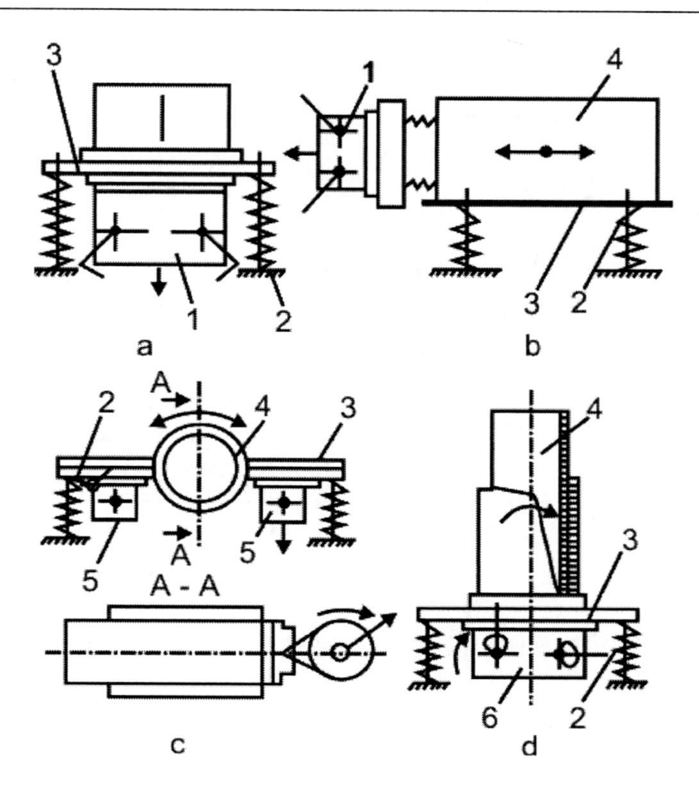

Figure 3.11. Charts of vibroplatforms: a - with rectilineal vertical oscillations; b - with rectilineal horizontal oscillations; c - with angular oscillations; d - with assembled vibrations 1- vibrator of the directed vibrations; 2 - spring; 3 - frame of vibroplatform; 4 - form with element; 5 - vibrator of angular vibrations; 6 - vibrator of the combined vibrations.

3.3. Thermal Treatment

The shaped elements usually go on thermal treatment as a result of which they acquire necessary physical and mechanical properties.

The basic varieties of thermal treatment are drying, burning (firing) and steam curing.

The process of evaporation of moisture from material with the subsequent removal of the steam formed in the environment is called *drying*. Raw meals, semi-finished and finished products are subjected to drying in the production of construction materials and elements. For example, in ceramic industry, drying is conducted to produce the clayey and other components, and also all-ready shaped elements with humidity, required for the normal behavior of burning process.

Drying is possible when the pressure of water steam on the surface of the dried out material is higher than in the environment. Pressure of water steam in material increases with the increasing of humidity and drying temperature and diminishes with strengthening of moisture interaction with a material.

There are three forms of moisture interaction with material: physical-mechanical, physical-chemical and chemical. The physical-mechanical water is contained in capillaries due to capillary pressure and surface tension. Water steam in microcapillaries is taken in due to the effect of capillary condensation, as if pressure over the concave meniscus is less than, than partial pressure in the environment. Water in macropores is absorbed at the direct contact with material and is free.

Physical-chemical water in materials can be adsorptive, osmotic and structural. Adsorptive water forms connections at the excessive energy level of surface of solid phase that is representative for the materials with the developed interphase. At the osmotic form of connection, the moisture is hold out due to an osmotic pressure which arises up mainly in organic materials cellular by origin. The closed cell is an osmotic cell, and diffusion of liquid is caused by the difference of osmotic pressures of soluted fraction inside and outside the cell. Structural moisture is contained in the closed pores. It is characteristic, for example, for the fine bentonite clays.

Chemical water reacts with material in exact stochiometric ratios and forms new substances. The examples of such formations are minerals: kaolinite $Al_2O_3{\cdot}2SiO_2{\cdot}2H_2O$, montmorillonite $Al_2O_3{\cdot}SiO_2{\cdot}5H_2O$, gypsum stone $CaSO_4{\cdot}2H_2O$, etc.

Free physical-mechanical water can be removed by mechanical method (by pressure, vacuumizing, filtration and centrifugation).

It is possible to delete physical-mechanical and physical-chemical moisture due to drying. Chemical water is removed only by high temperature treatment or due to the action of chemical reagents.

It is possible to dry materials in normal conditions that means outdoors without the additional heating and at heating in the special dryers. Heat to the material can be lead by: convectively - washed heat-transfer agent (hot air or flue gases); due to thermal conductance; by emission - irradiation of material with infrared rays; at the action of currents of high frequency. Sublimation drying by moisture evaporation without to the liquid state transmission at low temperatures and deep vacuum is possible in some cases.

The process of drying of solid materials can be presented by curves which have general character and conventionally divided on three areas: 1 - the stage of heating; 2 - period of constant-rate of drying (first period); 3 - period of falling rate of drying (second period) (Figure 3.12).

Materials, different according to the character of the moisture connection, are characterized by different rate performance curve of drying. The fibrous materials (paper, thin cardboard) give the curves with linear areas which answer to the period of rate falling (Figure 3.12, 3a); curves, turned with the convexities to the ordinate axis are characteristic for fabrics, thin leathers (Figure 3.12, 3b); for porous ceramic materials - by the convexities to abscissa axis (Figure 3.12, 3c). More difficult according to structure moistened materials are characterized in the second period with more complicated drying curves.

The temperature curves have high significance for technology of drying, as the quality of the dried up material depends often on a temperature and duration of its action. Temperature of material in the process of drying is considerably lower than the temperature of ambient air; therefore the heat-transfer agent can be used with higher temperature and low humidity in the first period.

Drying of construction materials and elements is conducted mainly due to a convective heat transfer between the heat-transfer agent and material. Thus, it is possible to define stream of heat from heat-transfer agent to material by a formula:

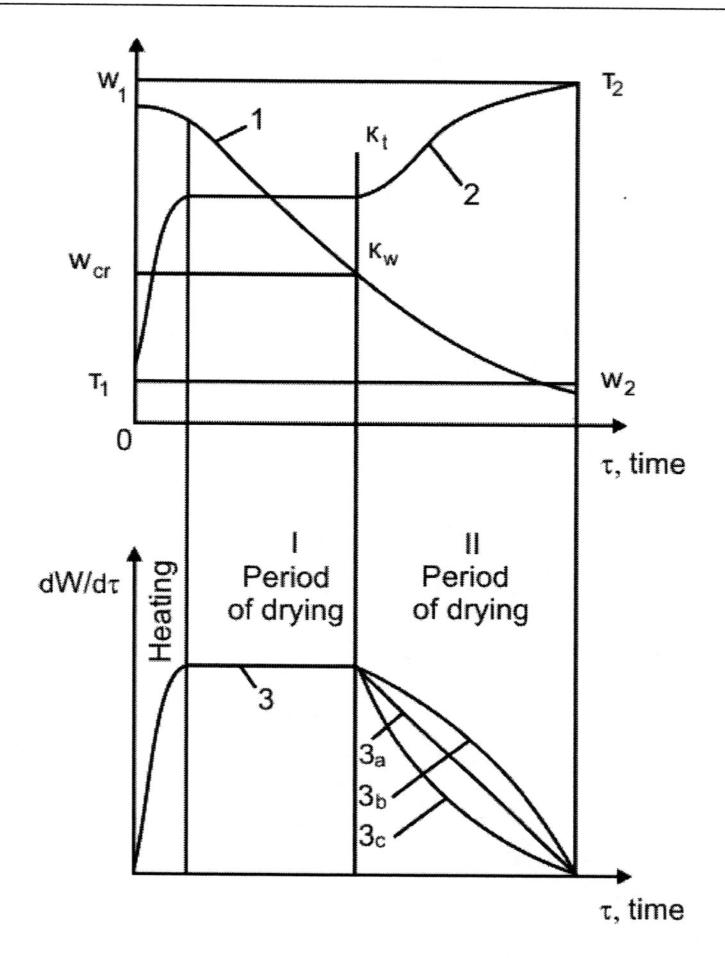

Figure 3.12. Typical drying curves: 1 - humidity; 2 – temperature; 3 -rate of material drying.

$$Q = \alpha(t_h - t_{m.s.}), \tag{3.8}$$

where α – a heat-transfer coefficient; t_h, $t_{m.s.}$ – medium temperature of the heat-transfer agent and material surface accordingly.

The heat-transfer coefficient depends on the aerodynamic conditions of flowing around of material by heat-transfer agent, its properties, surface of a material, from which the moisture is evaporating, and a series of other factors. The conditions of heat emission are studied with the use of law of similarity, applying the criterion dependences.

The drying units according to the operating regime are divided on periodic and continuously working, according to the method of transmission of heat on convective, contact, radioactive, with the use of currents of high-frequency, etc., to the traffic diagram of heat-transfer agent on counterflow, direct-flow, with recirculation, etc., by a construction on pulverizing for suspensions, with drying in the weighted and boiling layed, roll, screw, tunnel and conveyer.

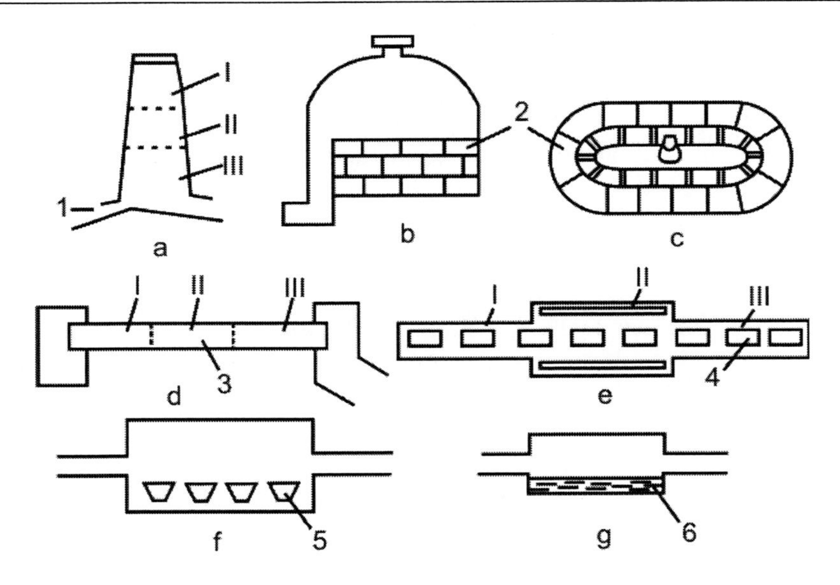

Figure 3.13. Furnaces with the different arrangement of working chamber: a- mine; b, c - chamber; d - rotating; e - tunnel; f - crucible; g - tank; I…III - accordingly areas of heating, burning, cooling; 1 - unloading; 2 - material; 3 - drum; 4 - trolley; 5 - crucible; 6 – bath.

Burning in technology of construction materials is the name of the high temperature treatment of raw materials and semifinished products, because of which there are irreversible physical-chemical processes in them, changing phase composition, structure and properties of materials. Burning is a basic technological operation of mineral binders and ceramic materials production. Depending on the type of materials, such processes as dehydration (gypsum, clay minerals), dissociation (carbonates), synthesis of silicates, aluminates, ferrites (cement clinker) prevail during the burning out.

Rate of interaction of components of raw meals at burning changes depending on the structure of crystals, (state of grate, type of admixtures and defects, type of connection, etc.), reduction size of grains, etc. The high-temperature burning of materials is finished by sintering and production of stony structure of a substance.

Table 3.3. Classification of industrial furnaces for construction materials burning

Classification attribute	Type of furnace
Technological assignment	For burning of lump and bulk materials. For burning of shaped elements Sintering (consolidation) of bulk material Silicate melts production
Operating regime	Periodically (cyclic) operating Continuously operating
Constructive diagram	Shaft, rotary, chamber, circular, tunnel, bath
Method of heat transfer	Direct fire, muffle
Origin of the heat generation	Flaming, electrical

The industrial furnaces are applied for materials and elements burning (Figure 3.13), which can be classified according to the technological assignments, operating regime, structural chart, method of heat transfer and sources of heat evolution (Table 3.3).

Along with physical-chemical transformations in the burned material there are complicated processes of burning and gasification of fuel, movements of combustion products in working space of furnace, heat- and mass transfer processes, related to the endothermic and exothermic effects. Achievement of required properties is determined by the optimality of temperature condition and chemical composition of furnace atmosphere (gas mode). The firing temperature is from 180°C for gypsum to 1450°C for Portland cement clinker. Chemistry of the furnace atmosphere is determined by surplus of oxygen in the kiln gases: to 1% in a reducing environment; 1.5-2% - in neutral one; over 2% - in oxidizing one.

The raw meal can partly melt during the sintering. At the production of a series of silicate materials and metals, the complete melting of mixes is required. Clean crystalline substances have strictly certain melting temperature, which corresponds to the thermodynamics equilibrium, when free energies of solid and liquid phases are equal. The melting temperature of minerals changes from 0°C (ice) to 3800°C (graphite). Raw meals contain the components with the different melting temperatures. They melts non-simultaneously, melts which correspond to the eutectic temperatures of reactive substances are produced at first. The most low-melt compounds are fluxes, which reduce the melting temperature of refractory components of mix according to its equivalent masses.

The amount of heat, required for melting of 1 kg of substance, preliminary heated to the melting temperature, is called the hidden melting heat. For the thermal designs of melting devices at the production of silicate materials the hidden melting heat is accepted approximately 335 kJ/kg of the melt.

Melting at production conditions is carried out in reducing, oxidizing and neutral environments, and also in vacuum.

Cast iron and also the most of ferroalloys are produced by melting in the reducing environment; steel in open-hearth and electroarc furnaces - in oxidizing ones, the special alloys (for example, in the environment of argon), glass - in neutral ones. Extra clean materials are produced by melting in a vacuum.

Melts are produced in *melting furnaces*. For the production of metals blast-furnace, open – hearth steel furnace, electric furnaces are applied; for mineral wool - cupola, tank and electroarc furnaces; for fritted glazes - the revolved furnaces of periodic working and tank furnaces; for glass - pot chambers and tanks.

Thermal treatment at which moisture is saved in the heated material is called *steam curing treatment*. Such treatment is a basic method of acceleration of concrete and reinforced concrete products hardening, silicate and other elements on the basis of mineral binders of hydration hardening. As heat-transfer agents at thermomoist treatment there are applied steam, electric power and combustion materials of natural gas, hot air, organic and inorganic oils.

The most widespread heat-transfer agent is steam. Combination of heat and moisture in it, required for hardening of hydraulic binders, has made the *steaming-out a* widespread method of heat-moisture treatment of concrete and reinforced concrete products.

The heat-transfer substance, contacting with the surface of material, exchanges a heat and mass with it. This process is called external heat and mass transfer. Heat and mass transfer between the surfaces and internal layers of material is called internal.

The mode of steaming-out is characterized with three periods: by the increasing of temperature, isothermal maturing and cooling of elements.

The increase of temperature in material causes temperature and moisture gradients. Temperature stresses from expansion of mix components and external layers of element along with moisture strains caused by swelling result in development of the complicated stress state in the material structure. In particular case expansion of air bubbles, which are contained in raw meals influences negatively. To remove the destructive phenomena, elements before a steaming-out should have required critical strength (0.6...0.8 MPa) at which they can withstand the appearing stresses.

The second period of steam curing is isothermal one, which characterizes by the intensive material setting of the strength, when the "thermal equilibrium" appears: the environment - element.

Cooling of elements is characterized by the moisture loss because of temperature increase and increase of steam pressure in material in relation to external environment. If elements are cooled extremely quickly, the cracks can appear in them because of development of stretching stresses, exceeding the strength of material.

For the steam curing of material heat apparatus of discontinuous working are applied (pit steam-curing chambers - Figure 3.14, mold batteries, hubcaps, and thermo-models) and continuous action (vertical, tunnel and gap chambers). The thermal-moisture treatment of materials at a steam curing takes a place in the conditions of the air-steam mixture saturated or near to it at a temperature to 100°C and relative humidity which is approximately 100%.

The methods of the steam curing treatment which enable to remove the part of water from concretes are effective for elements (for example, from the light-weight concretes) with raised initial water content. To that end thermal apparatus for *air-dry heating are* applied, the energy source in which is electricity, steam or gas.

It is possible to accelerate the hardening of concretes by electric curing with three methods: electric curing of elements in forms; preliminary electric curing of concrete mix (hot shaping) with the subsequent thermos maturing; heating of elements by electric heating elements. At the *electrical heating,* hardening of concrete is accelerated due to the heat released during the passing of electric current. The electro-heating of material at *the "hot shaping"* of elements is founded just on it. These methods of thermal treatment do not require the special chambers. During an *electric curing* the heat, required for the acceleration of concrete hardening, transfers from the electric heaters - light bulbs, spirals, pipe heaters, etc. embedded in a chamber, which are the sources of infra-red rays. Thus elements are exposed to the radiative-convective heating in dry air mixture. It is possible to replace electric heating units by gas infra-red radiants which work on natural or condensed gas. Thermal treatment with the combustion materials of natural gas is conducted in chambers, where air-gas mixture of the temperature arrives at.

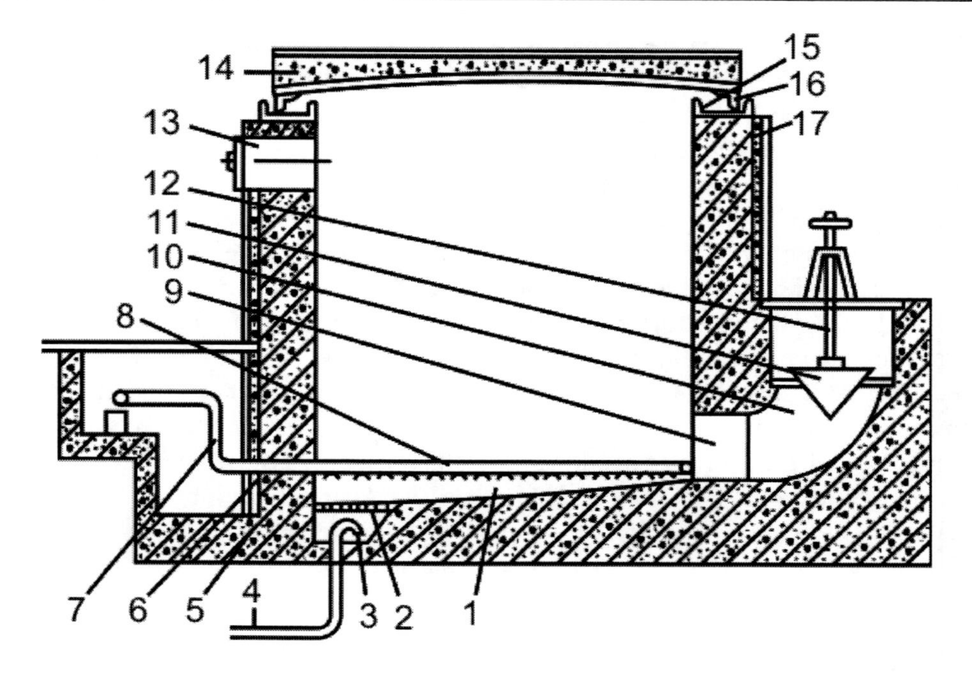

Figure 3.14. Steam-curing chamber of the pit type:
1- chamber slab; 2-trap for the condensation drainage; 3- hot-wells; 4- system of the hot-well ; 5- chamber walls: 6- orifice for the steam input; 7- pipe duct; 8-perforated pipe: 9- orifice for ventilation of chamber at cooling; 10- channel for the bleeding of steam-and-air mixture; 11 – pressurizing cone; 12-worm screw; 13- breech-block for the reception of air: 14- lid; 15- channel bar; 16- angle; 17- thermal protection.

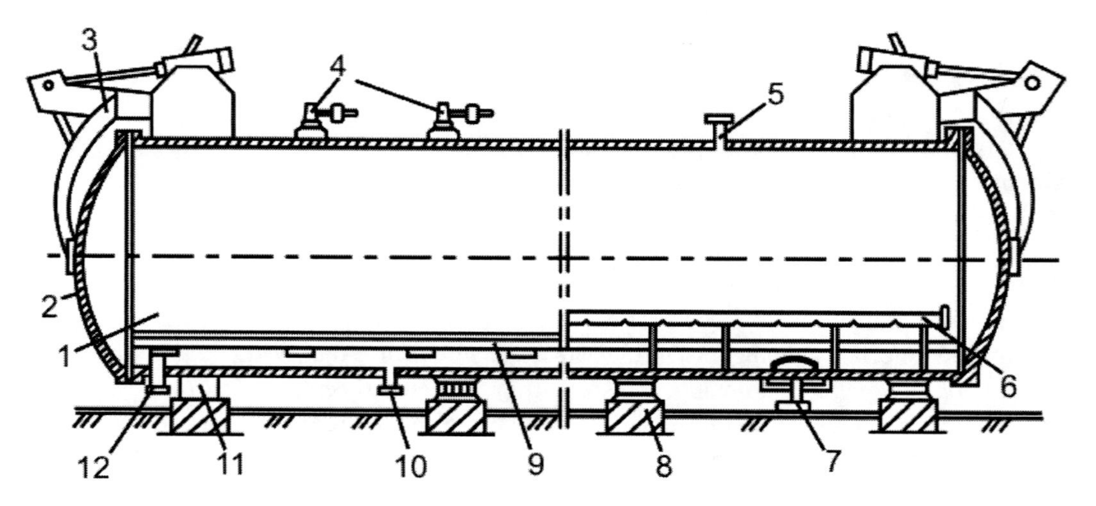

Figure 3.15. Scheme of the autoclave apparatus:
1 - envelop; 2 - lid; 3 - mechanism for closing and opening of lid; 4 – safety valve; 5 - the manifold for steam by-way; 6 - perforated pipe; 7 - the manifold for the condensate deriving; 8 - movable bearings ; 9 - rail track; 10 - vacuum-system; 11 - fixed bearing; 12 -the manifold for introduction of steam (13 - heat-insulation EXCISE!)

The currents of high-purity are intended for heating of materials practically without the gradients of temperatures and humidity (HDTV). Unlike the other this method of thermal treatment gives a possibility to transfer the large amount of the heat into material and to provide the uniform heating of materials in all of volume. For heating of HDTV material is placed in facings of condenser to which HDTV is tricked from a generator.

The methods of steam curing treatment considered above are carried out at atmospheric pressure. At considerable extra pressure of aquatic steam (0.9...1.3 MPa) thermal-moisture treatment is conducted in autoclaves (Figure 3.15).

Autoclaving for the elements based of lime-silica and other binders, which in ordinary conditions harden very slowly, are used. It is possible to use widely the various industrial wastes (slags, slurries, ashes, etc.) to produce the materials of the autoclave hardening.

Accelerated elements production, based on the different binders, foresees augmentation of its hardening at the simultaneous production of materials of required structure. An augmentation of hardening processes requires the accounting of all of the technological stages of material production. It is impossible, for example, to examine the acceleration of concrete hardening outside of complex of all influences on the concrete mix and shaped concrete.

The final result of the different technological influences on the material depends on a moment of its application, intensity and duration. Specifically, conformably to concretes, it is determined by the accordance of moment of imposition of influences and kinetics of structure forming of concrete mixture. The application of mechanical effects (reiterated vibration, activation treatment, etc.) during the optimum time enables to achieve, for example, increasing the concrete strength in 1.5-2 times, to improve substantially a series of properties.

Self-Assessment Questions

1. Describe methods of materials grinding.
2. What are the types of materials classification?
3. Tell about method of cleaning from dust.
4. Describe the basic methods of mixing and schemes of mixers.
5. What methods of shaping are used?
6. Tell about vibration method and parameters of vibrations.
7. What are peculiarities of drying process?
8. Tell about industrial furnaces for construction materials.
9. What are the peculiarities of steam curing and types of heat apparatus?

Chapter 4

CLASSIFICATION AND STANDARDIZATION OF MATERIALS

4.1. CLASSIFICATION OF MATERIALS APPLIED IN CONSTRUCTION

There is used a number of materials in the construction and their amount grows constantly. Uniform classification of construction materials is not developed; there are a lot of classification features and complex structural links between the separate groups of materials.

The classification features of construction materials are divided into physical, chemical, physical and chemical, mechanical, structural, technological and functional.

Except for natural stone and wood, construction materials are artificial products obtained by chemical-technological processes. The most important classification of construction materials is by assignment.

Structural materials which carry loads are selected taking into consideration the features of structural products and technical and economic assessment. Rolled steel and aluminium, concrete and reinforced concrete, brick, glued wood are widely applied to that purpose.

Materials for production of enclosing structures are self-supporting and do not experience influence of the large load. External non-load-bearing structures can perceive the snow and wind loads, and also should be proof to aggressive action of environment. Materials which occupy structural space between bearings have to be light and have a low heat-conducting.

Finishing materials give the surfaces of constructions of buildings and structures protective and decorative properties. They are divided into finishing and fitting materials. The first group is used for arranging of protective-decorative coverages on the structures surfaces (plaster, laques, paints, wallpapers), second one combines both decorative and structural functions at arranging of finishing (glass blocks, structural glass, particle boards and fibreboard, asbestocement and aluminium products).

The functional application *of heat-insulation materials* consists in diminishing of heat losses during buildings and structures performance, and also thermal aggregates and pipelines. Heat-insulation-structural materials are used for self-supporting structures of buildings and low-rise building structures (porous concretes, cement wood, fiberboard, etc.).

Acoustic materials are similar to heat-insulation materials, assigned for the decline of energy of voice vibrations (level of noises). They are divided into sound-absorbing and sound-insulating materials

For protecting of building structures and buildings from the harmful action of water and water solutions of aggressive compounds *water proofing materials* are applied. Such materials are divided by the purpose into antifiltrational, anticorrosive and sealing.

Upper water proofing layer of roof structure is roof coat. Some materials (roll mastics) can be used for roofing and for water proofing and others (asbestos-cement sheets, tile and roofing steel) - only for roofing.

Sanitary wares - baths, sinks, wash-stands, devices for kitchens heating, equipment of utility cores made of ceramics, polymers and metals are also construction materials.

Construction materials for special purposes - road, heat-resistant, acid-resistant, electrical engineering, bioprotective, pipeline and others can be referred to a separate group.

Construction materials are mostly composite materials (composites). *Composites* are natural or artificial heterogeneous materials, their common feature is the existence of boundary surface between components (phases), which they are formed from. There is differenced first phase, or matrix - a continuous binding component, that is in the solid crystalline or amorphous state, and second phase - one or several substances, dispersed in matrix, which can be in any aggregate state in a composite construction material (CCM).

CCM are divided by purpose into power, unpower and special. Power CCM (glass-fiber plastic, asbestos cement, concretes, etc.) must have high mechanical characteristics - strength, heat-resistance, durability. Unpower CCM perceive the insignificant mechanical loads. Heat-insulation materials made of different fibrous materials (fibrolite, mineral wool board, etc.), cellular concrete, foam glass, foam plastic are among them.

Special-purpose CBM can work under conditions of high temperatures (heat-resistant, fireproof), chemical aggression (alkali-resistant and acid-resistant), voltage (electrical insulating, electrical conducting). Soundproof and heat-insulating, decorative, unshrinkable, expansional and other CCM also belong to them.

CCM by matrix material are divided into cement, gypsum, ceramic, metallic, etc.

CCM fillers are various enough. Different fibrous and sheet CCM fillers produced of metals, glass, tree, asbestos, basalt and others are perspective by mechanical properties. Fiber reinforcement of CCM can be both oriented (reinforced concrete, glass cement, glass-fibre plastic) and dispersible (fibrous concrete). CCM with grainy fillers (concretes, mortars, mastics) are particularly widespread. At small content of filler (aggregate) CCM properties are determined mainly by matrix properties, and with growth of filler content their properties may substantially change, acquiring specific features, inherent only this type of CCM. For example, for cement concretes increase of filler content to the certain extent leads to strength increase at 20...30%, asphalt concrete - at 50...80%.

Artificial construction conglomerates are types of composition materials (ACC) in which aggregates are cemented by binders and initial bonds (chemical, electric, metallic, etc.) into a monolith. A series of rocks belongs to the natural conglomerates, to artificial ones - firstly various concretes and mortars. There is suggested to select two types of materials in classification of ABC - unfired, that appear as a result of low temperature physical and chemical processes of binders hardening and fired materials that form at cooling fusions or contact caking.

The raw material components of construction materials are solid, liquid and gaseous substances. Greater part of solid raw materials is presented by rocks, wood and industrial wastes (slags, ashes, screenings, etc.). Liquid substances are oil products, liquid wastes of chemical industry, water and water solutions. Recycling oil and coal, gaseous products are produced, which can be utilized for the production of polymers.

Among inorganic raw materials silicates are the most widespread, which take considerable place (66,5%) in the earth's crust. Clay, carbonate rocks, sands are multi-purpose mineral raw materials with the large consumption volumes. The products of silicate technology - cements, ceramic materials, glass, autoclaved concretes and other materials are widely used in construction.

By-products and production wastes, which are used as raw materials for construction materials production, should be classified depending on industry sphere, where they appear:

1. By-products of metallurgy: blast-furnace, ferro-alloy and steel-smelting slags, slags which formed at melting of colored metals ores, products of ore-dressing, nepheline and other slurries.
2. By-products of thermal power engineering and fuel industry: ashes, fuel slags, blast rocks, waste coal, etc.
3. By-products of chemical industry: ferrous and gypsiferous wastes, salts- and hydroxyl-containing slurries and potassium products, phosphoric slags, polymeric afterproducts etc.
4. By-products of mining industry.
5. By-products of processing of wood and other phytogenous raw material: bark, clips, lignin, sawdusts, cuttings, flax and hemps chaff, etc.
6. By-products of construction materials production: cement, asbestos cement, ceramic, polymeric and glass productions, non-metallic industry wastes.
7. Wastes of municipal services: waste paper, polymeric materials, tires, solid tailings of sewage, waste glass, etc.

4.2. General Information about Standardization of Construction Materials

The high volumes of the consumption of construction materials in the world dictate the need in creation and continuous improvement of standards to the construction materials. Standards to the materials in one or other form exist in all industrial and developing countries. The globalization of the world economy leads to the need in creating the internationally acknowledged documents, which contribute to the free displacement of products; to the propagation of the new technologies and methods of design. Several systems of the standardization of construction materials in the world are formulated. Largest of them became the system of the standards of International Organization for Standardization (ISO); and also regional systems - systems of European Committee for Standardization (CEN) and American Society for Testing and Materials (ASTM International).

Standardization is the process of developing and agreeing upon technical standards. Standardization helps to provide required quality of construction materials. Its aims to increase the requirements of the quality parameters.

Standard is a basic normative technical document, established norm or requirement. It is usually a formal document that establishes uniform engineering or technical criteria, methods, processes and practices.

There are different types of standards depending on their content:

- *Standard of specification* is an explicit set of requirements for construction, materials, products or structures. It is often used to formalize the technical aspects of procurement agreement or contract. For example, "EN 197 -1 Cement. Composition, specifications and conformity criteria for common cements".
- *Standard of testing method* describes a definitive procedure of materials, products or structures testing which provides a test result. It may involve making a careful personal observation or conducting a highly technical measurement. For example, a physical property of a material is often affected by the precise method of testing: any reference to the property should therefore reference the test method used; or in the case of product e.g. "EN 1344 Clay pavers. Requirements and test methods".
- *Standard of practice* or procedure gives a set of instructions for performing operations or functions. For example, there are detailed standard operating procedures for operation of a nuclear power plant.
- *Standard guide* is general information or options which do not require a specific course of action.
- *Standard of definition* is formally established terminology in specified area e.g. "EN 12670 Natural stone. Terminology".
- By geographic levels standards are divided into:
- *International standards* developed by international organizations; for example, ISO (International Organization for Standardization) and ASTM International (American Society for Testing and Materials).
- *Regional standards* developed and valid at the European Union countries by CEN (European Committee for Standardization) standards;
- *National standards* developed by National standards organizations and valid at the territory of specified country, but they also can be approved in other countries of certain region. Thus, BSI (British Standards Institute) in Great Britain, DIN (Deutsches Institute für Normung) in Germany, JISC (Japanese Industrial Standards Committee) in Japan, etc.

International standards is a one way for overcoming technical barriers in inter-local or inter-regional commerce caused by differences among technical regulations and standards developed independently and separately by each local, local standards organization, or local company. Technical barriers arise when different groups come together, each with a large user base, doing some well established thing that between them is mutually incompatible. Establishing international standards is one way of preventing or overcoming this problem. For example at present CEN is working on the harmonization of all the EU-standards according to the Directive 89/106/EEC for construction products.

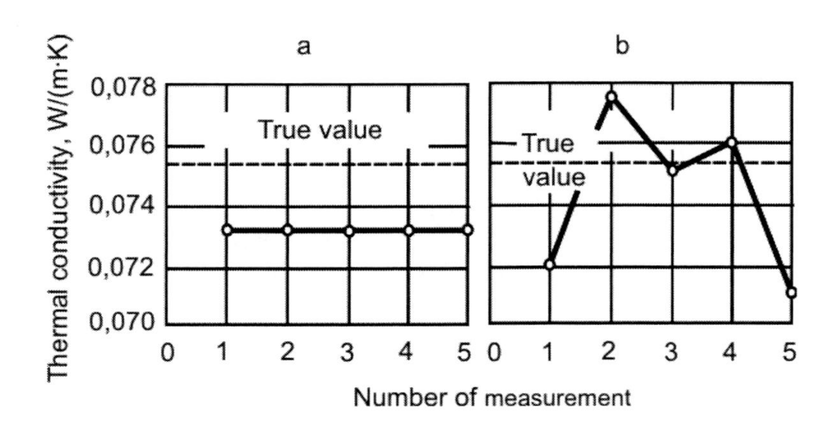

Figure 4.1. Types of measurement error:
a – systematic; b – random.

At making of construction products application of the *module sizes co-ordination system* enables to unify the quantity of dimension types, provide interchangeability of products made of different materials, to use products with equal dimension types in buildings of the different assignment.

The basic module in Europe (M) is 10 cm. The derivative modules that exceed by value basic one are called large-sized. They are recommended to apply such large-sized modules, mm: 20 (2M), 30 (3M), 60 (6M), 120 (12M), 300 (30M), 600 (60M).

Space-and-planning parameters of buildings (width of flights, step of columns, frames, floors height) and nominal sizes of large construction products (wall panels and blocks, panels and flags of ceilings, etc.) are assigned the multiple large-sized modules.

The derivative modules which are the parts of basic are called fractional. There is recommend applying such fractional modules, cm: 0.1 (1/100 M), 0.2 (1/50M), 0.5 (1/20M), 1 (M), 2 (1/5M). 5 (1/2M). Basic and fractional modules are used for assigning comparatively small sizes of structural products and details (cross area of columns, beams and partitions, tiles and sheet products thickness).

The nominal sizes of construction products, which include additionally, unlike structural sizes, guy-sutures and gaps are appointed by modular system. Deviation of actual-size of products (model size) from structural and project must not exceed possible. At making of precast structures admittances depend on the sizes of product and grade of accuracy.

Standardization of materials and products is connected with the necessity of the exact and reliable measurings, application of the unique facilities and methods of measurings and tests.

4.3. Estimation of the Materials Quality

The quality of materials characterizes the degree of their correspondence to the demands of user. The diverse methods of determining the qualitative indexes are used: *instrument* - measurement of properties by instruments; *organoleptic* - estimation of properties by the analysis of the sensations of man, by the comparison of the materials with the standard being investigated; *expert*, based on the estimations of the experienced specialists; *sociological*, based on the analysis of the estimations of the production users; *calculated* - calculation of the qualitative indexes depending on the parameters of its composition and structure taking into account the special features of technological process.

The instrument methods of evaluating of the materials quality, which have the prevailing scientific base, are applied most widely. Instrument methods are based on *metrology* - the science about the measurements, methods and means of the guarantee of their unity and methods of the achievement of the required accuracy. The results of all measurements contain the errors, caused by the imperfection of instruments and methods, by the inconstancy of observation conditions, by the insufficient experience of observer or by the special features of his sensory organs. Systematic errors in the process of sequential measurements remain constant (Figure 4.1) or change according to the specific law. For example, at the determination of the cement strength a systematic error introduces the use of forms with the rough surface. This error can be removed by additional polishing of forms. Systematic errors in the case of the impossibility of their elimination can be studied and taken into account as the corrections. In contrast to the systematic the random errors at the repeated measurements of one and the same value take different values. They cannot be excluded from the results of measurements; they study and consider them, processing the results of repeated experiences with the use of the mathematical statistics and probability theory.

The errors, which considerably exceed those objectively permitted (oversights), appear for a number of reasons and they are evaluated with the aid of the statistical criteria.

The values of the measured indices depending on the method of obtaining divide into four types: direct, indirect, joint and gross.

Indirect measurements are obtained on the basis of the direct, connected with the measured value by the known dependences (density - mass of unit volume, the ultimate strength - ratio of breaking stress to the cross-sectional area of specimen, etc.).

Joint measurements are the simultaneously produced measurements of two or several dissimilar factors for finding the relationship between them. In this case the values of the measured factors find according to the data of repeated direct or indirect measurements. Thus, for determining the modulus of concrete elasticity (E_{ls}) are measured stresses (σ) in the concrete with the different values of relative deformation (ε). With the stress equal to 0.2 of ultimate strength (R) the initial modulus of elasticity they, for example, calculate according to the formula:

$$E_{ls} = 0.2R / \varepsilon . \tag{4.1}$$

At the determination of temperature expansion it is also required to carry out the joint measurements of temperature (t) and to calculate the appropriate values of the relative elongation of (ε_t). For calculating ε_t it is required by the system of equations to calculate coefficients in the formula:

$$\varepsilon_t = at + bt^2 + ... \tag{4.2}$$

In contrast to the joint *gross* measurements are conducted for several similar values.

The methods and the means of the measurements of all basic properties, which characterize the quality of products, are standardized for the industrial production. Standardized methods and means of the measurements of the state parameters and characteristic of structure, physical, mechanical and chemical properties are used in connection to construction materials .

Quality level of production is evaluated by the system of the indices of designation, reliability and longevity, technological effectiveness, ergonomics, efficiency and others

The indices of designation characterize efficiency of the use of materials according to the designation and region of their application. They include predominantly the technical properties of materials, the indices of their composition and structure, transportability, etc.

The indices of reliability characterize the stability of the properties of materials in the assigned boundaries, which ensure their normal operation (fitness for work).

The state, with which the material in full or in part loses fitness for work they call failure. By failure-free performance is understood the ability of materials to remain operable under specific conditions during the specific time without the repair. The ability of materials to remain operable to the limiting condition with the required stops for the maintenance is characterized by durability. Durability is measured by the period of the materials service. In practice the concepts of reliability and durability are identified frequently.

The indices of technological effectiveness characterize the ability of materials to be processed in the products and structures. To such indices belong the ductility, weldability, etc.

The ergonomic qualitative indices unite hygiene, anthropometric, psychological and a number of other indices in the system “man - environment - product”.

Technical and economic indices characterize expenditures for development, production and operation of materials. Materials’ consumption for production which is determined by the ratio of quantity or cost of the spent on its production material resources to the volume, energy, labor, metal content and other, relates also to them.

The classification of the qualitative indices is conditional: one and the same index can belong to the different groups and the subgroups. On the basis of qualitative indices there are determined the types, the classes, the groups, the grades and other qualitative gradations of materials.

During the estimation of quality level of materials differentiated, complex and mixed methods are used. With the *differentiated* method the indices of materials compare with the base indices, given in the standards. For example, if at the required (base) average strength of concrete 20 MPa it was possible to ensure (without the overexpenditure of resources) strength 25 MPa, then quality level of concrete on strength 25/20 = 1.25. Similarly there are calculated the relative qualitative indices of other standardized properties.

If quality level should be described by one generalized index, *complex* method is used. In this case the integral qualitative indexes, for example, efficiency (E_f) per the monetary unit of the expenditures are used:

$$E_f = T_\Sigma /(E_m^o + E_m^u), \tag{4.3}$$

where T_Σ- total efficiency; E_m^o - expenditure for obtaining of material; E_m^u - expenditure for the use of material.

Table 4.1. Production control at the plant of the precast reinforced concrete

Stage of the production process	Object of the control	Stage of the control
Obtaining of the materials	Cement, admixtures, aggregates The reinforcement bars	Determination of the physico-mechanical properties Checking diameter, determination of the strength of the reinforcement bars
Production of the semifinished products	Concrete mixture	Control of the accuracy of dosage, time of mixing and level of workability
Forming of the products	Forms and shutters Preparation for the concreting Concreting	Checking the correctness of assembling form, quality of shutters and lubricant of the forms Testing the position of reinforcement bars, inspection of the degree of the reinforcement Control of the placing, duration and the degree of compaction of the concrete mixture
Heat treatment	Mode of the heat treatment	Control of temperature, humidity and duration of the heat treatment
Forms removal	Finished products	Control of form, dimensions and fineness of products
Inspection of products to the storage of finished products and delivery to the users	Control specimens Finished products	Determination of the ultimate strength of concrete, waterproofness and frost resistance Determination of the strength of concrete by instruments without the destruction; the thickness of the protective layer

Quality control is an important part of the production process of obtaining of construction materials and their application. An example stages and subjects of control at the plant of precast reinforced concrete are shown in Table 4.1.

Input, technological, final and receiving control is distinguished.

Input control consists in checking of correspondence to the established requirements of materials and products, which are supplied to the enterprise or the building site. Operating mode and other indices of technological process are subjects of the *technological control.* Finished products are subjected to the *final and reception controls*.

Control by the use of means of automation and statistical control is one of the most effective methods of control. At statistical control not all products are checked, but their specific part - specimen, which it should be sufficiently to define the quality of entire batch (general population). Batch is considered as the specific quantity of material, which entered to the storage or the manufacture simultaneously. The amount of batch for each material is determined by the appropriate standards.

So that sample would reflect the properties of the entire batch of material, i.e., it was representative, they select it according to the specific rules. For example, it is possible to compose specimens according to the table of random numbers; to select the part of all objects from the general population through the specific interval; to divide general population into the identical series to select from each series one or several objects for the testing. During the receiving monitoring of construction materials they indicate the volume of specimens. For example, for quality control of ceramic tiles from the batch they select for the visual control 50 tiles, dimensional control - 10 tiles, the determination of water absorption and frost resistance - 5 tiles. For the receiving control of gypsum paperboards from each batch 0.5% of sheets, but not less than 3 sheets are selected.

At the definition of the properties of construction materials the distribution curves approach in the nature, as a rule, to the normal Gaussian curves, which corresponds to the equal probability of the appearance both of positive and negative deviations from the center (Figure 4.2). For the characteristic of specific specimen of the numerical values of the indices of materials quality, including observations, the average values - arithmetic mean and the root-mean-square deviations are used.

Arithmetic mean of the deviation:

$$\overline{X} = \sum_{i=1}^{n} x_i / n \tag{4.4}$$

where $\sum_{i=1}^{n} x_i / n$ - the sum of the obtained values; n- the number of observations.

The root-mean-square deviation (standard) shows the limit of changeability, which is determined, i.e., the degree of the spread of its separate values relative to the average:

$$S = \sqrt{\frac{\sum_{i=1}^{n}(x_i - \overline{x})^2}{n-1}}, \tag{4.5}$$

where (n -1) = f - the number of degrees of freedom, i.e., the number of values in the final calculation of a statistic that are free to vary.

The root-mean-square deviation characterizes the absolute changeability of property. The *coefficient of variation* is intended for the expression of relative changeability:

$$C_v = (S/\overline{x}) \cdot 100\,\%. \tag{4.6}$$

The mean error of arithmetic mean can be calculated by formula:

$$m = \pm S/\sqrt{n}. \tag{4.7}$$

The ratio between the mean error and arithmetic mean is called *the index of the accuracy*:

$$\varepsilon = (\pm m / \overline{X})100\% . \tag{4.8}$$

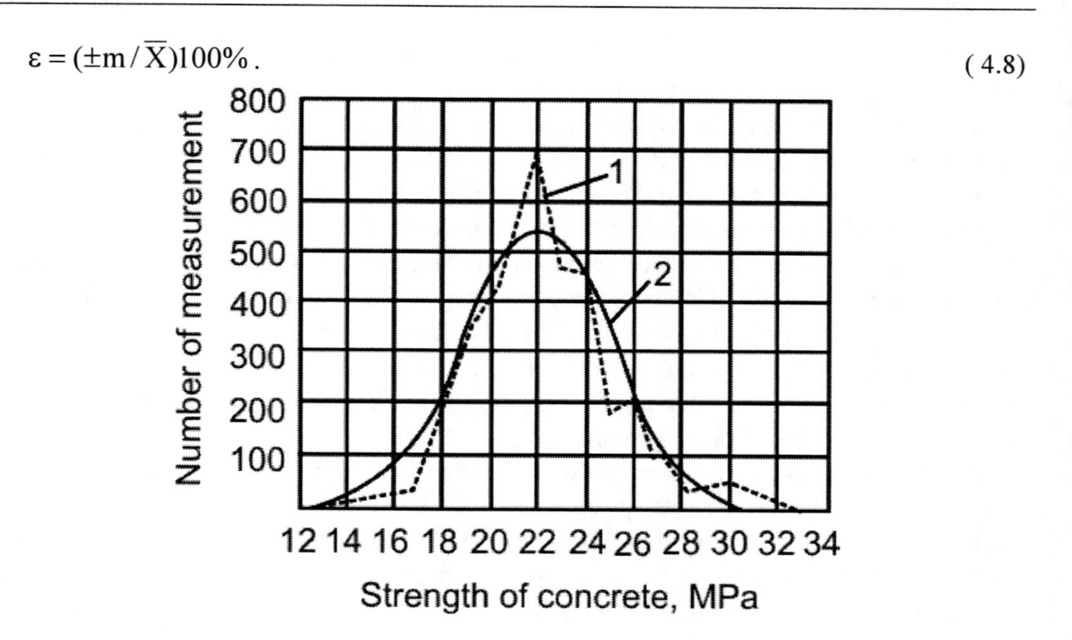

Figure 4.2. Experimental (1) and theoretical (2) curves of concrete strength distribution.

Statistical processing of the results of tests, besides the determination of the changeability of the measured index of quality and accuracy of research, assumes the evaluation of confidence coefficient or significance level of the obtained result.

The significance level is the quantity (or percentage) of the improbable cases, during exceeding of which it will possible commit error, after recognizing the obtained result as correct. A quantity or percentage, the reliable values of property, which is determined, is called the *confidence coefficient*. In the practice of the tests of construction materials are used two values of confidence coefficient – 0.95 and 0.99 and the corresponding to them significance levels -0.05 (5%) and 0.01 (1%) are used The value of confidence coefficient or the significance level is established in the dependence on the degree of accuracy, with which the tests are carried out.

Number of observations which necessary for obtaining of the reliable results can be calculated as:

$$n = C_v^2 t^2 / \varepsilon^2 . \tag{4.9}$$

where t - criterion, which is determined at the appropriate confidence coefficient by reference data.

For the correct application of statistical evaluations it is necessary to exclude possible gross errors during the experiment, i.e., to verify the uniformity of observations. For this it is possible to use *the maximum relative deviation*:

$$\tau = |x - x_{ext}| / S \tag{4.10}$$

where x_{ext} - extreme selected element, which must be verified.

Self-Assessment Questions

1. How are the construction materials classified?
2. What are the basic forms of standards?
3. What are the main systems of standardization?
4. What are the forms of production control?
5. What are the average statistical indices of quality of materials?
6. What methods of determining of the indices of quality of materials are used?

PART II. CHARACTERISTIC OF BASIC TYPES OF CONSTRUCTION MATERIALS

Chapter 5

NATURAL STONE MATERIALS

Natural stone materials are the materials which are obtained both directly during output, and at the subsequent mechanical treatment of rocks.

Natural stone materials due to high mechanical strength, durability, decoratively are widely used since the ancient times as walling and facing materials in construction for paving, hydraulic and other buildings, for strengthening and cladding of slopes, embankments, etc. At present they are about 50% of all of the materials mass applied in the construction. Especially requirement in such construction materials as crushed stone, gravel and sand is large.

Natural stone materials are divided into two groups - regular and irregular shape. Sawn, sledged, rubbed, polished products belong to the first group, and mainly quarry, crushed and graded materials belong to the second one.

Basic ways of application in construction of different products from natural stone are shown in Table 5.1.

Raw materials for the production of natural stone materials are *rocks* - mineral aggregations of certain composition and structure, which are the products of geological processes in the earth's crust.

Table 5.1. Application different kinds of stone in construction

Setting	Materials and products
Basements	Rubble, sawn and sledged stone
Walls	Wall (sawn) stone, large wall blocks, cut stone
External cladding	Facing slabs and stone, profile elements
Internal cladding	Facings slabs, profile elements
External stairs and grounds, parapet walls and fencings	Steps, slabs for grounds, pillars and walls, facing slabs
Internal stairs and grounds, floors	Steps, slabs for grounds, stair and floors
Highways coatings	Edge stone, pavestone, sledged stone and crushed stone
Hydraulic facilities	Angular, sledged and cut stone, boulders and crushed stone
Concrete aggregates	Angular, sledged and cut stone, boulders and crushed stone

The problem of the careful use of natural raw material, utilization of by-products of its processing such as screenings, stone powder, oversize and others is gained the greater value.

5.1. Rocks

Types, Composition and Structure of Rocks

About thousand types of rocks are known for today. By formation conditions (by genesis) they are divided into three classes:

- *Magmatic, or igneous rocks*, formed as a result of magma cooling in the bowels of the earth or on its surface, e.g. silicate melts;
- *Sedimentary rocks,* formed on the Earth surface are as a result of accumulation and transformation of destruction products of rocks formed before, remains of vegetable and animal organisms and products of their vital activity;
- *Metamorphic rocks,* formed on high deepness as a result of sedimentary and igneous rocks change under the action of high temperature and large pressure, caused by the influence of gaseous substances, escaped from magma, and hot solutions.

Slowly cooled *hypogene (intrusive)* rocks (granites, syenites, diorites, gabbro, etc.) and formed in overhead horizons of earth's crust, *(effusive)* rocks (basalts, andesites, porphyrites, pumice stones, etc.) are outpoured rocks.

Chemical composition of igneous rocks can be expressed by content of oxides of silicon, aluminum, iron, magnesium, calcium, sodium, potassium, hydrogen. The major components of igneous rocks, called *minerals*, are quartz and silicates. It is calculated, that average minerals content in igneous rocks is following %: feldspars - 60, quartz - 12, amphiboles and pyroxenes - 17, mica - 4, other silicates - 6.

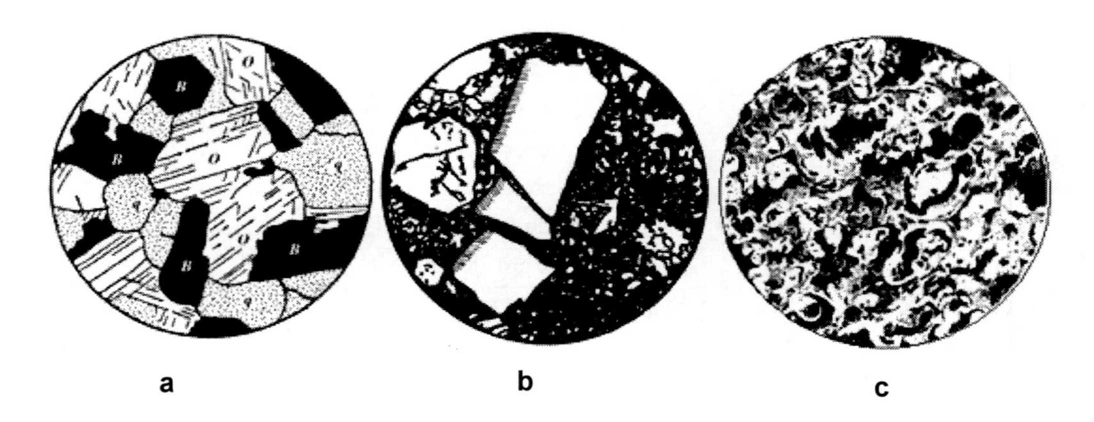

Figure 5.1. Structures of igneous rocks:
a- granular-crystalline structure of granite: q- quartz; O- orthoclase; B- mica; b- porphyritic structure; c- porous structure (pumice stone).

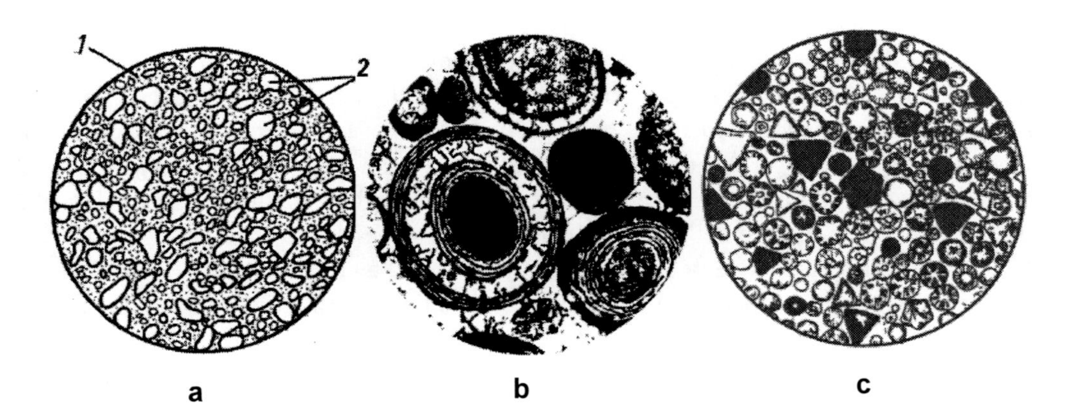

Figure 5.2. Structures of sedimentary rocks:
a- sandstone (1- cement; 2- sand grains); b- shell limestone; c- diatomite.

Sedimentary rocks by the way of their formation are divided into three groups:

Fragmental - the sediments mechanical by origin (boulders, pebble, sand, clay);

Chemical – the sediments chemical by origin (sulfates, carbonates, halogen compound);

Organogenic - the sediments biochemical by origin (carbonate, siliceous, carbonic rocks, etc.).

Fragmental rocks are appeared as a result of destruction (weathering) of igneous rocks influenced by temperature, water, glaciers and other external agents, chemical ones are appeared at sedimentation mineral substances from water solutions and organogenic ones are the products of sedimentation of weeds wastes and animal organisms at the bottoms of reservoirs.

Chemical and mineralogical composition of sedimentary rocks is more diverse, than igneous rocks. Along with minerals of initial mother rocks they can include also a series of other ones, formed in consequence of sedimentation (carbonates, mineral salts, etc.).

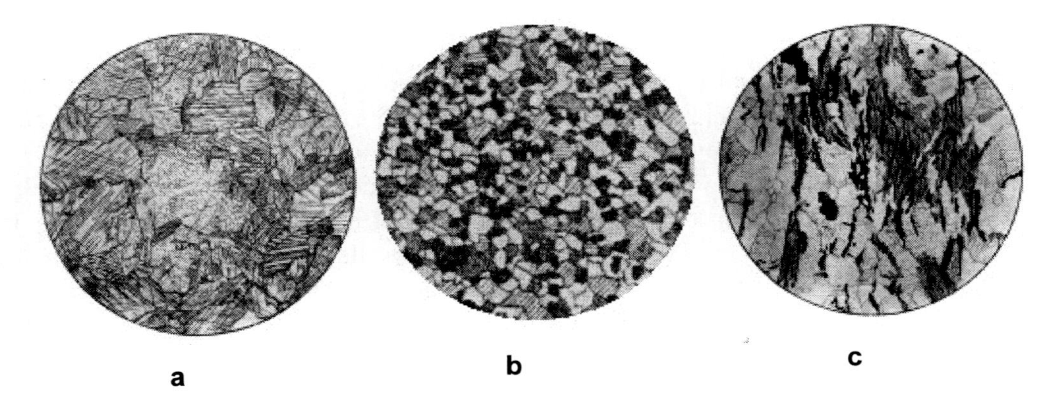

Figure 5.3. Structures of metamorphic rocks:
a - marble; b - quartzite; c – gneiss.

Among metamorphic rocks two groups are selected: modified changed igneous (gneiss) and modified sedimentary (for example: quartzite, clay shale). In the process of transformation, or metamorphism of rocks chemical and mineralogical composition undergoes certain changes depending on the actual values of temperature and pressure. Thus, at temperature up to 500°C and at pressure 40-90 MPa the various *stratified* silicates containing hydroxyl groups are appeared. If a temperature is higher, and pressure is lower, the crystals of amphibole, different micas and other similar minerals are appeared.

Rocks differ by the mode of occurrence, structural and textural features (Figure 5.1-5.3).

The peculiarities of their internal structure, predefined by the crystallinity degree, sizes and form of crystals are meant by a *structure* of rock, and the complex of signs determined by relative location of rock components in the space they occupy is meant by a *texture*.

By crystalllinity degree there are differed шт holocrystalline, semi crystalline and glassy structures. Holocrystalline structures are representative for the hypogene rocks, formed as a result of slow magma cooling, and also for the most of the metamorphic rocks. Semi crystalline and glassy structures are representative for outpours rocks.

Depending on the grains sizes the structures are divided onto coarse-grained (sizes of grains over 5 mm), medium-grained (2-5 mm) and fine-grained ones (less than 2 mm). If minerals grains which compose a rock are close by sizes, the structure is evenly grained, in opposite case — unevenly grained. The typical varieties of unevenly grained structures are *porphyritic* structures, which are characterized by a presence of glassy or fine-grained great bulk in the rock, where large crystals - inclusions are dissipated.

Basic types of textures are uniform, or massive, ordinary texture in igneous rocks, nonuniform (taxite) and gneissoid (foliated structure).

Granites are the most important in the natural stone materials manufacturing among the igneous rocks, carbonate rocks among sedimentary ones (limestones and dolomite), metamorphic ones - gneisses, quartzites and marble.

Granites are acidic (SiO_2>65%) hypogene rocks which are easy to distinguish (Figure 5.1), typical minerals: quartz (20-40%), feldspar (40-70%) and mica - muscovite or biotite (5-20%). Among other minerals amphiboles and pyroxenes occur. Grainy-crystalline structure is typical for granite. The color of granite is determined the colors of feldspars entering into composition, and more often grey, bluish-grey, dark red.

Syenites, diorites, gabbro are close to the granites. Unlike granites they practically do not contain quartz and consist mainly of feldspars and dark colored minerals - hornblende, augite, etc. These rocks are mostly grey or dark-green in colour.

The rocks of granite group belong to the strongest and most resistant rocks, they are well polished. They are applied in construction for cladding of constructions and structures, in particular structures subjected to impact and wearing loads (floors, starlings, etc.). The rocks of granite group are processed also on crushed stone, used as concrete aggregate.

Basalts are the most common outpoured rocks. Feldspar and large amount of dark-colored minerals are included in their composition. The structure of basalts is glassy or cryptocrystalline. High hardness and brittleness complicates their processing. The color of basalts is darkly grey or black. Basalts as well as granites are widely used as a building stone. They are applied in the industry of stone casting, as electrical insulation and acid-resistant materials. In those industries, that basalt, other outpoured rock *diabase is* applied.

Andesites and *porphyrites* are the widespread outpoured igneous rocks. Basalts and diabases belong to the high basic rocks (40-50% SiO_2); andesites and porphyrites to the low basic rocks (52-65% SiO_2). Porphyrites and andesites are applied for pavestone making, acid-proof elements, etc.

Carbonate rocks - are mainly *limestones* and *dolomite*. Limestones are formed mainly of calcite $CaCO_3$, dolomite $CaMg(CO_3)_2$, clay and other minerals are as admixtures in them. Structure and properties of limestones are predefined by terms of their formation. As a result of $CaCO_3$ precipitation calcareous tuff is formed from the sources of carbonate waters; it is a soft easily-sawn porous rock. A type of calcareous tuff is a *travertine* which is the result of $CaCO_3$ precipitation from thermal springs. Travertine has dense, fine-grained structure and is applied as cladding stone.

Some types of limestones are organogenic by origin. They are formed as a result of compression and cementation of skeletal residuals of elementary animals (shells, shellfishes, etc.). *Shelly limestone* and *chalk* are the limestones organogenic by origin.

Dense limestones are applied for obtaining cladding details, crushed stone for heavy-weight concretes, and porous ones for wall stone and blocks. Limestones widely are applied as raw materials for obtaining lime, Portland cement and other artificial construction materials.

Along with limestones there are also used conglomerate rocks - *sandstones* which consist of quartz sand grains, consolidated by clayey, siliceous, calcareous and other substances. Siliceous and calcareous sandstones are the most strongest and resistant.

Marbles, which are applied mostly as cladding materials, are formed as a result of limestones and dolomites recrystallization.

Typical metamorphic rocks - *gneisses by* mineral composition correspond to igneous rocks of granite type. For them, as well as for other metamorphic rocks, crystalline-grainy structure and foliated (gneissic) texture are typical. Gneisses are applied mainly as crushed stone for highways and for ballasting of railway bed.

Quartzites are formed as a result of metamorphization of quartz sandstones. Depending on the admixtures they can be white, yellowish and reddish colors. Quartzites are used as cladding, acid-proof material, for refractory materials production.

Rocks Properties

Application area of rocks is determined by their physical-mechanical properties, predefined by peculiarities of formation, chemical and mineralogical composition, structure and texture. The values of basic properties of a series of construction rocks are shown in Table 5.2.

Compressive strength is the most important property of natural stone. This parameter is highest for rocks which have a homogeneous crystalline structure. If glass prevails in rocks, their strength decreases, they are more affected by temperature changes. Fine-grained rocks, composed of grains of the irregular, ragged form, have higher strength. Porosity makes significant influence on compressive strength of rocks as it diminishes contact area between. If, for instance, porosity of limestones diminishes from 40 to 2%, their compressive strength grows from 5 to 180 MPa. Flexural strength of rocks is in 10-20 times lower, than compressive strength.

Table 5.2. Basic properties of rocks

Rocks	Average density, kg/m^3	Compressive strength, MPa	Module of elasticity, 10^4 MPa	Frost-resistance, cycles
Igneous:				
granites	2500-2700	100-260	5-10	100-300
gabbros	2800-3000	100-350	9-11	100-300
porphyrites	2500-2700	60-150	6-8	59-200
basalts	2200-3100	110-500	8-8.3	50-200
Methamorphic:				
gneisses	2000-2500	10-200	6-7	25-200
quartzites	2500-2700	100-250	7-9	100-300
Sedimentary:				
carbonates	1700-2700	5-200	0.2-9	3-300
sandstones	2000-2500	10-250	1.4-5	15-300

In rocks, especially sedimentary rocks, weak forms can be found, contained as separate layers. Homogeneity of rock properties is the most important indicator of their quality.

While selecting the type of the stone its average density is in great importance; for the rocks of certain mineral composition and structure it can describe their strength and durability. As far as chemical basicity of rocks increases (ratio of basic oxides content to acidic ones) the density of rocks grows.

Frost resistance of the most rocks can be defined approximately by water absorption. Water absorption of dense igneous rocks not subjected to the weathering does not exceed 0.7%, water absorption of sedimentary rocks is 10% and more.

Weathering resistance of minerals containing in the rock has substantial value for durability of natural stone. Quartz is resistant to weathering; orthoclase and microcline are low-resistant; plagioclases, amphiboles, olivine, calcite, dolomite, gypsum and others are nonresistant. Chemically active minerals like sulfides, sulfates and others can negatively influence on the stone preservation. For example, sulfides oxidization on the polished surface of cladding stone causes foxing and results in destruction, oxidization of sulfides in crushed stone - volume increasing and reducing in concrete strength.

In the case of application of natural stone as walling material thermal and sound conductivity, air permeability are very important properties along to porosity, frost-resistance, weathering resistance. All of these properties are correlated and predefined mainly by porosity of rock. Effective wall materials are elements from such high-porous rocks, as tuff, shell limestone.

For cladding natural stone their decoratively and processability are of important value.

The color of the rock is determined by the color of minerals, that it is consisted of. Igneous rocks have the most stable coloring, sedimentary and metamorphic rocks are less proof.

Workability of rocks (abilities to smoothing, polishing) reduces as far as their strength growths, at a coarse-crystalline structure, also at the presence of clots.

Soils as Natural Materials

Rocks which are the objects of structural engineering activity are called *soils*. Soils are natural basis, environment and construction material for various constructions. Soils are divided onto:

- Rocky - consolidated rocks, which are practically uncompressed and have compressive strength in the water-saturated state more than 5 MPa (granites, basalts, some types of sandstones, etc.);
- Semirocky or loose rocks with compressive strength in dry or water saturated state less than 5 MPa (gypsum conglomerates, etc.);
- Earth – coarse fragmental (uncemented rocks which contain more than 50% of particles more than 2 mm by mass), sandy (friable in the dry state rocks which contain less than 50% of particles larger than 2 mm by mass) and clayey (for which certain plasticity is typical).

Soils are divided also by softening coefficient, weathering degree, by solubility in water. Humidity has large influence on strength properties and especially on the soils cohesion. Dry clayey soils at considerable moistening become fluid. At the soil compression the most its density at minimal work is reached at ring optimal humidity.

In construction process of earthworks soil consolidates due to approaching of particles between each other under load action. Compressibility of soils is characterized by reduction of porosity coefficient at increasing in compressive stress. Determination of dependence between the porosity coefficient and compressive stresses is defined by compression equipment.

Strength of soils is characterized their ability to resist to sliding stress. Cohesion c (MPa) and angle of internal friction φ (degree), applied for the calculations of load-carrying capacity and resistance of soils are rated to a number of parameters of soil strength. Density and humidity of soil, loading rate influence significantly on the values c and φ .

Important parameter of soil at its irrigation estimation is filterability. Depending on the type of soil, volume of its pores and structure of pore space soils have different filtration coefficient (Table 5.3).

Table 5.3. Filtration characteristics of soils

Degree of water permeability	Filtration coefficient, m/days	Type of soil
Practically waterproof	0.01	Clay
Low waterproof	From 0.01 to 0.1	Loam
Waterproof	From 0.1 to 1	Loamy sand
High water permeable	From 1 to 10	Fine sand
Very high water permeable	Over 10	Coarse sand, gravel, pebble stone

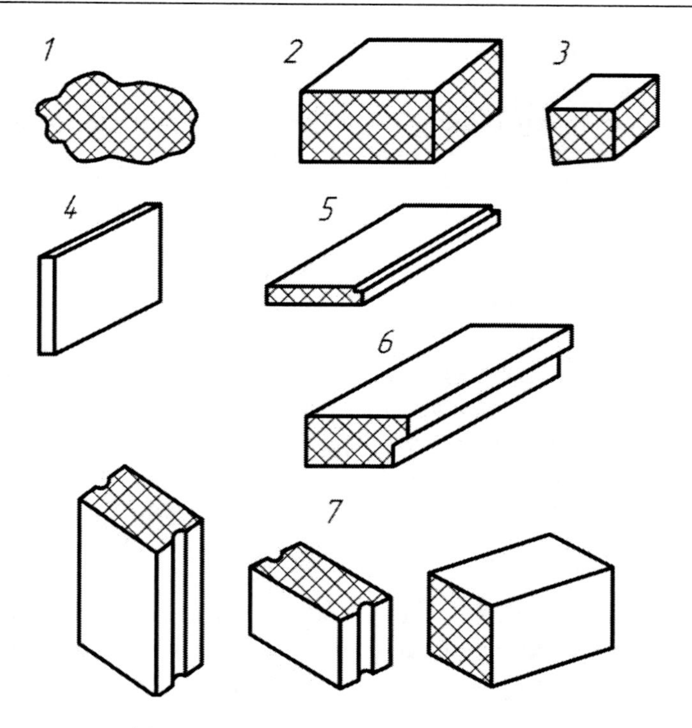

Figure 5.4. Types stone materials and elements:
1- rubble stone; 2- walling stone; 3- paving stone; 4- facing slab; 5- window sill; 6- step; 7- walling blocks.

For earthworks construction all the types of soils are accepted, with the exception of pulverescent sands, soils which contain the water soluble inclusions, and also soils which contain the decayed organic substances, etc. Clays are used for construction of waterproof elements of buildings, where they are at permanent humidity. Hydraulic earth-fill dams are built of gravel sandy, loamy sand soils. At the construction of earth-to-rockfill and rock-fill dams rocks are used with softening coefficient equal 0.9 (for igneous and metamorphic rocks) and 0.8 (for sedimentary). Stones should have sufficient strength and frost resistance.

5.2. Types of Natural Stone Products

Natural stone materials are produced by industry of construction materials. The basic types of products are crushed stone, gravel, sand, rubble stone, cladding elements, wall sawn stone, paving and other elements of the special assignment (Figure 5.4).

Rubble stone is a construction material which consists of pieces of rock irregular in shape 70-1000 mm in size. It is used for laying of foundations, underground parts of buildings, walls of nondomestic, subsidiary and production buildings, for strengthening of earthen slopes, in road construction, and also as an aggregate for rubble concrete. In hydrotechnical construction fill dams are also erected of rubble stones. Stone composition by coarseness is assigned from getting minimum porosity taking into account the peculiarities of dams.

Rubble stone is divided into self-faced (with one uneven face) and slabby (with two approximately by parallel faces). Cobble stone is a type of rubble stone, round inshape.

Strength of rubble stone is estimated by compressive strength of initial rock in the water saturated state. Ultimate strength of rubble stone is in the range of 10-140 MPa. Stone made of metamorphic rocks should have strength not less, than 40 MPa, but the stone made of igneous rocks - not less, than 60 MPa. Frost resistance of rubble stone is approximately equal 15-300 cycles of freezing and thawing. There are accepted not more than 15% fragments with dimensions, different from limit ones in rubble stone batch. Clay content in stone should not exceed 2% by mass, softening coefficient should be not less, than 0.7.

Facing and Walling Stone

Facing slabs and stones are produced by sawing of natural stone blocks. In the case of application of off-quality blocks there are obtained sledged slabs and stones.

For obtaining facings elements rocks of granite group and metamorphic rocks like marble and colored variations of quartzites are commonly used. Sedimentary rocks - limestones, sandstones, dolomites, travertines and gypsums are applied mainly for internal facing. Quartzites and granites have the highest durability; marbles have considerably lower durability, and can quickly to collapse at outdoor conditions.

Rocks which are used for the application of facings elements should have compressive strength not less than 5 MPa, frost resistance - not less than 15 cycles. Depending on blocks dimensions they are divided into some groups. There is accepted in a block no more than one crack, which is visible at two adjacent surfaces, less than one third of measuring long. Cracks on slabs less than one third of width length are allowed in that case, when slabs are made from the colored marble. Facings elements have abrasive (polished, rubbed, honed) or impact (tooled, boasted, pointed) finishes. The most common are sawn slabs 6-80 mm thick, 200-400 mm width.

Decorative slabs with mosaic and ornamented surfaces are produced utilizing separate pieces of natural stone, crushed stone and sand, and also inorganic or synthetic binders. Slabs are marked by grades, type, kind of face and overall dimensions are specified in which.

Due to high durability of natural stone facing their maintenance costs are considerably less than, at finishing of buildings by paints, coloured mortars and concretes.

As walling materials *sawn work stones* and *large-size blocks* are used. For walls light-weight porous rocks are used which have average density 900-2100 kg/m^3, compressive strength 0.4-50 MPa and frost-resistance not less than 15 cycles. Minimal value of softening coefficient of walling materials and blocks is 0.6, water absorption for tuffs is 50%, for limestones and other rocks it is 30%.

Walling stone and blocks often have following basic dimensions, mm:

Stones – length 390, 490, width 190, 240, height 188, 288;
Blocks - length 400, 500, width 820, 1000, height 2360, 2520, 2860, 3020.

By assignment wall stones are divided into facial and ordinary ones. Ordinary stones are used for walls erection with subsequent plastering.

Ultimate compressive strength of wall stones is 0.4-40 MPa, blocks - 2.5-40 MPa.

At symbolic notation of walling stone they specify type of stone by dimensions, its kind by assignment and compressive strength grade.

Walling elements made of natural stone have beautiful surface finish (rose, blue, yellow tuffs; white, yellow light grey limestones), they are processed easily.

From separate blocks, sawn from light-weight rocks, it is possible to make on mortar-based large *composite blocks* and *panels* the application of which enables to promote productivity of building.

It is effective to utilize composite blocks "floor" in high, which are made of 6-8 ordinary blocks, gaps between them are filled with cement mortar. Large panels, glued from separate elements by polymers are perspective products made of natural stone.

Border stone, pavestone, sledged stone and cobblestone are used for road construction as a workstone. These elements are served for detaching the traffic ways and highways from pedestrian ways (sidewalks), for paving roadways, tram-car ways, strengthening the slopes of earthen coatings and subbases. Raw materials for them are high-strength rocks with strength not less than 60 MPa, frost resistance not less than 25 cycles, with softening coefficient 0.6-0.9, water absorption 1-4%, impact resistance 15 kPa.

Piece natural stones are transported on pallets or separately, by purpose, by types and grades.

Stone materials and elements are widely used in water industry construction. At construction stone irregular in shape: rubble stone, cobblestone, sometimes boulders, gravel and crushed stone are utilized for laying of foundations of hydraulic structures, erecting the earth-and-rock-fill and rock-fill dams, preparation of rubble concrete for the erecting of interior zone of concrete and reinforced-concrete dams, foundations of the hydro systems of the pumping stations and hydroplanes, consolidation of hay-crops of canals and ponds, arranging of stoning round concrete hydraulic structures to protect concrete from corrosion. The stone of regular shape from igneous dense rocks and sandstones are used for dams stonework by the method of dry rubble (without mortar) by regular rows with bonding; arranging of finishing stonework of underwater and above-water parts of structures (embankments, pumping stations, hydroplanes buildings); masonry work of footings of aqueducts, bridges and retaining wall; architectural elements production (columns, semi-columns, basement parts of buildings). A cobblestone made of dense rocks is used for arrangement of fortifications from wave and speed influence of water, hay-crops of earthen dams, canals, water storages, at roadway pavings.

Hewn and sawn slabs made of dense water resistant rocks are used during the external cladding of underwater part of hydro systems, embankments, sometimes facing of hay-crops of the hydrosystems, floors arranging in buildings of hydroplanes and pumping stations. As facing material inside the pumping stations and hydroplanes buildings different types of marble, conglomerates and breccia are used. To improve architectural expressiveness of hydraulic structures there are applied different architectural compositions, produced of water resistant rocks (granite, gabbro, sandstone).

Especially wide application in water construction has clastic rocks such as sand, crushed stone and gravel, gravel-pebble and gravel-sand mixtures. These materials are used for concrete mixtures production for hydraulic and normal concrete; arranging the underlayers under monolithic and precast reinforced-concrete canal linings, roads, constructions; preparation of the drainage filings ups; at consolidation of hay-crops of canals and hydro systems, water storages.

Taking into account that stone materials used in hydraulic construction practically have permanent or periodical contact with water, which often is aggressive one, they have to meet raised requirements not only at density, strength but also in frost-resistance, resistance to aggressive environment.

5.3. Production of Natural Stone Materials

Production technology of natural stone materials is determined by their peculiarities, rock properties and terms of occurrence. A deposit, prepared for development of rocks, is called *quarry*. Preparation of a deposit starts with the removal of upper rocks and baring of minerals. Quarries are developed by one (or few) benches the height of which does not exceed 20 m.

The most mass produce material is crush stone from hard rocks. It is produced by drilling-and-blasting, related to the blasthole drilling or borehole located along ledge front; placing explosives inside (ammonite, ammonium nitrate, ammonal, etc.) and separation by the explosion of rock pieces which are ground down after that.

On crushing and screening factories and plants (Figure 5.5) there is made one-, two- or multistage crushing of initial raw material. Multistage crushing is the most prevailing; it enables to obtain both coarse and fine fractions of crushed stone. One-stage scheme is used at the low productivity plants at maximum dimensions of rock pieces 400-450 mm.

Crushing equipment can have periodical (jaw crusher and cone crusher) and continuous pressing of crushing surfaces (roll crushers and impact crushers - the hammer and rotor ones).

The crushed material is assorted on fractions by mechanical, hydraulic and air methods. At the mechanical method stone is assorted by screens. At the enterprises of construction materials most widespread screens are vibrocribble screens, in which one or two screens are set.

The stone can be crushed by opened and closed cycle. At open-cycle, material one time passes through a crusher, goes to assorting and then on finished storehouse. At the closed cycle stone which did not pass top screen, again goes to crushing, that enables to attain the greater productivity of crushers, regulate an output and coarseness of ready-made products.

Dusty and clayey admixtures which contaminate crushed stone are withdrawn by washing, for that blade flushing, drum or vibratory machines are used. Vibration machines are less metal- and energy- consuming, they require less water.

Along to washing special methods of cleaning of stone materials are used. Processing low-grade raw material in crushed stone, the method of selective crushing is applied in particular, based on more intensive destruction of pieces of weak rocks. The gravity concentration is based on dependence between strength and density of rock. Grains of material, heterogeneous by density, are divided in the special settling machines in the alternately ascending and descending streams of water or in dense media (water suspensions of magnetite or ferrosilicon), the density of which has an intermediate value between the densities of dividing fractions.

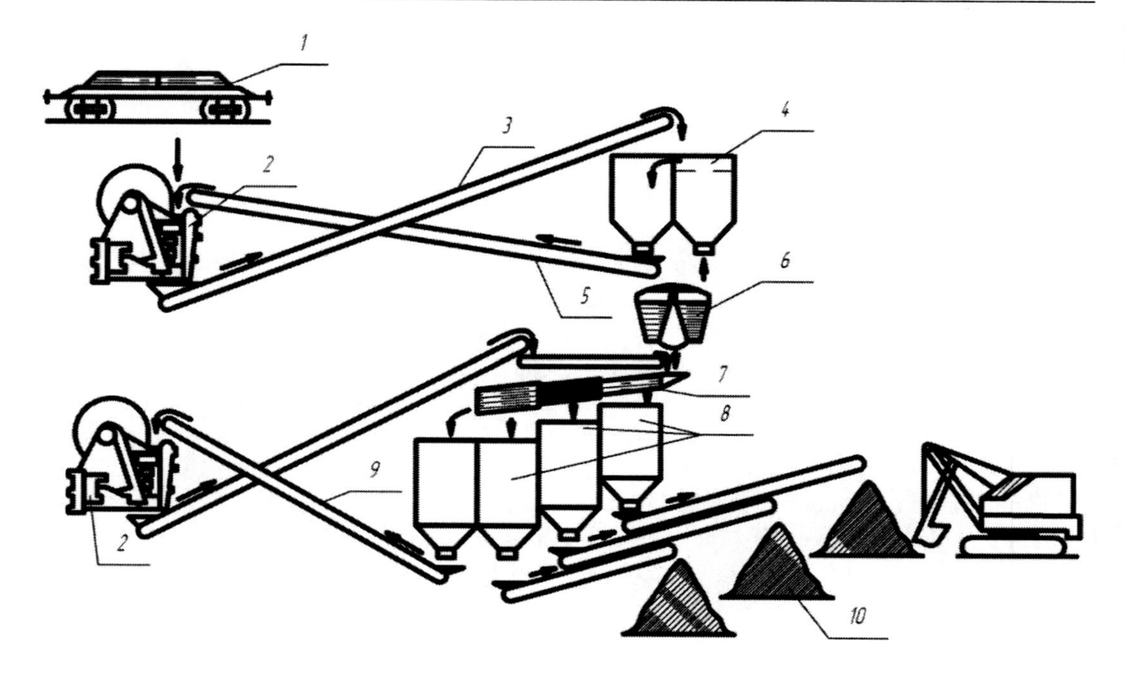

Figure 5.5. Chart of crushed stone manufacture:
1- stone; 2- jaw crusher; 3, 5, 9- conveyers; 4, 7- screens; 6- cone crusher; 8- hopper; 10- storages of crushed stone fractions.

Loose rocks (gravel, sand, gravel-sandy mixtures) are yielded by the opened method, mainly by one- or multibucket excavators. Along to dry-process on sandy gravel careers hydromechanical extraction is applied. Rock developed by this method is separated by the dynamic action of water, is loosened and is transported by dredging equipment as a pulp in the storehouse or on processing. Specific capital investments on extraction of stone materials by hydromechanization in comparison with other methods are usually lower.

Work stone from light-extracted rocks - marble, limestone, tuff is obtained by rock-cutting machines, the operating element of which are horizontal and vertical disk saws, set on a cart which moves in a quarry by rails along the working face. Rock-cutting machines enable to mechanize both cutting of stone and secondary operations, related to its stacking and transporting from a working face.

Except for machines with disk cuttings instruments machines with chainsaws are applied. The productivity of machines on the medium hardness rocks is increased in 4-5 times, if diamond cuttings nozzles are used.

Facing elements from hard rocks are obtained by separating from massive of large monoliths which are divided into blocks and slabs. For the separation of monoliths there are used drill and wedge, drill, fire and other methods. For quartz-containing rocks of granite groups thermal method is used. It is based on the destructive actions of internal tensions which arise up at thermal deformations of minerals under the action of high temperature.

5.4. LONGEVITY OF THE NATURAL STONE MATERIALS

Longevity is the generalizing (complex) construction property of natural stone. It serves as one of the important criteria in selection and evaluation the material, and combines a number of the physical, mechanical and chemical properties, which directly influence on the effectiveness of its use, namely: frost resistance, strength, porosity, water absorption, wearability, corrosion resistance. According to the modern classification the longevity is evaluated by the lifetime of stone (in the years) before the appearance of the first signs of destruction (initiation of destruction) and up to the final destruction. Depending on the value of these indices natural rock materials are classified into four groups of longevity (Table 5.4). According to the indices of longevity the most of the varieties of natural stone relates to the most stable rock materials. The fine-grained quartzites and quartz-like sandstones, and also fine-grained granites, which have been preserved in the earliest monuments of human culture, are especially long-lived.

The destruction of stone in the process of operation can be appeared in the different forms: peeling of surface, appearance of cracks, crumbling of the separate particles and fragments of constructions, appearance of plastic deformations.

Climatic action (the so-called physical erosion) is one of the main reasons, which reduce the period of the service of stone: that is displacement and the evaporation of moisture with the dissolved in it substances and sharp temperature variations with the passages through the zero temperature. Negative action on the water-saturated stone has not so much the value of temperature of freezing, as frequency of transition through 0°C, i.e. alternation of the cycles of freezing and thawing. This action on the stone leads to the development of cracks in it, the break of the cross connections between the separate voids, slackening of the connection between the crystals, reduction in the mechanical strength. In this case the basic reason for destruction is the action of the freezing water. The freezing water is enlarged and exerts the strong pressure on the walls of voids. This pressure can be reached up to 210 MPa at a temperature - 20°C. As a result of this process a large mechanical stresses can be acted on the sections of stone in the closed space, which can lead to its destruction.

Table 5.4. Classification of the rocks depending on the longevity

Categories the longevity	Designation of the rocks	Longevity in the years	
		initiation of the destruction	final destruction
1. Very long-lived	Quartzites, the fine-grained granites	650	–
2. Long-lived	Coarse-grained granites, gabbro, labradorites	220-350	1500
3. Relatively long-lived	White marble, dense limestones and the dolomites	75-150	1200
4. Short-lived	Gypsum stones, anhydrite, the porous limestones	20-75	100-600

If the rock has predominantly large voids, which are not connected, the periodic action of minus temperatures leads to the breakage of the walls between the voids, reduction in strength, and sometimes even to the disintegration of the stone. However in the case if rocks have voids, surrounded by the capillaries which being communicated with it, then part of the water during the freezing wrings out into the small voids and capillaries. Such small voids and capillaries are reserve capacities, and have not harmful mechanical influence upon the skeleton of rocks. In that case the stone can be enough long-lived even even at low mechanical strength.

Although the temperature coefficient of linear expansion in natural stone is relatively low, the stone can be destroyed because the crystals of different rock-forming minerals have different coefficients of linear thermal expansion. As a result rocks under the action of sharp temperature drops can be covered with the network of the small cracks, which facilitate further destruction.

Significant negative effect on the longevity of stone have the mechanical factors, such as: impact actions, compressive and bending loads, abrasion, action of abrasive particles.

There are also important chemical factors, which in a number of cases can sharply reduce the period of the service of elements made of the natural stone (so-called chemical weathering). In the case of the insufficient waterproofing of rock constructions the infiltration of ground water into the stone with the subsequent deposit of salts is possible, which leads to the appearance of bloom and saltings, and sometimes to the destruction of insufficiently steadfast stone. The chemical factors, adversely affecting at the longevity of the stone, also include the significant atmospheric content of oxide SO_3, which during the moistening leads to formation of the sulfuric acid or sulfates in the stone, which are expanded during the hydration.

The process of the gradual destruction of rock materials in the structures can be prevented by the different design and chemical methods of protection, which facilitate the reduction in the action of factors enumerated above.

Design methods are evinced by the making of the smooth or polished surfaces of materials, which can not detain rain and melt water and pass aggressive matters inside the rock materials.

The chemical methods of protection provide for the conservation of stone by the sequential impregnation of its upper layer with some substances. The undissolved substances, which clogged the voids and microscopic cracks, are formed as a result of the reaction between these substances. The silicatization, fluating and also modification by polymeric compounds are belonged to such methods.

Silicatization consists in the impregnation of the upper layer of stone with the solution of liquid glass, and then - by the solution of chloride calcium. Formed as a result of reaction silicate of calcium $CaO \cdot nSiO_2$ is the practically undissolved compound.

The process of floating consists in impregnation of the limestones and other rocks, which contain calcite, by the water solutions of the fluorosilicate salts. These salts (fluates) can be added into chemical reactions with the soluble components of stone with the formation of fluoric salts of calcium, magnesium and silica, which not dissolved in the water. They condense at the surface of stone and make its inaccessible for the aggressive matters. For instance, silica and fluoric salts are formed at processing of limestone with fluosilicate magnesium:

$MgSiF_6 + 2CaCO_3 = MgF_2 + 2CaF_2 + SiO_2 + 2CO_2\uparrow$.
fluate limestone undissolved salts

Process of fluating can be used for the noncarbonate (acidic) rocks. For this purpose they preliminarily are impregnated with the solution of lime salt, which subsequently forms with the fluate protective layer from the compounds undissolved in the water.

Also it is possible to use of water solutions and emulsions, polymeric substances and water-polymeric dispersions. For instance, silicone resin and also the water solution of urea-formaldehyde resin are used for the hydrophobization of surface and voids of stone. Reliable waterproofing is created by some monomers with their subsequent polymerization by the thermocatalytic processing. This method of modification is most effective for processing of porous rocks.

Combination of design and chemical methods leads to a considerable increase of the longevity of natural rock materials in different structures.

Self-Assessment Questions

1. What materials are called natural stone materials? What groups they are divided on?
2. What are known classes and groups of rocks depending on the terms of their formation?
3. Compare the concepts "structure" and "texture" of rocks. Give examples of basic types of structures and textures of rocks.
4. Describe the basic types of igneous rocks.
5. Describe the basic types of sedimentary and metamorphic rocks.
6. Give a description of basic properties of rocks important for obtaining and service of natural stone materials.
7. Give a description of rubble stone and specify requirements to it depending on the terms of application.
8. How facing elements from a natural stone are obtained? What are the requirements to them?
9. What products from natural stone are used at hydraulic and road construction?
10. What are the peculiarities of production technology of natural stone materials depending on the types of products?
11. What are basic ways of rising natural stone materials longevity?
12. What is the essence of natural stone weathering and what are the methods for weathering prevention of?

Chapter 6

METAL MATERIALS

Metals are the substances which have a number of specific features: brilliance, high thermal and electrical conductance.

The iron-carbon alloys - *steels* (up to 2.14 % C) and *cast-irons* (2.14-6.67 % C) are widely used in modern technique, particularly in construction. Their specific weight is up to 95-97 % in the general volume of metals used in metal structures. Cast-irons and steels are the basic representatives of the ferrous metals. Aluminium, magnesium, zinc, as well as their alloys belongs to the nonferrous metals.

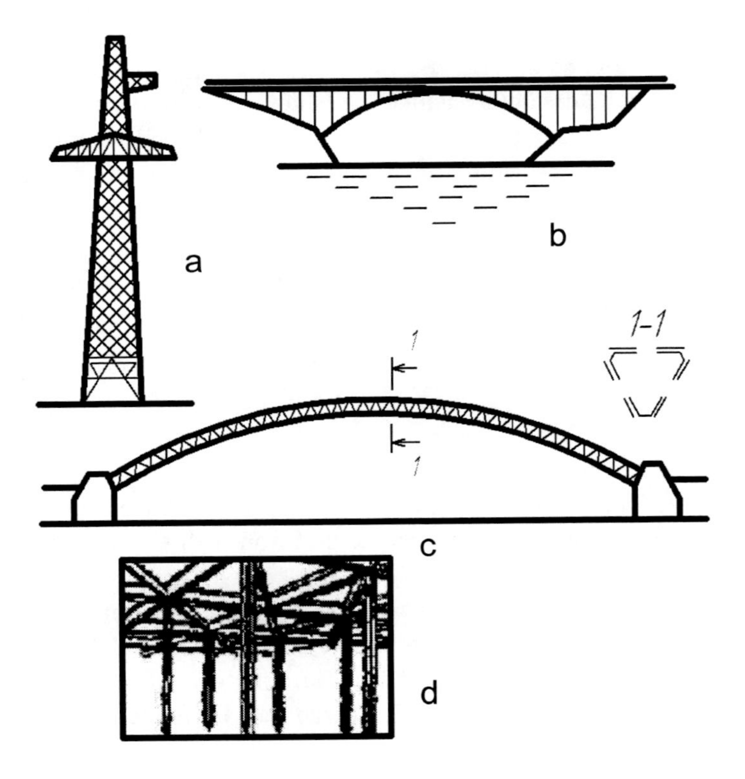

Figure 6.1. Aluminium framed structures:
a- power transmission tower; b- bridge frame; c- arch frame; d- building frame.

Metal materials in construction are mainly used as various products: plates, angled, double-T and channel-shaped sections (shapes, bars), pipes.

The production of the light steel structures made of the economic sections of rolled metals and low-alloy high-strength steel is increased. Mass of the light structures is less in 4-6 times than the ordinary steel structures; working hours diminish approximately at 30 %.

The application of the alloys based on light nonferrous metals, especially aluminum is enlarged (Figure 6.1). The external wall panels, structures of counter ceilings and partitions, window frames, doors, stained glasses, jalousies and others are made on the base of aluminum alloys. Aluminium alloys structures have less weight (2.7-2.9 t/m^3), better fire-resistance, seismic resistance, cold resistance and durability in comparison with structural steel and approach to its strength.

6.1. Metal Materials in Construction

Iron-carbon Alloys

The composition and structure, assignment and the most characteristic technical properties are the basic classification peculiarities of iron-carbon alloys.

Depending on composition and structure, cast-irons are divided into two basic groups: grey and white. In *grey cast-irons* a carbon is contained mainly as graphite, in *white* – as cementite (Fe_3C). The fracture dark grey in color is characteristic for the first one, and lusterless white – for the second one.

Grey cast-iron has good castability unlike the white cast-iron that has high hardness and fragility. The sanitary products, pipes, floors, tubing for tunnels and other products are made of grey cast-iron. The grey cast-iron can be graded depending on tensile strength. For example, its grades accordingly ASTM is given below.

ASTM Grade	Class 20	Class 30	Class 35	Class 40
Tensile Strength (PSI×10^3)	20-27	24-34	30-40	40-45
Brinell Hardness	201 Max	223 Max	241 Max	262 Max

Note: 1 Pa= 145,04×10^{-6} PSI.

Addition of the special admixtures to the liquid grey cast-iron before pouring, allows to get modified, in particular, *high-strength cast-iron*. It is alloyed by chrome, nickel, molybdenum, titanium, aluminum to produce the cast-iron with the special properties: heat-resistant, frictionproof, anticorrosive etc.

White iron is assigned, mainly, for its processing into steel and malleable cast-iron production. *Malleable cast-iron* differs from the grey one with the enhanced plasticity, good processability. It is manufactured by the protracted heating (annealing) of white cast-iron at the temperature 760-980°C. The malleable cast-iron can be graded depending on tensile strength, yield strength and elongation. For example, grades accordingly ASTM is given below.

ASTM Grade	60-40-18	60-45-12	80-55-06	100-70-03	120-90-02
Tensile Strength (PSIx10^3)	60 Min	65 Min	80 Min	100 Min	120 Min
Yield Strength (PSIx10^3)	40 Min	45 Min	55 Min	70 Min	90 Min
Elongation (%)	18 Min	12 Min	6 Min	3 Min	2 Min

Cast-irons with the special properties are used in such cases, when the cast, except of strength, should be resistant to the wearing, corrosion etc.

Steels are divided according to their composition into carbon and alloyed steels. *Carbon steels*, except of iron and carbon, contain the admixtures of row of chemical elements which are brought with the raw materials or predefined by the features of production. There are distinguished low (to 0.3% C), intermediate-alloy (0.3–0.5% C) and high-alloy (over 0.5% C) steels. Alloyed steels contain along with ordinary admixtures some alloying ones, which are specially included elements - chrome (Cr), nickel (Ni), molybdenum (Mo), tungsten (W), silicon (Si), manganese (Mn), copper (Cu), phosphorus (P), titan (Ti), vanadium (V), nitrogen (N), niobium (Nb), aluminum (Al).

According to the assignment steels are divided into structural, tool and special steels.

Structural carbon steel which contains 0.65-0.70% of the carbon is mainly applied for the building structures. At increasing the carbon content from 0.65 to 1.35% and manganese content up to 0.4% tool steel is obtained; at diminishing carbon to 0.2% - steel for the deep-drawing; with raised sulphur and phosphorus content - automatic steel which is used mainly for fixings production (bushings, nuts, screws etc.).

Strength and hardness of steels grow with the increasing of carbon content, but plasticity and weldability reduce (Figure 6.2). Carbon steel, fully deoxidized after melting, is called killed steel, partly deoxidized – semi-killed steel and unkilled steel. Killed steel hardens without noticeable gas emission. It have better strength properties, but at the same time less output of metal at rolling and higher cost in comparison with semi-killed and unkilled steels…

By quality there are carbon steels of ordinary quality and high-quality steels. Steel of ordinary quality is applied for production of building structures, fittings, pipes etc. There are applied steels with normalized composition and mechanical properties.

There are different grades systems for steel in different countries. Definition of steel grade normally represents combination of letters and numbers, representing the composition and the most important steel properties. For example, in USA there are used several definitions of metals and alloys related to the standardization organizations: AISI, ACI, ANSI, ASTM, etc. In the wide-spread systems of definitions AISI (American Iron and Steel Institute) carbon and alloyed steels, as a rule, are determined by four numbers. For carbon steels there are 4 groups of grades: 10XX, 11XX, 12XX, 15XX. The first two numbers determine the number of steels group, and the last two – average content of carbon in the steel, multiplied per 100. Thus steel 1045 relates to the group 10XX of qualitative structural steels (unsulfonated, with Mn content less than 1%) and contains about 0.45% of carbon. Along with four numbers in the names of steels there also can occur the letters. Letters M and E are placed before the name of the steel, which means that the steel is purposed for production of irresponsible section steel (letter M) or smelting in the electric furnace (letter

E). In the end of the name there can be present letter H, which determines that hardenability is the characteristic feature of the steel.

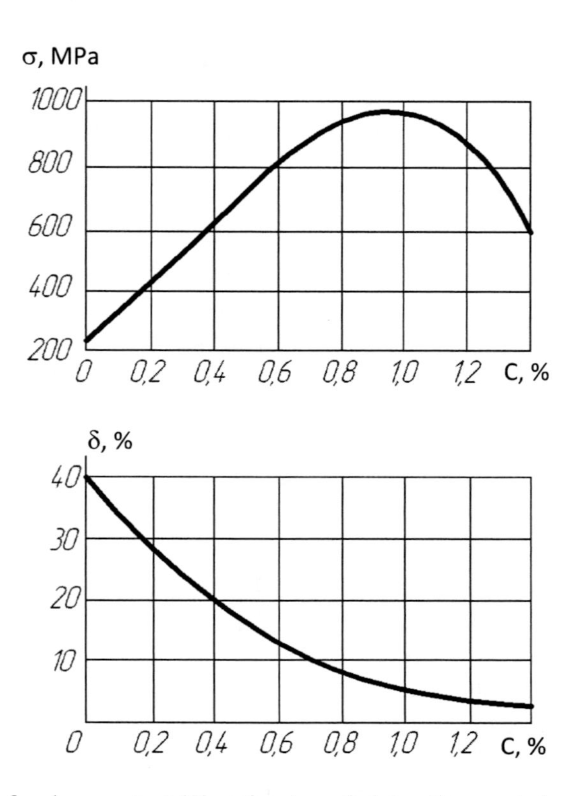

Figure 6.2. Influence of carbon content (C) at the strength (σ) and percent elongation (δ) of steel.

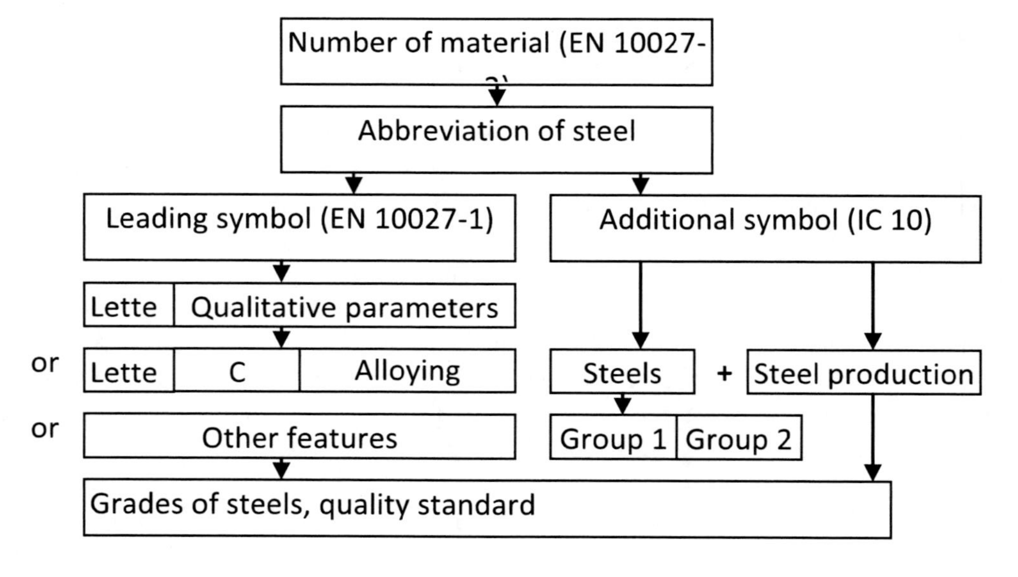

Figure 6.3. Designation systems for steel.

Table 6.1. Group 1. Mechanical properties of steel

<table>
<tr><th rowspan="2">Letter</th><th rowspan="2">Leading symbol</th><th colspan="2">Additional symbol for steels</th></tr>
<tr><th>Group 1</th><th>Group 2</th></tr>
<tr><td>S

G -
cast steel
work
placed
before, if
required</td><td>Structural steels

E.g S355J0

Properties:
Minimal yield stress (N/mm^2) (three numbers).</td><td>Fracture energy (in Joules) tests
27 J 40 J 60 J °C
JR KR LR +20
J0 K0 L0 0
J2 K2 L2 -20
J3 K3 L3 -30
J4 K4 L4 -40
J5 K5 L5 -50
J6 K6 L6 -60

M= thermal mechanical deformed
N= normalized
Q= heat-treated
G= other properties, if required with 1 or 2 numbers</td><td>C= with special ductility in cold state
F= for forging
L= for low temperatures
M= thermal-mechanically deformed
N= normalized
Q= heat-treated
S= for shipbuilding
T= for tubes
W= weather resistant</td></tr>
<tr><td>B</td><td>Reinforcing steel

E.g: B500N
Properties: Yield stress (N/mm^2) (three numbers).</td><td>N= normal elongation
H= high elongation
G= other properties, if required with 1 or 2 numbers</td><td rowspan="2">Letter and number, if required</td></tr>
<tr><td>Y</td><td>Hot-rolled steels for prestressed structures

For example:
Y1770C

Properties: Minimal temporary resistance (N/mm^2) (four numbers, zero before is possible).</td><td>C= cold-drawn wire
H= hot-rolled or prestressed bars
Q= heat-treat wire
S= thin cable
G= other properties, if required with 1 or 2 numbers</td></tr>
</table>

Grading system of steels in Europe is given in Figure 6.3 and Table 6.1, 6.2.

Carbon high-quality steel have an enhanced cleanness, the limits of carbon content are restricted, sulphur content should not exceed 0.04%; the phosphorus content should be the same. Depending on deoxidation degree these steels permanent can be killed or unkilled.

Table 6.2. Group 2. Chemical composition of steels

Letter	Leading symbol	Additional symbol for steels	
		Group 1	Group 2
C G - cast steel work placed before, if required	Alloy-free steels with average content of Mn < 1%. E.g.: C35E4 Carbon content: (up to three numbers = 100 × average content of C of the assigned range	E= specified maximum content SR= specified range of content SD= for wire drawing C= specific ductility in cold state W= for welding wire G= other properties, if required with 1 or 2 numbers	Additional symbols are not provided
G - cast steel work placed before, if required	Alloys-free steels with content of Mn > 1%, E.g.: 28Mn6 Carbon content: (up to three numbers = 100 × average content C of assigned range.	Alloying elements: Letters = symbols for characterization of alloying elements. Numbers = separated, correspond to the average content of element, multiplied on the following factor. Element Factor Cr, Co, Mn, Ni, Si, 4 Al, Be, Cu, Mo, Nb, Pb, Ta, Ti, V, Zr 10 Ce, N, P, S 100 B 1000	

For low-alloy structural steels depending on the assured characteristics, categories are set which differ by the testing conditions on the impact strength.

Selecting the degree of responsibility and conditions of performance of steel structures the grades of steel and category of their delivery are taken into consideration.

Nonferrous Metals Alloys

For the production of aluminum alloys, foil, cable and electrically conductive elements aluminum of technical cleanness, containing 0.15-1% admixtures is used. Aluminum alloys are divided into two groups: 1 - alloys which are deformed, from which by rolling, pressing, dragging, forging and punching the various products and mouldings are obtained; 2- alloys which are assigned for making of production. The most typical aluminum alloys which are deformed (as *duralumin*) contain 2.2-5.2% Cu, up to 1.75% Mg, up to 1% Si, up to 1% Fe and up to 1% Mn. The grades of duralumin are signed with the letter D with a subsequent digit - conditional number of alloy with the increasing of which its mechanical strength increases. A commonly duralumin contains 4.4% copper, 1.5% magnesium, 0.6% manganese and 93.5% aluminum by weight. Typical yield strength is 450 MPa, with variations depending on the composition and temper. Although the addition of copper improves strength, it also makes these alloys sensitive to corrosion. For sheet products, corrosion resistance can be greatly enhanced by metallurgical bonding of a high-purity aluminum surface layer. Among castings aluminum alloys *silumins* are the most widespread alloys of aluminum with silicon, which is added in amount of 4-13%. Among the advantages of silumin is its high corrosion resistance, making it useful in humid environments.

Except of the aluminum alloys, *copper alloys* – brass and bronze, are applied in construction. Zinc is a basic alloying element (to 45%) in *brass*, and tin, aluminum, silicon and other elements are in *bronze*. At marking bronze and brass their composition is specified.

Magnesium alloys belong to the lightest structural materials (magnesium density is 1.7 g/cm^3). The same as aluminum alloys, they are divided into alloys which are deformed and casting. Low density in combination with high strength and corrosive resistance are also characteristic for *titanic alloys*.

Metal Products

Steel structures are produced of the rolled products and also of the formed and welded sections (Figure 6.4).

Rolled products are the most widely used, divided into four groups: steel shape, sheet steel, special types of rolled metal, pipes. Various reticular and whole structures are assembled from the rolling sections: columns, beams, bunkers, masts, towers, pipelines, reservoirs etc.

The steel shape includes the sections of popular demand (round, square, L-steel, channels, I-beams) and sections of the special assignments (rails and others). The L-steel, I sections and channels are widely used in construction.

The *angle sections* are divided into two types: equal and unequal. The lightest angle sections have sizes 20×20 mm and thickness 3 mm (20×3), and the heaviest - correspondingly 250×250 and 30 mm (250×30).

The double-T sections and channel are selected by the numbers which correspond to their height in millimeters. The numbers of I sections vary from 10 to 60, channels - from 5 to 40. The I sections are rolled with long up to 19 m, channels – up to 18 m. They are applied mainly as beams which work on a bend and axle loading. Channels differ from the I sections with the moved wall to the edge of shelves. The form of channels simplifies assembling to their walls of other elements. Channels are widely used as the roof beams of industrial buildings.

In the last years metallurgical industry started to supply ight-weight I-beams, the main distinguishing feature of which are thinner walls.

Steel depending on the thickness of sheets is divided onto the plate steel (4-16 mm), sheet steel (0.2-4 mm), universal wide strip steel (4-60 mm), slabs (60 - 200 mm), roll and corrugated steel.

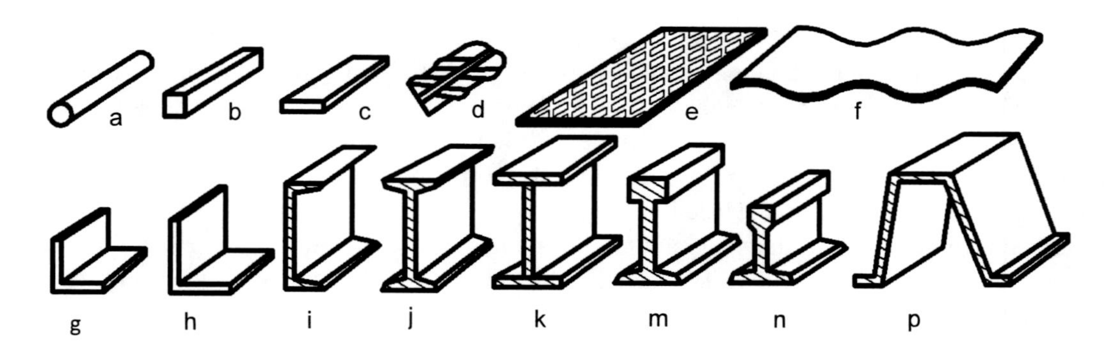

Figure 6.4. The rolling shapes:
a- round; b- square; c- strip line; d- die-rolled section; e- channeled; f- corrugated; g - L-equilateral; h – L-scalene; i- channel; j - double-T; k- welded double - T; m - crane rails; n- railway rails; p - grooved pile.

The sheet steel is in the form of the plates 8500 mm in width and up to 12 m in length. Steel sheets up to 40 mm thick are the most widely used in construction. Steel plates have length up to 4 m, width – 600-1400 mm. They are applied at production of bent slender sections and formed construction bent slender sections production. Manufacturing of bent sections enables to simplify substantially production technology of details and save up to 10% of metal. Shaped boarding is used for arranging of light and comfortable at the assembling roof.

The basic type of the rolled metal for construction is *reinforcing bar (rebar)*. Rebar, or reinforcing bar, is a common steel bar, and is commonly used in reinforced concrete and reinforced masonry structures. It is usually made of carbon, and has ridges for better mechanical anchoring into the concrete. (Figure 6.5).

Rebar is available in different grades and specifications that vary in yield strength, ultimate tensile strength, chemical composition, and percentage of elongation.

The grade designation is equal to the minimum yield strength of the bar in ski (1000 phi) for example grade 60 rebar has minimum yield strength of 60 ski. Rebar is typically manufactured in grades 40, 60, and 75.

Common specifications (accordingly ASTM) are:

- A 615 Deformed and plain carbon-steel bars for concrete reinforcement
- A 706 Low-alloy steel deformed and plain bars for concrete reinforcement
- A 955 Deformed and plain stainless-steel bars for concrete reinforcement
- A 996 Rail-steel and axle-steel deformed bars for concrete reinforcement

Historically in Europe, rebar is made of mild steel material with yield strength of approximately 250 N/mm². Modern rebar is made of high-yield steel, with yield strength more typically 500 N/mm². Rebar can be supplied with various grades of ductility, with the more ductile steel capable of absorbing considerably greater energy when deformed - this can be of use in design to resist the forces from earthquakes for example.

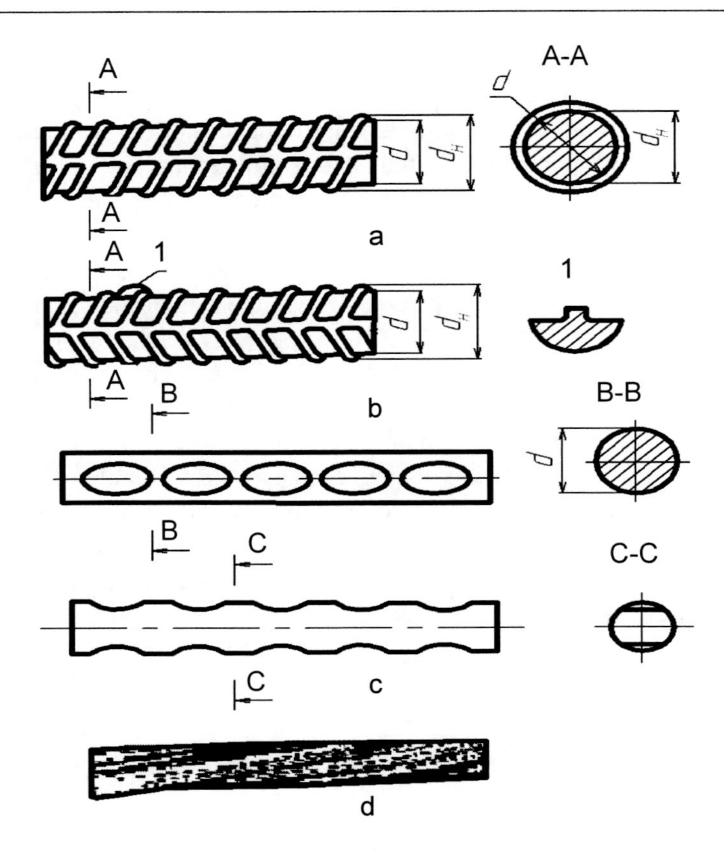

Figure 6.5. Reinforcing steel:
a, b- varieties of rebars; c- wire of die-rolled section; d- seven-wire bundled yarn.

Most grades of steel used in rebar are suitable for welding, which can be used to bind several pieces of rebar together. However, welding can reduce the fatigue life of the rebar, and as a result rebar cages are normally tied together with wire. Grade ASTM A706 is suitable for welding without damaging the properties of the steel. Besides fatigue concerns welding rebar has become less common in developed countries due to the high labor costs of certified welders. Steel for prestressed concrete may not be welded absolutely.

When welding or wire-tying rebar is impractical or uneconomical a mechanical connection or rebar coupler can be used to connect two or more bars together. These couplers are popular in precast concrete construction at the joints between members and to reduce rebar congestion in highly reinforced areas.

The metal pipes are widely used for the erection of towers, masts and other grillworks. The metal pipes are applied for the pressure pipings, to the pipelining in out-of-the-way places, and also when the use of pipes made of other material is impossible. The steel pipes are applied in pipelines with the power pressure more than 1.5 MPa, cast-iron - more than 1.2 MPa. Stand-pipes are subjected to the hydraulic pressure (P), MPa which is calculated by a formula:

$$P = \frac{20S\sigma_q}{D_{int}} \tag{6.1}$$

where S– minimum thickness of wall, mm; σ_q – permissible stress; D_{int}– internal diameter of pipe, mm.

Depending on the production method they are divided into the seamless hot-drawn and cold-drawn and electric-welded with longitudinal or spiral seam pipes.

The steel electric-welded pipes with internal cement-sand and external bituminous or other coating are widely applied for water transporting at the operating pressure up to 2 MPa for the reclamation work. The type of the steel pipes with anticorrosive coating is galvanized thin-walled pipes. They are applied for placing of the pipelines of the irrigative systems with pressure 0.8-1.5 MPa.

Along with the round welded pipes for steelworks square and rectangular welded pipes are applied.

Besides the steel, elements made of aluminum alloys as sheet products, bent and pressed sections are utilized in construction. Pressing enables to get aluminum sections, not only similar to steel ones but also other including comparatively complicated by shape.

Metal, assigned for the erection of building structures, is kept in stacks accordingly to its type, size and grade. It is palletized on the metal or wooden plates. For the aluminum products storage only the wooden plates are used. The wooden plates are saturated by the special mixture and dye with the paints to prevent the electrochemical corrosion of metal.

For storage of pipes, round and square steel the metallic shelvings are set which prevent the stack breakage.

6.2. Brief Technology of Metals

Cast-iron and Steel Production

Technological process of ferrous metal production includes melting of cast-iron from iron ores with the next processing in steel.

The basic method of the cast-iron production is *blast-furnace*. Blast-furnace process consists of three stages: iron reduction from the oxides containing in ore, iron carbonization and slag-forming. The iron-stones and fluxes are the raw materials.

Iron ores are subjected to the previous processing and crushing, to concentration and pelletizing before the melting. The crushed iron-stones are often concentrated by the magnetic separation. For the exception of sandy and clay particles ore is cleansed by water. Fine and dusty iron ores are pelletized with the agglomeration - baking on the fire grates of the sintering machines or running in the granulator with the subsequent drying and burning-out. The basic fuel at melting of cast-iron is cokes which is the source of heat and directly participates in the iron carbonization. Fluxes (limestones, dolomites or sandstones) are applied to reduce the temperatures of melting of spoil and binding it with the ash of fuel in slag.

The blast furnace is a vertical mine stack with walls made of fire-brick, placed in a steel jacket (Figure 6.6). The prepared raw materials are layerwise stoked from above in a furnace. As a result of coke burning at presence of oxygen from the air which is heated in the underbody of furnace, the carbon oxide forms which reducts iron from ore and can chemically react with it. At the same time the silicon, phosphorus, manganese and other admixtures are reduced.

Molten at a temperature of 1380-1420°C the cast-iron and slag is passed through the notch. Cast-iron is teemed into forms, and a slag goes on the reprocessing. The cast-iron, smelted in the blast furnaces intended for a steel production, foundry-iron – for obtaining various cast-iron elements. Special types of cast-iron (ferrosilicon, ferromanganese) are used in a steel production as the deoxidizing agents or alloy additions.

Steel is smelted from the iron by the oxidation with the open-hearth, converter and electrosmelting methods. The open-hearth method is basic method, but oxygen-converter method which has substantial technical-economic advantages in the last years become widespread.

At *open-hearth* method steel is smelted in open-hearth furnace, where gas or fuel oil is burned, and in the special chambers- regenerators air and gas fuel are heated due to the accumulated warmth of rejected burning products. The charge contains iron and metal scrap - debris and iron ore. Steel is produced as a result of charge melting, at which appears large amount of protoxide iron; products of oxidization of carbon and other admixtures and deoxidation- reduction of iron from a protoxide by additions of ferrosilicon, ferromanganese or aluminum.

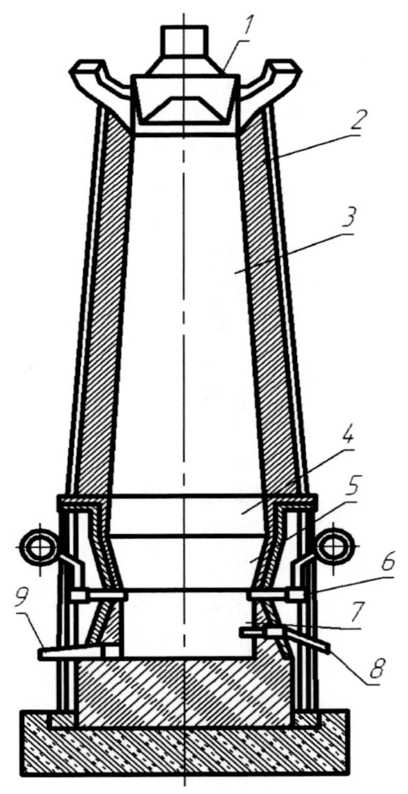

Figure 6.6. Scheme of the blast furnace:

1- charging equipment; 2- bell and hopper; 3- mine; 4- bosh; 5- shoulders; 6- blower nozzle; 7- hearth; 8- spout for the slag output; 9- spout for the cast iron output.

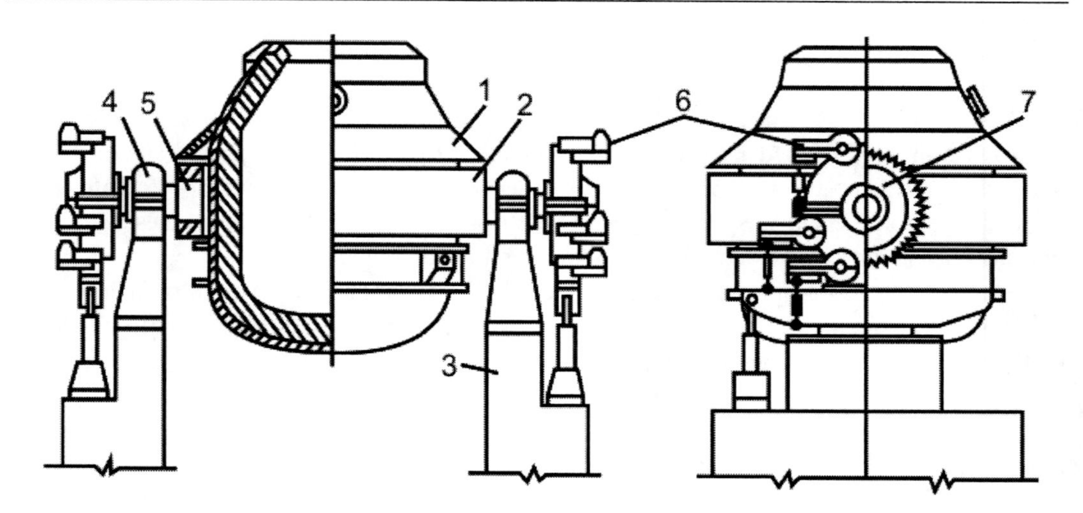

Figure 6.7. Chart of the converter:
1 - protective case; 2 - mantle ring; 3 - support frame; 4 - bearing; 5 - journal; 6 - mounted electromotor; 7 - follower gear.

Converter method of steel production consists in blowing out air or oxygen through the iron, poured in a converter (Figure 6.7). Upon the termination of process the converter is returned horizontally, the blowing is stopped, steel composition is checked up and outpoured in a scoop. At present oxygen-converter method at which blowing is carried out with the clean oxygen is widespread. Oxygen-converter method is characterized with a high productivity, does not need large capital investments and fuel. Coefficient of heat use at this method is about 70 %.

Electro-smelting method of the steel production is the most modern one. In the electric furnaces the high-quality steel is smelted which has a high purity and good deoxidation. The steel scrap is a basic material for the electro-smelting of steel.

Electro-smelting is carried out in arc (Figure 6.8) or induction furnaces. In arc furnaces metal is heated due to the heat of the voltaic arc, which appears between an electrode and molten metal, in induction one - due to the high frequency currents.

Steel production by *duplex process* is a progressive method, when liquid cast-iron is processed into steel in the converters, and then in electric furnaces steel is taken to the set chemical state.

The *electroslag and plasma-beam remelting* are the modern methods of steel production. High-quality alloyed steel is manufactured from the bars of ordinary steel at the electroslag remelting, which melts due to a heat which is generated at passing of the electric current through them. A molten metal passes through the layer of liquid slag and clears up from harmful admixtures and gases. At the plasma-beam melting plasma arc is a source of the heat, and at electron beam melting – electrons flow which is emanated by the cobalt unit with forming of deep vacuum in the melting space.

The steel production from ore by the *direct reduction* without blast-furnace is an effective process. The sponge iron which is semi-finished product at this method is obtained in rotary or blast furnaces and subjected to the crushing and separating from a barren rock. In the case of direct reduction of iron from ore there is no need in by-product-coking production - the basic one at a blast-furnace process that substantially promotes the productivity of steel production.

Steel, produced in the furnaces, is inundated in the special cast-iron patterns – casting molds or supply on the continuous casting. In the case of continuous casting the stream of molten metal goes at first to the crystallizer which is cooled down with water, and then to the area of aftercooling and automatically is cut on the ingots of required length.

The continuous casting, comparatively with an piece one, enables to promote the high labour productivity, quality of steel and to decrease the amount of the recrements of the casting industry.

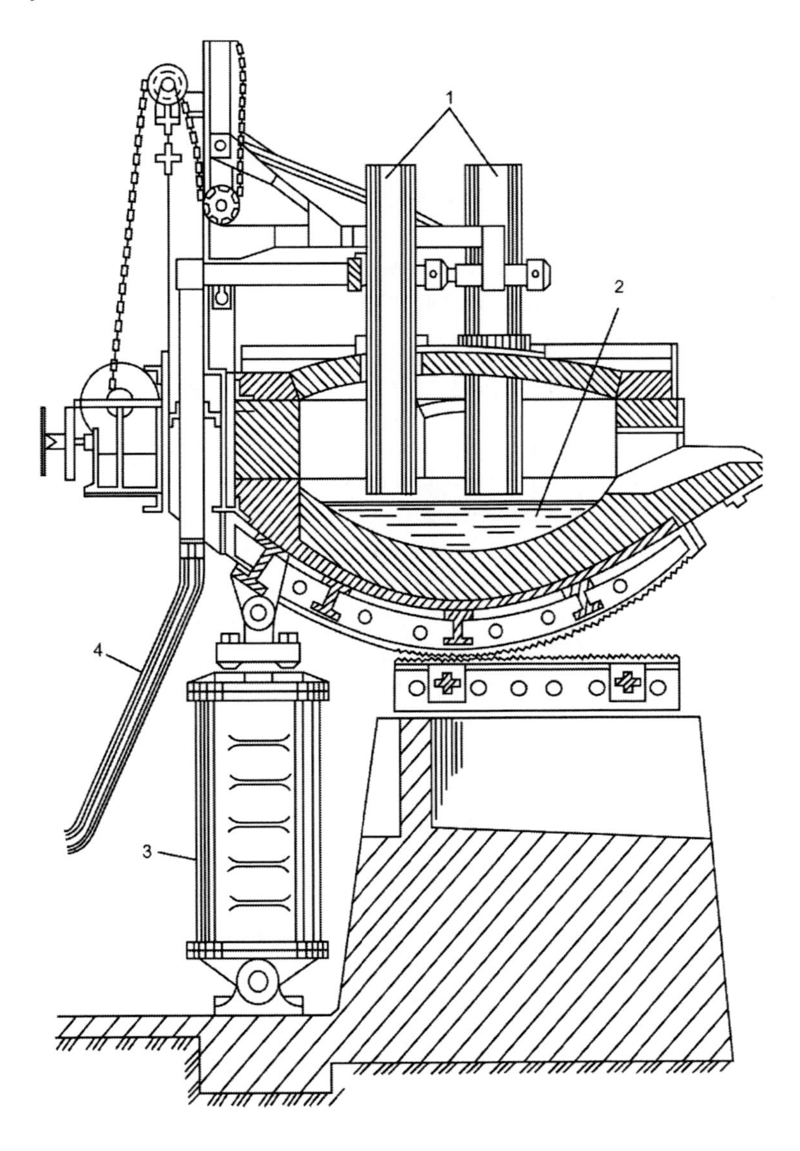

Figure 6.8. The electroarc furnace for the steel smelting:
1- electrodes; 2- bath; 3- mechanism for the bath overturn; 4- electricity cable.

Production of Aluminum

The rocks with high content of Al_2O_3 - bauxites, nephelines, alumstone and others are the raw material for the production of aluminum. The process of aluminum production consists of three technological stages: manufacturing of pure alumina, its electrolysing and extraction of original aluminum and finishing (that is cleaning). For the production of the alumina aluminum ore is processed in alkalis or acids depending on the composition of admixtures. Water solution of sodium aluminate forms at the alkaline method of admixtures extraction. It is easily decayed with formation of the aluminum hydrate and then Al_2O_3 formed after burning. Aluminum salts, which also consistently transform in $Al(OH)_3$ and Al_2O_3, appear at a acid method.

Metallic aluminum is manufactured by the method of electrolysis of alumina in a liquid electrolyte - molten cryolite Na_3AlF_6. Electrolysis is carried out in electrolysis baths which are consistently connected in series of 80-100 units. The liquid aluminum is cleared or refined in the closed scoop by the blowing out with the chlorine at a temperature 750-770°C.

6.3. Structure, Composition and Properties of Metals

Structure of the Metal Alloys

Metals are typical crystalline substances properties of which are predefined by peculiarities of their crystalline structure. The certain crystalline lattice is representative for every metal. Adhesion between the elements of metal crystalline lattice is a result of physical and chemical forces action, the most substantial among which are the forces between the positively charged ions, forming the lattice, and free electrons, which gather round them (metallic bonds). The theoretical strength of metals is comparatively high and approximately equals, for example, for iron 10^4 MPa. However, the different metals have actual strength in oftentimes lower than theoretical one because of the defects of crystals, microcracks, different impurities and especially dislocations.

The properties of metals are usually predefined by the peculiarities of their crystallization which takes place in transition metals from the liquid state to solid. The metals are stronger and more plastic the finer the metal grains. Each metal crystallizes at strictly designated temperature. In the case of the heat removal of the melt a lot of crystallization centers appear and their intensive growing takes place. Crystallization process of metals is exothermic. The metals at a different temperature can have a different crystalline structure. These transformations which are called polymorphic, also substantially affect their properties.

Alloys as mechanical mixtures, solid solution or compounds appear at the combined crystallization of several elements.

The properties of alloys which are the mechanical mixture of accrete crystals are medium between the properties of elements, forming them.

Solid solutions and compounds can have the properties which substantially differ from the properties of elements forming them. The characteristic samples of compounds among the iron - carbon alloys are ferric carbide or cementite Fe_3C, and among the alloys of aluminum copper - $CuAl_2$.

Unlike the pure metals, crystallization of alloys is carried out not at a strictly designated temperature, but in some temperature interval between the beginning and ending of crystallization. The equilibrium state of alloys, depending on their composition and temperature, is studied in accordance with the phase diagram (Figure 6.9). There is a row of characteristic lines and points on the phase diagram. The line of start of alloy consolidation is called the *liquidus* line, end of consolidation – *solidus* line. A point at the diagram at which the lowest temperature of alloy melting is achieved is called an *eutectic.*

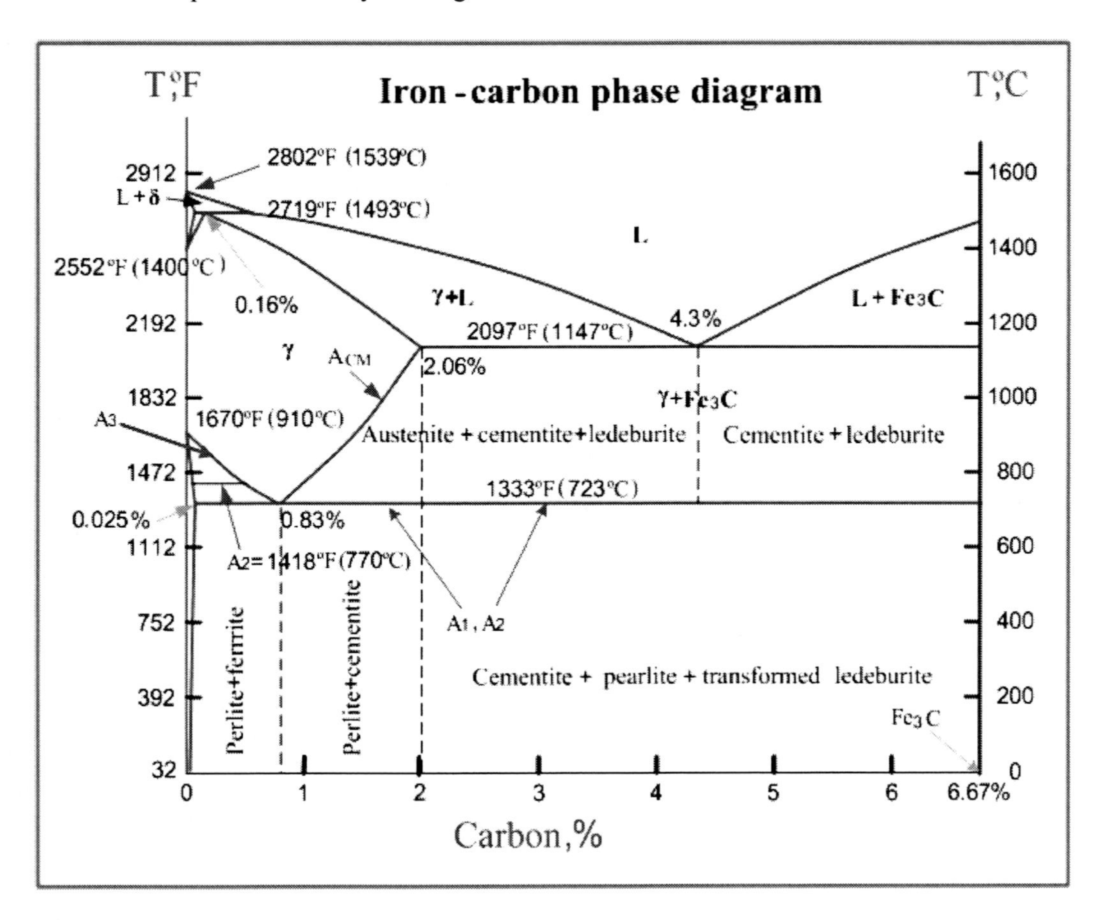

Figure 6.9. Phase diagram of iron-carbon alloys:

- L - Liquid solution of carbon in iron; - δ - ferrite – Solid solution of carbon in iron;
- Upper critical temperature (point) A_3 is the temperature, below which ferrite starts to form as a result of ejection from austenite in the hypoeutectoid alloys;
- Upper critical temperature A_{CM} is the temperature, below which cementite starts to form as a result of ejection from austenite in the hypereutectoid alloys;
- Lower critical temperature A_1 is the temperature of the austenite-to-pearlite eutectoid transformation. Below this temperature austenite does not exist;
- Magnetic transformation temperature A_2 is the temperature below which α-ferrite is ferromagnetic.

Table 6.3. Influence of basic components on some properties of steel

Element	Ultimate tensile strength	Yield stress	Percent elongation	Firmness	Impact elasticity	Weldability	Cold brittleness	Hot-brittleness	Corrosive resistance
Carbon	++	+	--	++	-	-	0	0	0
Manganese	+	+	-	+	0	0	0	-	+
Silicon	+	+	-	+	--	-	0	-	-
Nickel	+	+	0	+	+	0	0	0	+
Chrome	+	+	0	+	+	0	0	0	+
Copper	0	0	0	0	0	0	0	0	++
Vanadium	+	+	-	+	0	+	0	--	+
Molybdenum	+	+	-	+	0	+	0	-	+
Titanium	+	+	0	+	-	+	0	0	0
Phosphorus	++	+	-	+	-	-	++	0	+
Sulphur	-	-	0	-	-	0	0	+	0

Note: "++" - increases highly, "+" - increases, "--" - diminishes highly, "-" - diminishes, "0" - no noticeable influence .

In the iron-carbon system following basic phases form: liquid solution of carbon in the iron; *ferrite* that is a solid solution of carbon in α - or δ - Fe with the body-centered cubic lattice; *austenite* that is a solid solution of carbon in γ - Fe with a face-centered cubic lattice; *cementite* that is ferric carbide with a rhombic lattice.

Ferrite is similar by its properties to the pure iron, it is ductile, hardness equals HB = 80-100, elongation is 30-50 %, ultimate tensile strength – 250-300 MPa. Austenite has higher hardness and ductility than ferrite (HB = 170-200), though the strength is lower. The cementite - one of the hardest (HB = 800) and brittle components of iron-carbon alloys.

The solid solutions can decay (eutectoid decay) at the alloy cooling. The product of such decay of austenite at the temperature 727°C and carbon content 0.81 % is particularly *pearlite* - mixture of ferrite and cementite. *Ledeburite* - eutectic mixture of austenite and cementite forms also at the process of crystallization of iron-carbon alloys. As a result of austenite decay at a temperature 250-450°C, *bainite* - the superfine mixture of ferrite and ferric carbide crystallizes.

Phase transformations in the process of alloys crystallization are the basic phenomenon which is used for the change of their structure and properties. The variety of properties of steels is mainly determined by the transformations of austenite.

By the phase diagrams there are selected the alloys of such composition and structures to which are inherent the required properties.

Properties of Alloys

The choice of alloys for building structures is mainly based on the evaluation of four main descriptions: yield stress, ultimate tensile strength, percent elongation and impact elasticity at a temperature below zero.

With the increasing of carbon content the amount of friable and solid cementite grows in steel, ultimate tensile strength is increased (at C < 1%) and the percent elongation diminishes. Character of influence of basic components on the steel's properties is resulted in Table 6.3.

Steels for the metal constructions are welded well enough, if the carbon content does not exceed 0.17-0.18% in them, and total content of alloying elements is equal to 4-5%.

The low-alloy steels with the content of alloying elements to 3%, which are used in construction, have higher yield stress, less sensible to the deterioration and less capable to cold brittleness than ordinary carbon steels. The heat-resistant, corrosion-proof, wearproof and magnetic steels are the special alloyed steels.

6.4. Heat Treatment and Metalforming

Heat Treatment of Metals

The effective method of purposeful change of the metal structure to obtain required properties is heat treatment. Heat treatment of metal materials consists in heating of them to the required temperature, curing and cooling with specified rate to the certain temperature. The processes of heat treatment are divided into heat treatment itself under the action of the heat only - heat hardening, annealing, tempering, normalization; thermo-mechanical treatment at the combined action of heat and plastic deformation; physic-thermal treatment at combination of action of heat and change of chemical composition of metal.

Heat hardening of metals consists in their heating to the temperature not below the critical points that are the points on the phase diagram, at which the phase state of alloys quality changes, their next curing and fast cooling. The heat hardening carbon and low-alloyed steels have an aim to get the needle-shaped structure of martensite. *Martensite* - is the oversaturated solid solution of carbon in a - Fe. At cooling in water austenit is saved to the temperature approximately 200°C, and then instantly transforms into martensit. Structure of martensite is the most solid and brittle structure of steel.

To remove the internal stresses which arise up at heat hardening of alloys, and achievement of the best combination of the strength and plasticity they are subjected to *tempering* - that is to the heating till temperature below the lower critical points. There are distinguished low, middle and high tempering.

Temperature of heating at low tempering is 150-200°C, and at high tempering – 600-650°C. During the tempering of steel the martensit transforms into more stable structures.

If it is required to reduce brittleness, to increase plasticity and viscidity of metals, their workability, *annealing* is applied. The characteristic feature of annealing is slow cooling, which is achieved under the layer of sand, ash, slag etc. At annealing of steels after the previous heating austenite structure forms, which slowly cools down, transforms to the equilibrium structure in accordance with the phase diagram.

For achievement of fine-grained homogeneous structure with lower plasticity, but with higher brittleness, than after annealing, *normalization* of alloys is conducted, the peculiarity of which is air cooling. Normalization is more simple type of treatment comparing with annealing. It promotes brittleness and enables to get the cleaner surface of steel at cutting.

The *thermomechanical treatment* of metals includes heating, plastic deformation and cooling of metal, combined in the single technological process. The essence of combined process consists in the hardening of part blanks just after the end of hot-plastic working (forging, rolling). At the same time part blanks are specially not heated, but remaining heat is utilized after a hot molding. Due to that fuel is saved for heating at hardening, demand in heater furnaces diminishes, time on making the details shortens and mechanical properties are substantially improved.

The chemical thermal treatment of metals consists in the saturation of products surfaces with the carbon, nitrogen, aluminum and other elements. Products, assigned for the high weathering in combination with the shock loadings, are subjected to this type of treatment. Such wares must have high hardness of surface coatings and enough viscid core. Depending on the peculiarities of the deformed product it is possible to influence also on the fatigue strength, to promote resistance of material surface to the action of external aggressive environments etc.

More frequently *cementation* - saturation of steel with a carbon - is used. Surface coating of low-carbon steels during the cementation carbonizes to 0.8-1.1% C and is subjected to the heat treatment, resulting martensite structure forming. At the carbonization the products are placed in the steel boxes, filled with the grout mixture which consists of absorbent carbon and carbonate, heated to the temperature 900-950°C. Oxygen of air, heating with carbon, forms the carbon oxide which in the case of presence of iron dissociates with formation of carbon.

The nitriding, cyanidation, aluminizing, chromium coating, and other types of the chemicothermal treatment of metals are used except of the cementation.

The surface treatment of products by the optical quantum generators (lasers) spreads at the last years. Comparatively with other kinds of treatments the laser heating has such advantages: possibility of treatment of the out-of-the-way places and surfaces of details of the complicated configuration; absence of warping and deformation of details; possibility of ray energy passing on the large distances and treatment of details, making from any materials (irons and steels, non-ferrous and solid alloys, powdered metals); high speed of the process and others like that. Application of modern laser devices provides possibility of complete mechanization and automation of the process.

Metalforming

Metalforming is carried out by rolling, dragging, pressing, forging and punching. These methods are based on the use of plastic deformation of the cold or heated metal under the action of rollers, stamps, firing-pins and others.

Not only their form but also structure changes at the plastic deformation of metals - there is stretching of grains, grinding of them. Treatment of metals in the cold state results in peening - the increasing of ultimate strength and hardness and the declining of ductile and shock viscidity. The temperature considerably influences on the strain hardening. If temperature rises, ductility grows continuously, and deformation resistance diminishes.

Within reasonable limits of heating there is the recrystallization, the previous structure of material recommences. The deformation resistance of metals grows if the speed of deformation increases.

Rolling - metalforming, at which the deformation is carried out with squeezing between the cylinders (rollers) of rolling mill, which are revolved (Figure 6.10, a). The initial materials for the production of rolled products are casting blocks, and final products are various sections.

Dragging is a treatment of metals by pressure by pulling of wire, stick or pipe through the opening of matrix with a section, smaller than initial section of blank. The products obtain the specified geometrical shape, exact sizes, clean surface by dragging. Dragging of metals is carried out mainly in the cold state in the special drawbenches (Figure 6.10).

The compacting (pressing) is punching shear of ductile materials through the opening of matrix. This method of ductile working is applied mainly for manufacturing products of various types from the nonferrous metals and in some cases also from steel.

Forging and *punching* are processes of plastic working of metals, executable on the special equipment — hammers with mass of falling parts to 5000 kg or hydraulic presses, which developing efforts to 200 mN. Forging is produced at heating of metal to the so-called forging temperature with the purpose of increase of its ductility and decline of resistance to deformation. The temperature interval of forging depends on chemical composition and structure of the processed metal, and also from the type of technological operation. For steel temperature interval 800-1100°C and for aluminum alloys — 420-480 °C. At punching a metal is limited from every quarter by the walls of stamp. During deformation it takes the shape of this cavity.

6.5. Corrosion of Metals

Chemical corrosion of metals is observed in dry gases and nonelectrolytes. In the first case the metals collapse, reacting with gases and vapour at a temperature over 100°C (oxidization of metal during heating, corrosion of the furnace accessories, blades of steam and gas turbines). In the second case - corrosion takes place in petroleum, petrol, oils etc.

The metals mostly collapse in result of *electrochemical corrosion*, i.e. under the action of electrolytes - water solutions of salts, acids and alkalis. This type of corrosion takes a place in atmospheric conditions, at the action of sea, river, underground and other types of water.

In the iron-carbon alloys mostly, ferrite is an anode, and the cementite or nonmetallics are cathodes (Figure 6.11).

Especially intensively electrochemical corrosion flows at periodic influence of electrolyte on a metal, for example, in the case of the previous moistening and drying of metal constructions of hydrotechnical structures.

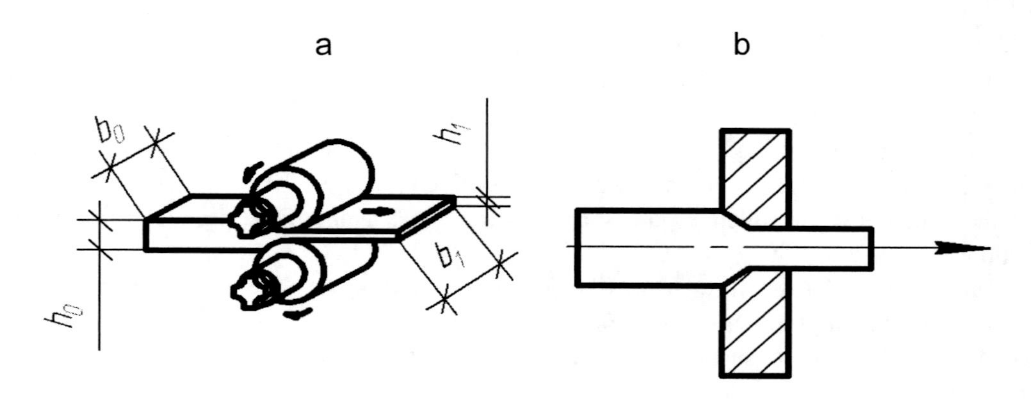

Figure 6.10. Plastic working:
a- rolling; b- dragging.

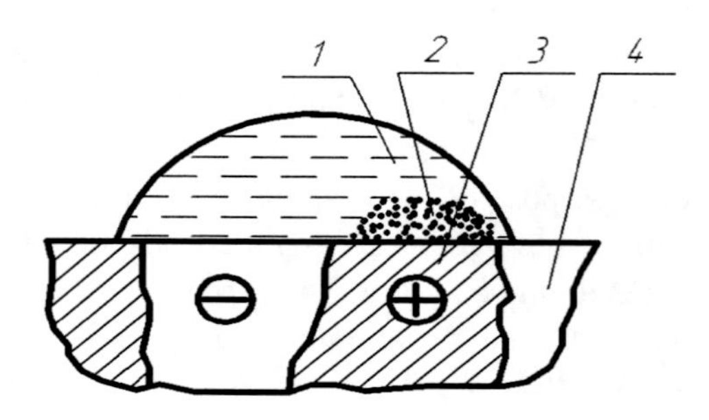

Figure 6.11. Scheme of electrochemical corrosion of steel:
1- water; 2- products of corrosion; 3- cementite; 4- ferrite.

Corrosion of metals is accelerated by the action of electric current. *Electrocorrosion* of metals takes place during the connecting of construction to the source of current, for example, if the pipeline is located near-by tram-car rails. The carbon and sulphur dioxides, chloride of hydrogen and chlorous salts are instrumental in the active passing of corrosion process.

The corrosion resistance of different metals is compared with the unique scale (Table 6.4).

The corrosion of the metal products control is a very important task. Annually, about 10% of melted metal are lost in result of corrosion in the whole world.

Table 6.4. Scale of corrosion resistance of metals

Group of resistance	The corrosion rate, mm per year	Points	Group of resistance	Corrosion rate, mm per year	Points
High resistant	Less than 0.001	1	Reduced resistant	0.1-0.5	6
Enough resistant	0.001-0.005	2		0.5-1	7
Resistant	0.005-0.01	3	Low resistant	1-5	8
	0.01-0.05	4		5-10	9
	0.05-0.1	5	Nonresistant	More than 10	10

Different methods are applied for the metal protection against corrosion. The additions of nickel, chrome, phosphorus and especially copper effectively promote the corrosion resistance of steels; manganese influences negatively. The widespread method of increasing of the structural steels corrosion resistance is their alloying with 0.2-0.4% Cu, at that the corrosion resistance grows on 20-30%.

Metals are protected from the corrosion by applying paints and varnishes, inorganic non-metal and metal coatings. Effective varnish coatings are developed on the basis of synthetic polymers. The most various protective films belong inorganic coatings of metal: oxide, phosphate, cement, asbestos-cement, enamel and others.

Metal coatings differ by the mechanism of protective action (anodal and cathode), and also by the method of application (galvanic, metallizative, sprayed from the fusion, etc.). Anodal coating is characterized by more negative electrode potential, than in a metal which is protected. In the case of damage, the coating collapses with greater speed, carrying out a protector role in relation to the basic metal. The zinc and aluminum are appointed mainly for anodal coatings of steel; the copper, tin, nickel and other-for cathode. If the anodal coatings are damaged, their protective action disappears.

Service life of coatings is 25-50 years. They enable to provide replacement of high-alloy steels with the low-alloy, to defense the steel in atmospheric conditions and at a high temperature, in salt water, etc.

Aluminum is the most perspective among the nonferrous metals for coatings. The protective films from aluminum and its alloys are coated by arc spraying pistols, by evaporation in a vacuum, electrolytic deposition, plasma and laser spraying.

The active method of corrosion protection is an electrochemical method. It is based on the change of electrochemical potential of construction due to polarization of them by a direct current from an external source or from a protector. At protector defense (Figure 6.12) to the metal, which is protected from corrosion, connect zinc anodes which enable to defend a valuable construction from corrosion. The cathode defense is use for the corrosion protection of marine hydraulic structures, main pipelines and other responsible constructions. In this case, the protected structure is connected to the negative pole of permanent current source, so the steel acts as cathode. The mechanical scrap, connected to the positive pole is an anode.

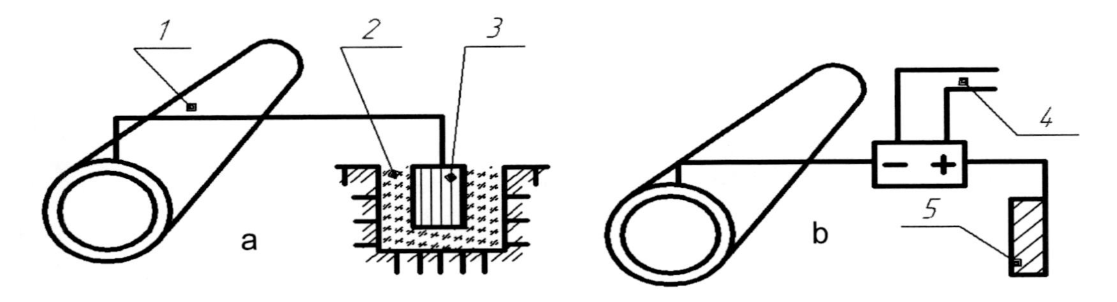

Figure 6.12. The corrosion protection of metal pipelines:
a- protector; b- cathode; 1- pipe; 2- filing up; 3- protector; 4- source of electric current; 5- metal.

The corrosion rate of metals is reduced by additions of the inhibitors and passivators. The salts of alkaline and alkaline-earth metals (chromites, nitrates and others) as the strong oxidants – successfully decelerate the corrosive process in neutral and alkaline environments. The using of additions (in particular calcium nitrite-nitrate) is an effective at the protection against corrosion of steel reinforcement in the concrete.

Self-Assessment Questions

1. What is the value of metal materials in construction? Give the definition of the ferrous and nonferrous metals.
2. Point out the classification of cast irons and tell about the peculiarities of their types.
3. Point out the classification of steels and tell about the peculiarities of their types.
4. Tell the peculiarities of cast iron and steel marking; give the examples of cast irons and steels grades.
5. Describe the basic alloys from the nonferrous metals which are used in construction.
6. Give the description of steel products used in construction.
7. Specify the basic classes of reinforcing steel and types of pipes, applied in construction, their peculiarities.
8. Describe the technological process of cast iron production in the blast furnace.
9. What are the peculiarities of technological process of steel production by the open-hearth, converter and other processes?
10. What basic phases form during the crystallization in the system of iron-carbon, their influence on properties of alloys?
11. Give a concept about the phase diagrams and their value in the metallurgy.
12. Which basic properties are taken into account at selecting steel for building structures? How does chemical composition of steels influence on their properties?
13. Describe the basic methods of heat treatment and its influence on properties of metals.
14. Describe the basic methods of plastic working of metals.
15. What are the peculiarities of the chemical and electrochemical corrosion of metals?
16. What are the methods of anti-corrosive protection of metals?

Chapter 7

CERAMIC MATERIALS

Materials which are obtained from mineral, mainly clay raw material by forming, drying and firing at a high temperature are called ceramic materiaals.

Ceramic materials – are the most ancient amongall of artificial stone materials. A word "ceramics" results from Greek "κεραμικός" - pottery art. Age of ceramic brick is over 5 thousand years. Remains of buildings and constructions from a ceramic brick are found on territory of Ancient Egypt (3rd - 1st Millennium BC). Ceramic brick was known also in India. The first temple of Gera in Olimpya (6th century BC) was covered with a roof tile and designed with the decorations from terra-cotta in Ancient Greece. 3-4 floor dwelling-houses, arches and bridges were built from a ceramic brick in Ancient Rome. Some buildings from ceramics were saved to our time on the territory of different countries and strike harmoniousness and beauty of architectural decisions.

Depending on application ceramic materials are divided into building, heat-resistant, electrical engineering, of the special destination (technical ceramics), chemically proof, service-utility.

There are distinguished *rough ceramics*, which have a heterogeneous structure, and *thin ceramics* with a fine-grained homogeneous structure. Greater part of construction ceramic materials and refractories, acid-proof brick, and others belong to the rough ceramics. The basic representatives of thin ceramics are the glazed faience - fine-pored ceramic material, covered by transparent fusible glaze, and porcelain - sintered ceramic material of white colour with water absorption less than 0.5%. Varieties of ceramics are *majolica* - products from the coloured fired clay, covered by glaze; *terracotta* - one-colour naturally painted unglazed products.

Construction ceramic materials are classified by a degree of sintering and destination. There are distinguished dense and porous materials depending on a degree of sintering. *Dense ceramic materials* have water absorption less than 5%. Flooring tiles, clinker brick, and sewage pipes belong to this group. *Porous ceramic materials* have water absorption over 5%. Construction brick, facade tiles, drainage pipes etc are included in the group of porous ceramics.

According to the destination construction ceramic materials are divided into such types: *masonry (wall)-* brick, ceramic stones and panels; *facing-* facing brick, tiles for facade and internal facing and so on; *roof-* roof tile; *road* - clinker brick; *sewage-technique* - washstands, water-closet pans, etc; *underground communications* - sewage and drainage pipes; *heat-*

insulation – diatomite and other light products; *light aggregates* for concretes - claydite and aggloporite; *special assignment* - heat-resistant, high acid-proof, etc.

7.1. Compositions of Ceramic Mixtures

Raw materials for ceramic mixtures are divided into plastic and inductile ones. The role of plastic materials in ceramic mixtures is carried out by clays. Kaolin (variety of clays), that consist mainly of mineral kaolinite is basic raw material for faience and porcelain mixtures.

Clays are sedimentary rocks which consist mainly of clay minerals with the characteristic stratified structure. Clays have a property to form plastic paste with water which after a firing has strength like a stone.

Clay minerals form the most micronized factions in clays - less than 0.005 mm, particles 0.005-0.05 mm make drank factions, and over 0.05 mm - sand.

All of clay minerals belong to aqueous aluminum silicate.

Kaolinite($Al_2O_3 \cdot 2SiO_2 \cdot 2H_2O$) has the stratified structure (Figure 7.1, 7.2), its crystalline lattice is composed from the double-layer flat packages, every package is included by the layer of oxygen-silicon tetrahedrons and layer, formed with the atoms of aluminum, oxygen and hydroxyl group. Size of particles of kaolinite hesitates from 1 to 3 μm.

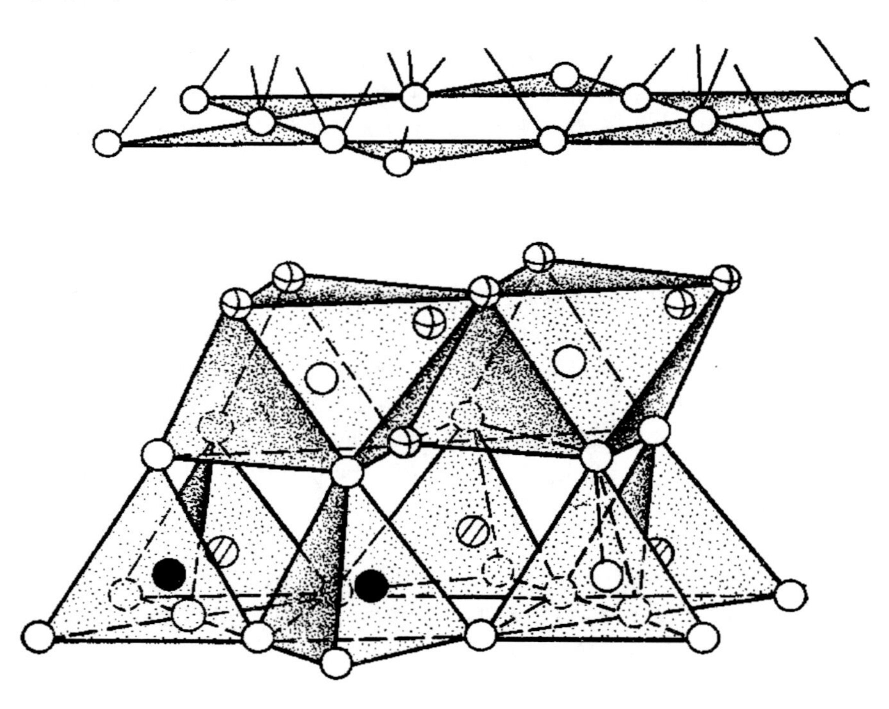

Figure 7.1. Schematic structure of kaolinite.
Circles indicate the centres of gravity of separate ions:
○ - O; ⊕ - OH; ● - Si; ⊘ - Al (Mg)

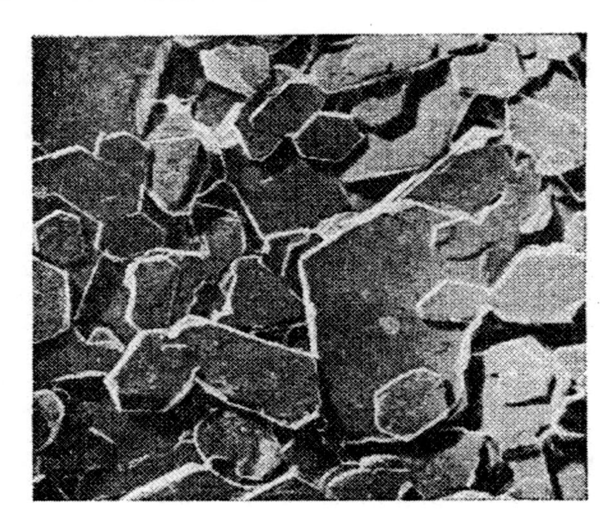

Figure 7.2. Electronic microphotography of kaolin (×17200).

Montmorillonite ($MgO·CaO·Al_2O_3·4SiO_2·nH_2O$) has the stratified crystalline lattice as well as kaolinite. Separate layers can be moved apart under the act of the wedged molecules of water. It is explained by montmorillonite clays ability to take in plenty of water intensively, firmly to retain it and hardly to give at drying. As a result of the strong swelling at moistening the volume of montmorillonite can be increased to 16 times. Sizes of particles of montmorillonite considerably are less than 1 μm.

Hydro micas ($K_2O·MgO·4Al_2O_3·7SiO_2·2H_2O$ and etc) are products of long-term hydration of micas. It occupies intermediate position between kaolinite and montmorillonite according to the intensity of connection with water. There is characteristic feature of separate cations to the isomorphic substitutions. So, Si^{4+} can be replaced by Al^{3+} and the last one with Mg^{2+}. Size of particles is about 1 μm.

Basic admixtures in clays are presented by quartz, by carbonates, ferrous compounds, feldspars, and organic impurities.

Clays are presented frequently in nature as polymineral systems, including two or three clay minerals. Kaolin clays, approaching to the monomineral clays, can be used also for ceramic materials manufacture. Clays are classified according to the large number of features. Classification of clay raw material, applied for the manufacture of ceramic construction materials according to the most characteristic features is resulted in Table 7.1.

Possibility of application of clays for the production of one or another ceramic product is determined foremost by their chemical composition (Figure 7.3).

Clays belong to the hydrophilous colloids and contain a plenty of water, considerable part of which is in a view of thin layers, dividing solid particles.

The ability of clays to retain the certain amount of water is called its *moisture capacity*. There is of, the grains of clay minerals having negative charges in the system clay-water, create the force field power. The dipole molecules of water which are in a direct closeness to clay particles form adsorbed multilayer. As far as a removal from solid surface diffuse layer of weakly bond water is formed.

Table 7.1. Classification of clays for the obtaining of construction ceramics

№	Classification feature	Value		Types of clays
1	Content of clay substances (less than 0.005 mm)	More than 60% 30-60 20-30 5-10		Heavy clays Clays Clay loam Loamy sand
2	Contents of disperse fractions (less than 10 and 1 μm)	<10 μm 1) More than 85% 2) 40-85 3) Less than 40%	<1 μm More than 60% 20-60 Less than 20%	1) Fine-dispersed 2) Dispersed 3) Coarse-dispersed
3	Plasticity (difference in humidity required for achievement of fluidity of clay and its rolling)	>25 15-25 15-7 >7		High-plastic Average-plastic Moderate-plastic Low-plastic
4	Sintering ability (water absorption by weight, %)	<2 from 2 to 5 >5		Strongly-sintering Medium-sintering Non- sintering
5	Content of Al_2O_3+TiO_2 in the fired state, %	More than 40% 30-40% 15-30% Less than 15%		Highly-basic Basic Semiacid Acidic
6	Refractoriness	>1580^{0}C 1350-1580^{0}C <1350^{0}C		Fireproof Refractory Fusible

Granulometric composition of clays depends on content of clay, dust and sandy fractions in it and determines its fitness as ceramic raw material (Figure 7.4). The moisture-capacity of clays increases with the increase of dispersion. It has the most value in the montmorillonite, the least at kaolinite clays.

The basic mechanical property of clays that is shown at influence of external forces on them is *plasticity* –an ability to acquire the set form and save it after the action of efforts stops without the break of continuousness.

Plasticity index of P_{lst} can be calculated by formula:

$$P_{lst} = \omega_{f.l} - \omega_{r.l}, \qquad (7.1)$$

where $\omega_{f.l}$ - humidity on fluidity, %; $\omega_{r.l}$ - humidity on rolling of clay specimen, %.

Plasticity of clays is determined by mineralogical composition and content of clay minerals and also by dispersion.

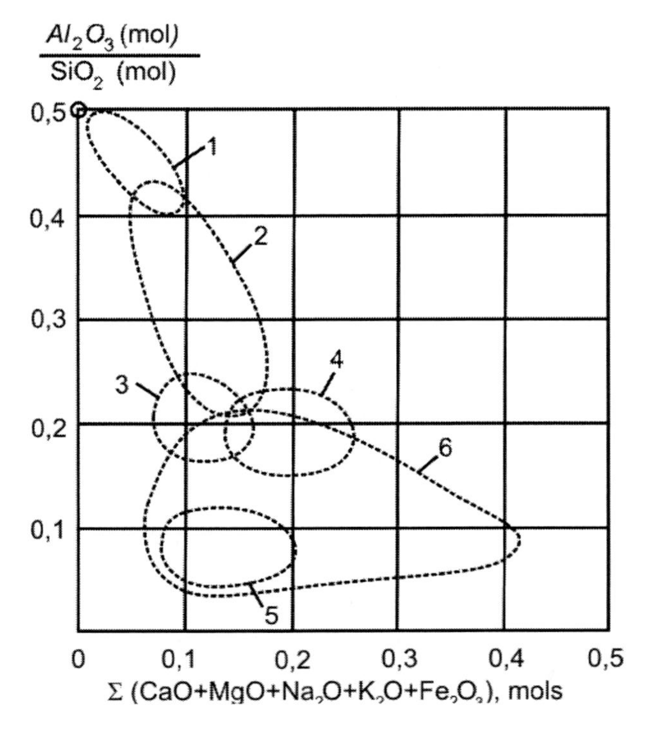

Figure 7.3. Industrial assiqnment of clays depending on its chemistry:
1 - fire-resistant chamotte products; 2 - tiles for floors, sewers, acid-resistant and stone products; 3 - pottery and terra-cotta products; 4 - roofing tile; 5 - paving clinker; 6 – brick.

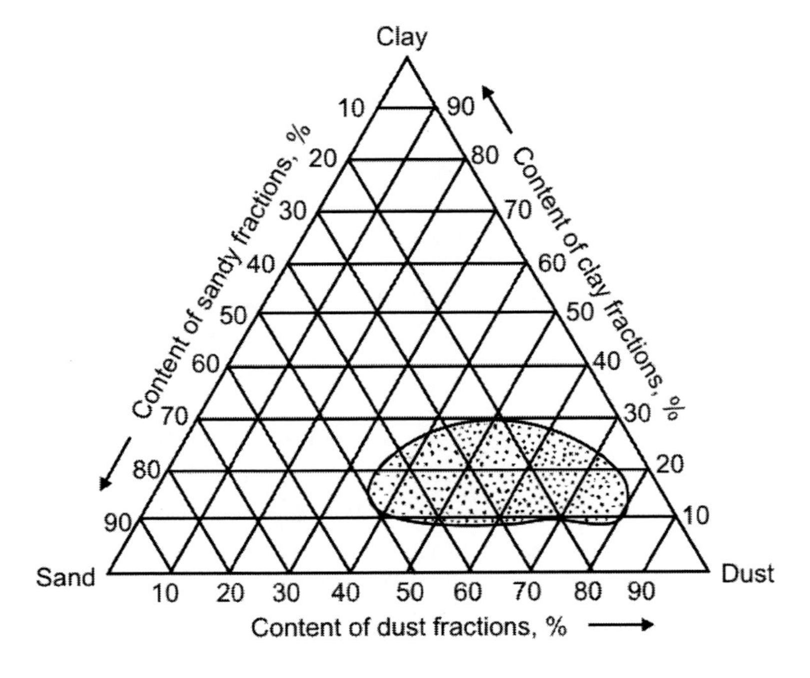

Figure 7.4. Graph of granulometric composition of clays, which are applicable for ordinary brick manufacture.

Changes, what be going on with clay at drying are characterized with a size of so-called air shrinkage of specimens, by a sensitiveness to drying, by the properties of hydraulic conductivity.

Shrinking deformations of clay at drying are depended on the forces of capillary pressure. Pressure arises up at deleting of water from capillaries, when the power equilibrium is violated between capillary and gravity forces.

The quantitative measure of shrinking deformations at drying of clays is a value of relative shrinkage (S_{hr}):

$$S_{hr} = \frac{l_0 - l_1}{l_0} \cdot 100\%, \quad (7.2)$$

where l_0 and l_1 - accordingly initial and final after drying length of specimen.

For the most of clays relative air shrinkage changes in the range of 2-8%.

Shrinkage is increased with growth of moisture capacity of clays; it depends also on the mode of drying, achieving at a greater value in the conditions of slow evaporation of moisture.

The different value of shrinkage of clay products at drying causes appearance of tensile stresses and cracks. The index of crack resistance of clays is its coefficient of drying sensitivity ($K_{d.s}$):

$$K_{d.s} = \frac{V}{V_0\left(\frac{m_0 - m}{V_0 - V} - 1\right)}, \quad (7.3)$$

where V_0 and m_0 - accordingly volume and weight of moist specimen directly after forming; V and m – the same, for specimen, dried up at 16 - 20^{0}C.

At $K_{d.s}<1$ clays are considered as not sensitive; $K_{d.s}=1\text{-}1.5$ – medium sensitive and $K_{d.s}>1.5$ - highly sensitive to drying.

Along with the shrinkage deformations, crack resistance of clay products at drying is determined by its strength, tensility and water conductivity.

Required properties of ceramic mixtures can be achieved by the use of different additional materials. For example, in raw mixtures, used for wall ceramics production can be added additional materials for achievement:

- Improvement of forming properties of mixture (high-plastic clay, surface-active substances);
- Improvement of drying properties (chamotte, sand, dehydrated clay, sawdust);
- Improvement of the terms of firing (ash of heat and electric power plants, slags, coal);
- Increase of the strength and frost-resistance (crushed glass, pyrite drosses, iron-stone);
- Special purpose - improvement of products color, efflorescence prevention, neutralization of harmful influence of the natural impurities in clays (dyes, liquid glass, calcium chloride, etc.).

Table 7.2. Compositions of the mixtures for the manufacture of sanitary products

Materials	Content, % for the mixtures		
	Faience	Semivitreous product	Porcelain
Fireproof clay	25	17-20	13-23
Kaolin clay	30	25-30	27-30
Pegmatite	-	15-20	-
Fieldspar	-	-	18-20
Quartz sand	33	24-26	22-27
Dolomite	-	1	-
Crushed finished product	12	4-6	8-12
Sulfurous cobalt (beyond 100%)	0.025	0.025	-
Soda	0.2	0.12	-
Liquid glass	0.1–0.15	0.2	-

Compositions of ceramic mixtures are determined by the required structure of products and their properties. It is necessarily apply fluxing agents - admixtures, formative with clay fusible compounds at firing and promoting the increase of sintering degree (feldspar, pegmatite, nepheline syenite, etc.) at a production, for example, of the faience and porcelain construction products along with clays and kaolin clays. Some compositions of the mixtures for the production of sanitary-building products are shown in Table 7.2.

Fireproof clays and kaolin clays are basic raw materials for the obtaining of the most widespread refractories - chamotte products. For the manufacture of high-aluminous refractories is applied corundum as basic raw material; for magnesite refractories - magnesite, talcous refractories - talc, etc. The oxides of some metals can be used as basic initial components for the manufacture of technical ceramics.

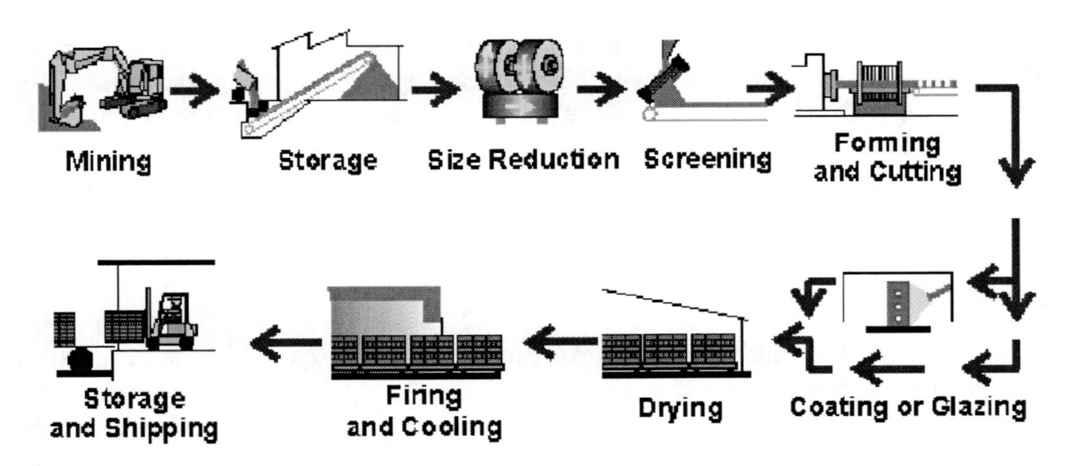

Figure 7.5. Diagrammatic representation of brick manufacturing process.

7.2. Basic Methods of Ceramic Products Manufacture

There are three basic methods of preparation of ceramic mixture – soft-mud (plastic), dry and slip.

At a soft-mud method ceramic mixture has humidity 20-30%. It can be obtained directly from clay with quarry humidity, mixing up with admixtures, or from dry powders with the next moistening.

At a dry method mixture has humidity of 6-12%, it is obtained by preliminary drying, grinding and mixing up the components.

At a slip method ceramic mixture in the form of suspension (slip) with the humidity 30-35% can be used.

A softmud method is the most effective at the using of clays with enhanced quarry humidity, which gets wet in water well and dry - at the dense structure of clay raw material and low initial humidity. Slip method is used when it is needed to attain the especially careful mixing of initial components (porcelain-faience production, production of facings tiles).

The *brick manufacturing process* has six general phases: 1) mining and storage of raw materials, 2) preparing of raw materials, 3) forming the brick, 4) drying, 5) firing and cooling and 6) de-hacking and storing finished products (Figure 7.5, 7.6).

Surface clays, shales and some fire clays are mined in open pits with power equipment. Then the clay or shale mixtures are transported to plant storage areas.

Continuous brick production regardless of weather conditions is ensured by storing sufficient quantities of raw materials required for many days of plant operation. Normally, several storage areas (one for each source) are used to facilitate blending of the clays. Blending produces more uniform raw materials, helps control color and allows raw material control for manufacturing a certain brick body.

To break up large clay lumps and stones, the material is processed through size-reduction machines before mixing the raw material. Usually the material is processed through inclined vibrating screens to control particle size.

Tempering, the first step in the forming process, produces a homogeneous, plastic clay mixture. Usually, it is achieved by adding water to the clay in a pug mill, a mixing chamber with one or more revolving shafts with blade extensions. After pugging, the plastic clay mixture is ready for forming. There are three principal processes for forming brick: stiff-mud, soft-mud and dry-press.

In the stiff-mud or extrusion process, water in the range of 10 to 15 % is mixed into the clay to produce plasticity. After pugging, the tempered clay goes through a deairing chamber that maintains a vacuum of 15 to 29 in. (375 to 725 mm) of mercury. De-airing removes air holes and bubbles, giving the clay increased workability and plasticity, resulting in greater strength.

Next, the clay is extruded through a die to produce a column of clay. As the clay column leaves the die, textures or surface coatings may be applied. An automatic cutter then slices through the clay column to create the individual brick. Cutter spacings and die sizes should be carefully calculated to compensate for normal shrinkage that occurs during drying and firing. About 90 % of brick in the United States are produced by the extrusion process.

The soft-mud or molded process is particularly suitable for clays containing too much water to be extruded by the stiff-mud process. Clays are mixed to contain 20 to 30 % water and then formed into brick in molds. To prevent clay from sticking, the molds are lubricated with either sand or water to produce "sand-struck" or "water-struck" brick.

Dry-press process is particularly suited to clays of very low plasticity. Clay is mixed with a minimal amount of water (up to 10 percent), then pressed into steel molds under pressures (3 to 10 MPa) by hydraulic or compressed air rams.

Wet brick from molding or cutting machines contain 7 to 30 % moisture, depending upon the forming method. Before the firing process begins, most of this water is evaporated in dryer chambers at temperatures ranging from about 100 °C to 200 °C .The extent of drying time, which varies with different clays, usually is between 24 to 48 hours. Although heat may be generated specifically for dryer chambers, it usually is supplied from the exhaust heat of kilns to maximize thermal efficiency. In all cases, heat and humidity should be carefully regulated to avoid cracking in the brick.

Hacking is the process of loading a kiln car or kiln with brick. The number of bricks on the kiln car is determined by kiln size. The brick are typically placed by robots or mechanical means. The setting pattern has some influence on appearance. Brick placed face-toface will have a more uniform color than brick that are cross-set or placed face-to-back.

Figure 7.6. Chart of ceramic bricks manufacture by the method of the dry forming:
1- excavator and trolleys for raw materials transporting;
2- batcher; 3- grinding mill; 4- mixer; 5- extruder with a cutting device; 6- trolley-platform for adoby bricks; 7- tunnel drier; 8- rotating device; 9- tunnel kiln; 10- finished-products storage; 11- transportation of wares.

Brick are fired between 10 and 40 hours, depending upon kiln type and other factors. There are several types of kilns used by manufacturers. The most common type is a tunnel kiln, followed by periodic kilns. Fuel may be natural gas, coal, sawdust, methane gas from landfills or a combination of these fuels.

In a tunnel kiln, brick are loaded onto kiln cars, which pass through various temperature zones as they travel through the tunnel. The heat conditions in each zone are carefully controlled, and the kiln is continuously operated. A periodic kiln is one that is loaded, fired, allowed to cool and unloaded, after which the same steps are repeated. Dried brick are set in periodic kilns according to a prescribed pattern that permits circulation of hot kiln gases.

Firing may be divided into five general stages: 1) final drying (evaporating free water); 2) dehydration; 3) oxidation; 4) vitrification; and 5) flashing or reduction firing. All except flashing are associated with rising temperatures in the kiln. Although the actual temperatures will differ with clay or shale, final drying takes place at temperatures up to about 200 ^{0}C, dehydration from about 450 °C to 700 °C, oxidation from 550 °C to 950 °C and vitrification from 850 °C to 1200 °C.

Bricks are sorted, graded and packaged. Then they are placed in a storage yard or loaded onto rail cars or trucks for delivery. The most of bricks are packaged in self-contained, strapped cubes, which can be broken down into individual strapped packages for ease of handling on the jobsite. The packages and cubes are configured to provide openings for handling by forklifts.

Drainage and sewage pipes are made as a usual by a soft-mud method. The pipes of small diameter are formed on horizontal band presses, and large - on vertical vacuum.

From the powdered mixtures products are formed on the presses of high-pressure (10-30 MPa and anymore). At the pressing drying of raw - protracted and complicated process is abbreviated. Wall brick products, refractories, and different thin-walled products are made of the powdery mixtures.

Slips are used for making of the thin-walled products with complicated configuration (technique, decorative, chemically proof ceramics and others like that). This method of forming is based on property of gypsum forms to absorb part of water from slip in itself inundated in it.

Formed products are dried, for giving required strengths to them at a firing. Distinguish the convection and radiation drying. At the convection drying warmth is passed from heat-carrier - smoke gases or air, and at radiation it emanates from the heated surfaces. For drying of products there is selected an optimum temperature condition for which a drying device has the best production, and the term of drying and amount of spoilage are minimal. The tunnel and conveyer dryers of continuous action, which provide a high productivity of manufacture, are widespread.

The determinative stage of ceramic technology, at which the properties of ceramics are formed, is a firing. Formation of liquid phase as fusions begins at a temperature close 700°C and intensively develops as far as the increase of temperature of firing. Glassy fusions are glued together separate grains of ceramic mixture in the unique monolith. Sintering of ceramic mixtures can be also due to reactions in a solid phase.

The most important crystalline compound at the firing of ceramic mixtures is a mineral - mullite $3Al_20_3{\cdot}2Si0_2$, which most intensively forms in the interval of temperatures 1000-1200°C.

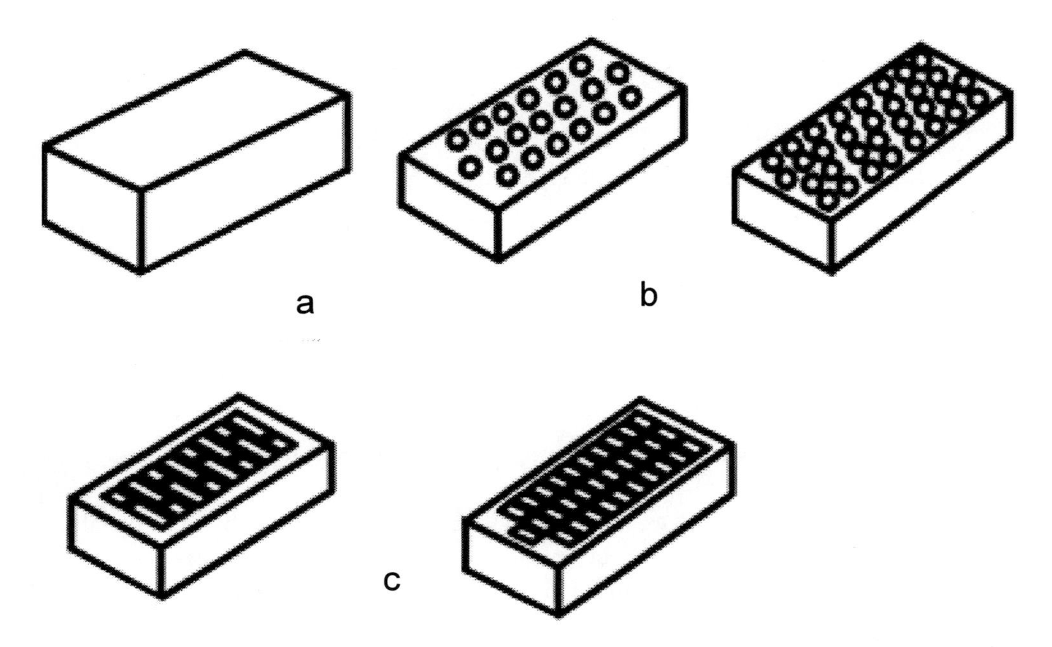

Figure 7.7. The variety of ceramic bricks: a -solid; b–round hollows; c –square hollows.

For the firing of products of every kind set a certain temperature condition. If ceramics are fired at a temperature, below than required, it has enhanced porosity and low strength, and at a higher temperature strength grows, but products can be deformed and melt.

Ceramic materials are formed as a result of high temperature processes, going at firing of various mineral materials, possessing with ability to sinter. Under *a sintering ability* in general case understand ability of matters at firing to make a more compact with formation of stony solids. Along with crystalline phases at sintering usually there is the partial melting, that results in formation of glassy phase. Correlation of crystalline and glassy phases determines physical - mechanical properties of materials in a great deal.

Sintering is carried out on air or in a protective gas environment at a temperature, as a rule, not below 0.6-0.7 from the temperature of melting of material.

Rough ceramic products are firing usually at temperature 950-1050^0. Noticeable formation of fusion takes a place at temperatures 850-900 and more high. From a physical side the action of fusion shows up in shrinkage of products, declining of its porosity. Such ceramics are characterized with a coarse-grained structure, often by high porosity (7-20%).

Firing of products of thin ceramics (facings tiles, sanitary ware) on traditional technology is conducted at temperatures to 1250-1280^0 C with formation of densely caked mass (water absorption to 4%). For firing on the speed conveyers of facings tiles at a temperature to 1100 ^{0}C it is required to utilize the proper compositions of raw material charge. The special types of hard porcelain are fired at temperatures to 1450^0 C. As a result of firing to sintering the structure of thin ceramics is characterized with sweating of superficial areas of grains of quartz, feldspar with formation of grains of mullite $2Al_2O_3 \cdot 3SiO_2$; voids of such ceramics are mainly closed, its size is about 10 μm, amount does not exceed 5%. The amount of glass mylite phase is 22-28%.

7.3. BASIC TYPES OF CERAMIC MATERIALS

Masonry Materials

Masonry materials are the most considerable for construction among ceramic products. They can be divided by the type of products at bricks, blocks and panels; by destination - at ordinary (for building of external and internal walls) and facing (for facing of walls); by the method of manufacture - on the products of the soft-mud and dry pressing; by heat engineering properties and density - at effective ceramics with average density ρ_0 <1400 kg/m^3; conditionally effective - with average density equals 1400-1600 kg/m^3 and ordinary - with average density $\rho_0 \geq 1600$ kg/m^3.

A ceramic brick (Figure 7.7) has a form of rectangular parallelepiped and different standard sizes (Table 7.3).

Bricks may also be classified as solid (less than 25% perforations by volume, although the brick may be "frogged," having indentations on one of the longer faces), perforated (containing a pattern of small holes through the brick removing no more than 25% of the volume), cellular (containing a pattern of holes removing more than 20% of the volume, but closed on one face), or hollow (containing a pattern of large holes removing more than 25% of the brick's volume).

Blocks may be solid, cellular or hollow. Blocks have a much greater range of sizes. Standard coordinating sizes in length and height (in mm) include 400×200, 450×150, 450×200, 450×225, 450×300, 600×150, 600×200, and 600×225; depths (work size, mm) include 60, 75, 90, 100, 115, 140, 150, 190, 200, 225, and 250.

The minimal compressive strength of bricks and blocks produced in the USA and European Union is about 7 MPa. Strength can be changed according to the field of application. Usually, in England common house brick has the strength in a range of 20–40 MPa.

House walls from the solid brick with density 1700-1900 kg/m^3, as a rule, have a thickness up to 2.5 bricks. If apply the hollow bricks with density 1300-1450 kg/m^3, thickness of wall is diminished on half-brick, weight on 35%, consumption of mortar on 45%. At application of ceramic blocks, consumptions of materials and labour are diminished at 20-30% comparatively with an ordinary ceramic brick.

Table 7.3. Standard sizes of brick

Standard	Metric sizes
http://en.wikipedia.org/wiki/File:Flag_of_the_United_States.svgUnited States	203 × 102 × 57 mm
http://en.wikipedia.org/wiki/File:Flag_of_the_United_Kingdom.svgUnited Kingdom	215 × 102.5 × 65 mm
http://en.wikipedia.org/wiki/File:Flag_of_South_Africa.South Africa	222 × 106 × 73 mm
http://en.wikipedia.org/wiki/File:Flag_of_Australia.Australia	230 × 110 × 76 mm
Russia	250 × 120 × 65 mm

Water absorption of solid bricks should be no less than 8%, hollow bricks no less than 6%. Frost-resistance of common house brick in the saturated water state can be in the range of 15-50 cycles of freezing and thawing.

For a bricks and blocks the row of other indexes are also standardized: size of hollows, deviation from sizes, number of the damaged corners and edges, amount of cracks, etc.

Overfired and insufficiently fired ceramics should be rejected as defective. Products, which have the lime impurities, should be also rejected due to ability of cracking at steam treatment.

Large wall blocks and panels belong to industrial ceramic masonry materials (Figure 7.8).

Large blocks are made from a solid or hollow bricks and blocks on the special plants with filling of seams by cement-sand mortar. In order that mortar hardened quickly, blocks are steamed. Application of wall blocks enables to decrease the total labour consumption on the erection of walls, and also to reduce the construction terms. Basic defects of the of large blocks use are plenty of nominal sizes, complication of installation and increase of cement content.

Most radically the task of industrial application of ceramic bricks and blocks decides making from them at the plant *of panels* from which then walls are installed. By construction panels are divided on one-, multi-layered, continuous and with cuts; by purpose - for external and internal walls; by the type of loadings, what they are taken up - on bearings, self-bearings and partitions. Nominal sizes of panels are determined by the construction of building for which they are intended.

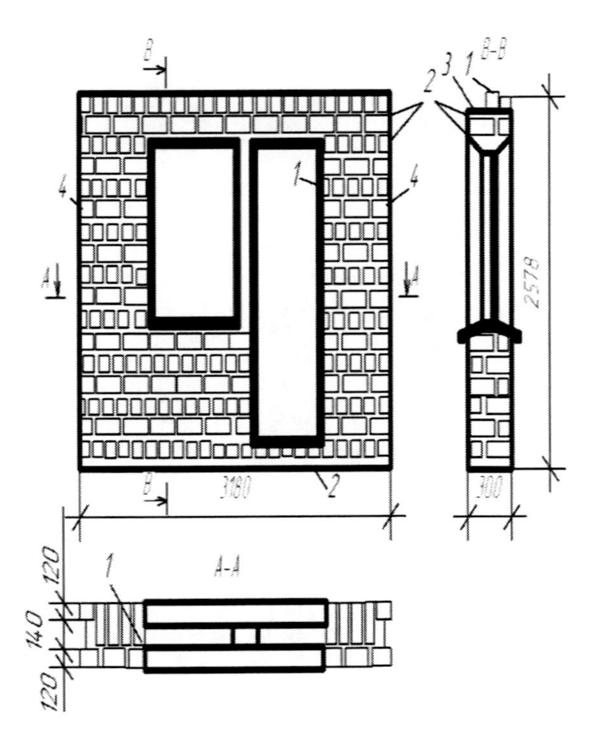

Figure 7.8. One-layer brick panel:
1– vertical welded framework; 2– horizontal welded framework; 3–fixed detail ; 4- seam.

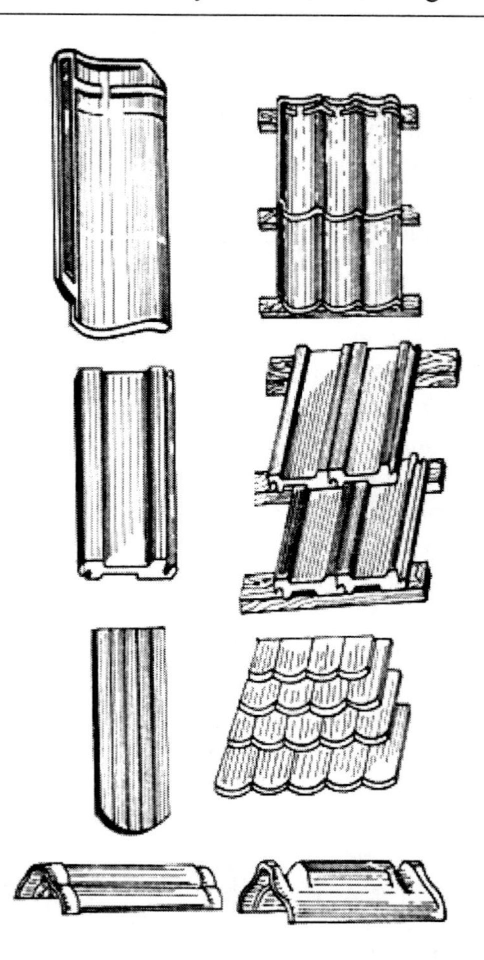

Figure 7.9. The variety of roof tiles.

Depending on the number of floors and loading which is perceived, they are designed by thick in one, one and a half and two bricks. Panels are reinforced by the horizontal and vertical welded steel frameworks.

The technological process of manufacture of ceramic panels includes preparation of mortar and metallic framework, forming, finishing of panels, heat and moisture treatment. It is applied technology of making of ceramic wall panels with the compacting of mortar by vibration and without vibration.

Labour consumption for the construction of walls from ceramic panels essentially decreases comparatively with the same at the monolithic bricking.

Ceramics for Roofs and Facings

Basic ceramic material for roofs is a *roof tile*; its main positive features are high strength and longevity. So, if the normative term of service of roll materials is 15 years, steel sheet 25, asbestos cement - 30 years, tiles - 80 years. The lacks of tile are its large weight (35-70 kg/m^2), fragility, necessity in large inclination of roof. Small-size and hand method of layering of tile predetermines large labor intensiveness of arranging of tiling. Due to the increasing of issue of industrial roofing materials application of roof tiles is diminished. The variety of roof tiles (Figure 7.9) is manufactured.

A large number of shapes (or "profiles") of roof tiles have evolved. Basic of them are:

- Flat tiles - the simplest type, which are laid in regular overlapping rows. An example of it is the "beaver-tail" tile. This profile is suitable for stone and wooden tiles, and most recently, solar cells.
- Imbrex and tegula, an ancient Roman pattern of curved and flat tiles that make rain channels on a roof
- Roman tiles - flat in the middle, with a concave curve at one end at a convex curve at the other, to allow interlocking.
- Pantiles - with an S-shaped profile, allowing adjacent tiles to interlock. These result in a ridged pattern resembling a ploughed field. An example of this is the "double Roman" tile, dating from the late 19th century in England and USA.
- Mission or barrel tiles are semi-cylindrical tiles made by forming clay around a curved surface, often a log or one's thigh, and laid in alternating columns of convex and concave tiles.

Roof tiles are 'hung' from the framework of a roof by fixing them with nails. The tiles are usually hung in parallel rows, with each row overlapping the row below it to exclude rainwater and to cover the nails that hold the row below.

There are also roof tiles for special positions, particularly where the planes of the several pitches meet. They include ridge, hip and valley tiles.

A roof tiles have to maintain the destructive loading 800-1400 N, its frost-resistance should be not less than 25 cycles of freezing and thawing.

Ceramic materials for facings are divided into products for facing of facades and internal facing. A face bricks and blocks, facade tiles, carpet ceramics, terracotta details belong to the products intended for facing of facades of buildings.

The most effective materials for facing of facades are *face ceramic bricks and blocks,* which execute the functions of masonry and facing materials at the same time. They are produced solid, perforated, cellular and hollow with the same strength as the ordinary masonry materials. White or ordinary clays are used for facing products making. Water absorption of products from white clay is 6-12%; from ordinary clay is no more than 14%. There are additional requirements in relation to homogeneity of raw material and color for the fired products.

Besides the one-layered a double-layered bricks are produced. They have facing layers in thick 3-5 mm. From white and colored clays the decorative facing surface of bricks and blocks is obtained by engobe coating or glazing. A glaze unlike engobe during a firing melts and forms a brilliant glassy layer. Engobed surface is dull.

Facade tiles depending on sizes are divided into large- and fine-size, on the methods of fixing - embedded, which are set simultaneously with construction of walls and attached, which are set on mortar after the building of walls.

Tiles are produced of many nominal sizes. The large-size tiles of square and rectangular forms can be manufactured. A facing surface of tiles can be smooth and shiny, rough or fluted, one- and multicolored, with a pattern, glazed and unglazed. The frost-resistance of tiles can achieved 50 cycles of freezing and thawing, water absorption - up to 4%.

By coordinating sizes (a coordinating size includes a total size of nominal size of tile and width of mortar seam which are 4-8 mm). Usually they can be from 50x50 mm to 300x150 mm (thickness 7 or 9 mm).

Fine-size tiles can be supplied also in carpets. In this case they are glued by a facing side on a paper. The sizes of carpets should be related to the sizes of constructions which are faced.

Tiles for internal facing are produced square (100× 100, 150×150 and 200×200 mm, etc), rectangular (150×25, 150×75, 150×100, 200×100 and 200×150 mm, etc) and shaped (for the corners of facing, cornices and plinths). Most widespread tiles are covered by transparent or opaque glaze. Tiles for internal facing are applied in dwelling houses, public-service centers and industrial buildings with large humidity or with enhanced hygiene and sanitary requirements.

Floor tile should have a high density, strength, abrasion resistance. The clays which are well sintering with admixtures of fluxing agents and pigments are used for their production.

Water absorption of the unglazed tiles should be no more than 3.5%, glazed – 4.5%, loss of weight at abrasion no more than 0.06 g/sm^2. Tiles for a floor by a form are square, rectangular, three -, five-, six-, octahedral and figured. A facing surface of tiles can be smooth and shiny, fluted, stamped and multicolored, figured, marble-like.

Ceramic tiles may be painted and glazed. Small mosaic tiles may be laid in various patterns. Floor tiles are usually set on mortar, consisting of sand, cement and often a latex additive for extra adhesion. The spaces between the tiles are filled with sanded or unsanded floor grout.

The unglazed, natural are painted ceramics with water absorption of 8-10% belong to terracotta. Bas-reliefs, cornices and other details of the architectural decoration are made of terracotta.

Ceramic Pipes

Drain-pipes are applied for arranging of the closed drainage-systems. Drain-pipes can be cylinder, six- and octahedral forms with different internal diameters.

Drain-pipes are produced unglazed, with smooth internal surface, water absorption 12-18% and frost-resistance (if Required) not less than 15 cycles. During a test pipes should maintain hydraulic pressure not less than 0.05 MPa and destructive effort not less as 1.5-2.0 MPa.

Efficiency of drain-pipe is determined by the amount of water which is taken by it for time unit.

Unlike drain-pipes, *sewage-pipes* belong to the dense sintered ceramics. They are produced as hollow cylinders with a bell, with an internal diameter 125-600 mm and length 300-1200 mm. For the increase of watertightness and chemical resistance, and also diminishing of resistance flowing of liquids, sewage-pipes are covered from outside and from within by a glaze.

Basic quality indexes of sewage-pipes are following: possible external loading 2 - $3 \cdot 10^4$ N/m; hydraulic pressure not less than 0.2 MPa, water absorption not less than 9-11%; acid resistance not less than 90%.

Heat-resistant and Acid-proof Ceramics

For ordinary refractories a fire-resistance should be 1580-1770°C; high fire-resistant refractories – 1770-2000°C; the highest fire-resistance refractories - over 2000°C. Fire-resistance is measured accordingly to the softening temperature of the pyroscopes, specially made from ceramic mixture, which have a form of the three-cornered cut away pyramid, and also by optical and electric pyrometers.

Typical refractoriness are dinas, quartz and chamotte ones. They differ by a raw materials and methods of making. *Dines refractories* are obtained by the firing of quartz materials, as a rule, at a lime on a clay copula. *Quartz refractories* are made by melting of natural quartz, *chamotte* - by the firing of fire-clays.

Singularities of dinas refractories are a heat-resistance and low firmness against the action of most slag. They are not used for the lining of stokers; and are applied for construction of glass-work, steel-smelting, coke and other furnaces.

Among the high fire-resistant materials alumina, dolomite, chromic, carborundum refractories are applied the widest; highest fire-resistance - magnesite, chromic-magnesite, zirconium, graphite are used. Products from clean oxides and unoxygen compounds - nitrids, borids, carbides, silicides, etc have a highest fire-resistance and other valuable properties.

Ceramic refractories are made as a bricks of different forms and sizes.

For protection of building constructions and equipment which work under the conditions of acid aggressive environments, in particularly at the lining of chimneys, an *acid-proof brick* is used. It is made rectangular, radial and shaped from plastic refractory or heat-resistant clays of the proper chemical composition, which do not contain the impurities of gypsum, sulphuric pyrites, and carbonates.

Ceramic products are transported in containers, which provide their safety, mechanized loading and unloading.

Self-Assessment Questions

1. What is the value of ceramic materials in construction? Tell about classification of ceramic materials.
2. Describe clays as basic raw material for the manufacture of ceramic.
3. Give examples of ceramic mixtures compositions.
4. Tell about the basic methods of the ceramics manufacture.
5. What are the features of ceramics firing?
6. Give characteristic of ceramic bricks.
7. Give characteristic of masonry ceramic materials.
8. What are the features of effective ceramic materials?

9. What are the features of roof and face ceramic materials?
10. What are the features of ceramic pipes?
11. Give characteristic of fire-resistant and acid-proof materials.

Chapter 8

GLASS AND GLASS-CERAMIC MATERIALS

Glass is an amorphous material which acquires mechanical properties of solid after cooling of mineral fusion.

The basic products of industry are sheet glass as: unpolished, polished, multi-layered, tempered, etc. Last years an assortment of glass products is considerably extended and mastered a number of new products from glass, particularly colored, reinforced, structural profile and plate. The increase in production of window pane 3...4 mm in thick was caused by the increase of area of the light openings. The various products of the architectural-construction assignment are widely used: glass blocks, structural glass, glass packs. They are applied as effective transluent materials for external and internal enclosing structures in housings, civil and industrial buildings.

The glass carpet tessellated tiles, marbled glass - coloured tiles from opaque glass mass, sheet glass coated by ceramic paints on one side, patterned stained glass are assigned for revetment of buildings.

Glass pipes, glass-fiber waterproofed and heat-insulating materials are widespread.

The production of sheet and pressed glass crystalline material - *slag glass-ceramic* is developed. The small cost of raw materials and highly-mechanized technology of production of slag glass-ceramic along with high operating qualities promotes this material to become one of the most effective construction materials.

Glass with overcoated on its surface thin transparent metallic or plastic film; sheet glasses, agglutinate transparent synthetic films belong to progressive composition materials. The production of structural profile glass box-like section, stained reinforced glass, glass with selective light penetration for all ranges of waves' lengths, stained rolled glass and other new perspective construction glass products is advanced.

Enamels and *glazes* are similar to glass by composition and materials structure, which are sheeted by thin layer on the surface of metallic and ceramic products. They are intended for a giving to the products decorativeness, protection of them against corrosion, improvement of mechanical and dielectric strength.

8.1. Vitreous State. Compositions and Properties of Glass

Vitreous State

A number of the theories, which explain the vitreous state of substance, is known at present. Crystallite theory and theory of continuous disordered lattice are obtained the greatest acknowledgement.

Before the *crystalline theory* there were ideas about the glass as about the completely amorphous supercooled liquid. According to the crystalline theory glass has the ordered zones of submicron sizes - crystallites. Crystallites consist of the tetrahedrons [SiO_4] and polyhedrons [MeO_n], their sizes are within the limits $(15...25)\cdot10^{-10}$ m.

The author of the *theory of disordered lattice* - American researcher W.H.Zakhariasen advanced it in 1932. He considered the glass- as continuous atomic three-dimensional lattice, deprived of symmetry and of periodicity. This lattice, according to Zachariasen, is the infinitely large cell, in the units of which the atoms or ions are located, not one pair of which is structurally equivalent.

The hypothesis of Zakhariasen was proved to be unable to explain many experimental facts, established late. Least convincing proved to be the assertion of Zachariasen about the chemical uniformity of multi-component glass.

Along with basic theories examined above others were proposed: polymeric, micro-heterogeneous and others.

Types and Compositions of Glass

Glass classifies depending on composition and assignment.

Oxide, chalcogenide and halide glasses on composition are distinguished. Silicate (quartz), aluminosilicate, borosilicate, alumino-phosphate and other glasses, which names are determined by the glass-forming oxides are included into the most numerous group of *oxide glass*.

In the construction silicate and aluminosilicate glass is used frequently united into one group - *silicate glass*, taking into account that in their composition SiO_2 predominates. The most of industrial glass include the silicates. *Phosphate glass* fusions are used for the production of optical, electric vacuum glass; *borate* - for the special forms of glass. The mixed borosilicate glass is used for manufacturing the optical and thermally resistant glass products.

The oxides of lithium, potassium, sodium, beryllium, calcium, magnesium, strontium and other elements can be added in the composition of many glasses for regulating the properties.

The chemical composition of construction glass is within the limits (%): SiO_2 - 71,5...72,5, Al_2O_3 - 1,5...2, Na_2O - 13...15, CaO - 6,5...9, MgO - 3,8...4,3.

The basic component of glass is silica which introduced into the glass charge with the quartz sand, ground quartzite or sandstone. Alumina is introduced in the form the feldspar, kaolin, the sodium oxide - soda and sodium sulfate, and the potassium oxide - in the form potash and potassium nitrate. The oxide of sodium accelerates the process of glass formation, reducing melting point and facilitating the clarification of glass mass, but increases the coefficient of thermal expansion and decreases the chemical stability. The oxide of potassium decreases the tendency of glass toward the crystallization, gives luster to it and improves light transmission. The oxides of calcium and magnesium are ensured in the glass charge by different varieties of the natural carbonates of calcium and magnesium. These oxides increase the chemical stability of glass, and MgO - decreases the tendency of glass toward the crystallization.

Boric anhydride is introduced into the charge with the borax and boric acid. It increases the melting speed, contributes to the purification of glass, it increases thermal and chemical stability, decreases the tendency to crystallization and reduces thermal expansion coefficient.

Basic requirement for all types of raw material is the absence of impurities and uniformity in the content of basic oxide.

Different auxiliary raw materials are introduced together with the basic into the glass charge: *lighters, opaquers, dyes*. Lighters contribute to removal of the gas bubbles from the glass mass (chlorides and sodium sulfates, fluorspar). As the opaquers, fluorine and phosphorus compounds and sometimes antimony of tin are used for obtaining the opaque glass. Dyes for the glass are divided into the molecular, which are dissolved in the glass mass, and colloidal, dispersed in the form of colloidal particles. The first include the compounds of cobalt (dark-blue color), chromium (green), manganese (violet), uranium (yellow), iron (brown and blue-green tones), and the second - gold (ruby), silver (golden yellow), selenium (pink) and other

The classification of construction glass according to the designation and the basic types of the glass products, produced by industry, are given in Table 8.1.

Glass with the special properties is obtained by the regulation of their chemical composition and, using the appropriate methods of treatment in the production process. Thus, *figure glass* is obtained from the fusion of metal or salts. In this case one side of glass is thermally polished, and deep relief with the intermittent pattern and the alternation of sections with the polished and lusterless surface is formed on another.

For obtaining the *uviol glass*, which has the ability to transmit not less than 25% ultraviolet rays, the raw materials with the especially high value of the degree of purity (less than 0.03% oxides of iron) are used. A *bsorbing heat* of the long-wave part of the spectrum of solar rays' glass with the lowered light transmission are made with the introduction the contribution of the oxides of cobalt, nickel and iron.

Table 8.1. Classification of construction glass and glass products

Form and the designation of the glass products	Types of the glass products
Sheet transparent	Window, plate glass, reinforced figured, heat absorbing, colored
Structural	Glass blocks, glass packet, profile the glass-, door leafs
Facing	Carpet- mosaic tiles, enameled tiles, glass-crystallite, smalt, glass crumb
Heat- and soundproof	Products from the glass staple fiber, mats construction, the foam glass
Non-woven fiberglass materials	Fiberglass roofing, moisture-proof, thermal insulating linens, glass paper, filtering materials

A number of special glasses are obtained during the putting at the surface of glass of the metal films and their oxides. Film coatings are brought by electrochemical treatment, by chemical precipitation from the solutions, by cathode sputtering, by evaporation in the vacuum. The transparent plastic films belong also to the surface of glass, which change the natural vibration frequency of glass. Such glass is used for the soundproof glazing.

The large group of glass products are based on the fiberglass. There are distinguished continuous and staple fibers. Fiberglass is made from the alkali-free aluminum-borosilicate glass, the alkaline and neutral glass. High-temperature-resisting fibers are made from the vitreous, siliceous and kaolin. Fiberglass differs of high tensile strength, relatively low density, heat- and chemical stability. They are used for preparing of fiberglass fabrics, electro and thermal insulation materials, sand paper, glass-fiber-reinforced plastics. Glass wool, which consists of short glassy fibers and has high thermal insulation properties are obtained by processing of mineral fusions and glass mass.

Properties of Glass

The most important properties of glass are optical properties. For ordinary structural glass the refraction coefficient, that is ratio of speed of light transmission in vacuum to speed of light transmission in glass, is 1.52...1.53. Depending on chemical composition, structure of glass, character of its surface, refraction coefficient hesitates from 1.47 up to 2.05. Optical transmission of ordinary window-pane is 83...90%.

Ability of glass of light passing is characterized by the ratio between amount of light energy which passes through the glass and total light energy.

Strength of glass is not equal at the different types of loading - at bending and tensile strength in 7...10 times less than at compression.

Ordinary glass has poor impact resistance, its strength at impact bend is only 0.15...0.20 MPa. The additive of oxides of magnesium, silica, iron increases the impact resistance at 5...20%, adding of boric anhydride - at 50%.

Fragility - main failure of glass, it is predefined by high relation of the modulus of elasticity to the value of tensile strength, and also by absence of plastic deformation of glass before bursting and high-rate of cracks spreading. Fragility of glass diminishes if content of B_2O_3, SiO_2, Al_2O_3 is increased and also at heat treatment.

Glass has comparatively low thermal conductivity [λ= 0,4...0,8 W/(m•K)], the temperature coefficient of linear expansion hesitates from $5 \cdot 10^{-7}$ to $20 \cdot 10^{-6}\ K^{-1}$, it diminishes at addition into the composition of glass SiO_2, Al_2O_3, MgO and rises due to alkaline oxides.

Glass is more proof to the action of the sharp heating, than sharp cooling, because in its superficial layers compression strains appear at heating, and at cooling – tensile strains. The most heat-resistant - quartz glass stands at cooling the overfall of temperatures to 1000°C, low alkaline borosilicate glass – 150...300°C, ordinary structural glass - 80...100 °C.

Glass is characterized by high firmness to the action of acids (except for hydrofluoric and phosphoric), neutral and acid salts. Chemical firmness of glass in 10...20 times goes down under the action of solutions of alkalis, phosphates, phosphoric and especially hydrofluoric acid. Chemical corrosion of glass sharply increases in the case of increase of temperature and pressure.

Ordinary construction glass well skips neutrons and gamma-radiation. Protective properties of glass in relation to gamma- and x-ray radiation, as well as other materials, rise with the increase of density. Heavy-weight glass is obtained by increasing the content of lead, boron and caesium in it. Flow of neutrons loss is provided by the oxides of boron, lithium, cadmium.

A number of the physical-mechanical properties of glass is calculated from the additive formulas, which assume the calculation of shared participation in the synthesis of the properties of separate oxides. Following formula can be used:

$$P = \sum P_i m_i, \tag{8.1}$$

where P - the property of glass; P_i - coefficient of shared participation; m_i - content of the i-th oxide in mass %.

The computed values of the indices of properties are averaged; they should be corrected taking into account the special features of technology, method of treatment.

The *density* of construction glass depending on composition varies in interval of 2.47…2.56, foam glass density of 0.15…0.80 g/cm^3.

The theoretical *compressive strength* of silicate glass is equal to 7000… 12000 MPa, quartz of 1200… 2500 MPa, actual 500… 800 MPa. This significant difference is explained by the presence on the surface of glass and in it volume the scratches and heterogeneities. Micro-scratch and microscopic cracks create in the glass the wedging efforts, which facilitate its destruction. The values of the strength of the glass specimens with a thickness of 5 mm in the dependence on the surface condition are given below:

- natural (fiery polishing) - 218 MPa;
- scratched by fine-grit emery paper - 131 MPa;
- scratched by coarse-grit emery paper - 41 MPa;
- slicked and polished - 71 MPa;
- polished and non-slicked - 215 MPa;
- slicked and polished with the subsequent hardening - 180 MPa.

With the decreasing of size of specimens and diameter of glass fibers the influence of defects decreases and the strength of glass grows. This effect of scale factor follows from the statistical theory of strength, in accordance which with increasing in the sizes of specimens grows the probability of the appearance in them of dangerous defects and microstresses.

Residual stresses in the glass are mainly removed by *annealing* - the heat treatment of material, i.e. its heating to the specific temperature and slow cooling.

Reduction in the strength of glass is conditioned the characteristic defects, caused by the insufficient uniformity of glass mass, by the disturbances of the technological parameters. Filament like cords in state of tension and located on the surface of products are especially dangerous.

For the glass the influence of "fatigue", caused by the lasting effect of loads is characteristic. Safe constant load for the glass taking into account fatigue is considerably less calculated.

To the development of microcracks surface-active media and, first of all, water contribute. Effect of the environment grows with an increasing in the duration of contact, temperature, value of the accompanying stress. Established, for example, if the strength of glass is in air 5.3 MPa, then in the water with 20°C it is reduced to 4.2 MPa, and with an increase of the temperature of water to 80°C to 3.9 MPa.

There are a number of methods of improvement of the mechanical and other properties of the glass: hardening in the air flow and in the special liquid media, etching in the hydrofluoric acid, ion exchange, crystallization of surface layer, reinforcement, coating glass with films and others. At the hardening the resistance of glass to bending rises in 4…5, etching and the coating with films - 5…10 times.

Hardening glass consists in its heating to 700… 900°C and subsequent sharp, but uniform cooling. They for the first time revealed the phenomenon of hardening glass in 17th century, but for strengthening the glass products in industrial scales began to use in the 30th years of 20th century. In the hardening operation on the surface of glass the evenly distributed compressive stresses, which increase its ultimate strength under the action of the external bending or impact loads, appear. The surface of glass is cooled by air or by some organic, for example, silicon liquids. Liquids with the high boiling point, which especially considerably increase strength as a result of the formation, on the surface of glass of durable films are used. The comparative characteristic of the basic properties of initial (burned) and hardened sheet glass is given in Table 8.2.

In the magnitude of the strengthened effect to hardening the method of the etching of the surface of glass by the solution of hydrofluoric acid in is dissolved the weakened surface layer not yield. The method of the three-stage treatment of glass - hardening in the liquid at the conditions of the ultrasonic field, etching and then application of protective coatings is also effective.

At the method of ion exchange in the surface layer of glass are created the stresses of compression as a result of the diffusion substitution of the ions Na^+ by the alkaline ions, which pass from the salt fusion.

Triplex method consists in the production of the three-layered sheet glass, which consists of two external sheets of glass, firmly glued between itself by the intermediate inner layer, which consists of the transparent filler plate of elastic organic material. The basic merit of triplex consists in its nonshatterability. The group of shatterproof glass includes also the hardened and reinforced by wire mesh glass.

The *thermal resistance* of glass characterizes its ability to maintain sharp temperatures without the destruction. It is measured by the temperature, to which it is possible to cool suddenly glass specimen without its destruction.

With the chilling of glass as a result of the unequal rate of cooling in the surface layers the stresses of tension appear, in the internal - compression, while with the heating - vice versa. Considering that the destruction of glass starts from the surface and compressive strength is much more than tensile strength, the chilling of glass products is more dangerous than sharp heating.

Table 8.2. Comparative characteristic of the basic properties of the initial (annealed) and hardened sheet glass

Property	Glass	
	Annealed	Hardened
Impact strength	The impact of sphere with a mass of 800 g withstands from the height 150 mm	The impact of sphere with a mass of 800 g withstands from the height more that 1200 mm
Bending strength, MPa	To 50	To 250
Elasticity	-	Bending deflection is 4…5 times more than in annealed
Heat resistance, °C	60...70	To 175
Electrical conductivity	-	It is 2…3 times more than in annealed
Density, g/cm^3	2.5	It is reduces insignificantly
Coefficient of the linear expansion, 10^{-6}/K	8.8…9.5	It is increases insignificantly

The thermal resistance of different glass is within the limits 90...1000°C. For example, it is not above for the sheet window glass 90°, chemical-laboratory - 120...140°C, and quartz - 800...1000°C.

8.2. Glass Products in Construction

Construction glass is divided into sheet glass and products of the architectural-construction assignment.

Table 8.3. Basic types of sheet glass

Type of glass	Thickness, mm	Recommended industry of application
Mirror improved glass	2…6	Producing of high-quality mirrors
Mirror glass	2…6	Producing of mirrors of the general assignment
Technical polished glass	2…6	Producing of decorative mirrors, safe glasses of transport vehicles
Window-pane glass polished	2…6	High-quality glazing of light transparent constructions
Window-pane glass unpolished	2…6	Glazing of light transparent constructions, safe glasses for agricultural machines
Plate glass polished	6,5…12	High-quality glazing of windows, stained-glass windows
Plate glass unpolished	6,5…12	Glazing of windows, stained-glass windows

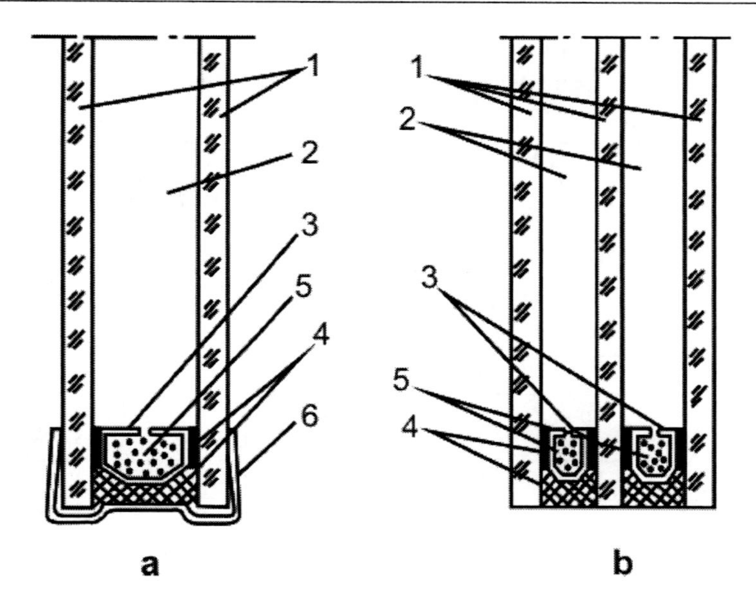

Figure 8.1. Constructions of glass packs:
a- double-layer (single chamber); b - three- layer (double chamber):
1 - glass; 2 - air space; 3 - metal frame; 4 - pressurizer; 5 - desiccant; 6- edging.

Sheet Glass

The range of sheet glass is various (Table 8.3): plate glass, figured and reinforced, with the special properties, that skips and takes in ultraviolet rays, with semi-transparent mirror coverage, which takes in or beats back warmth, electro- lead, treated by hardening.

Sheet glass has high coefficient of optical transmission, wide range of values of linear sizes and thickness.

A window-pane is divided into sorts depending on the presence of defects: striped, crinkles, bubbles, other impurities and scratches.

For glazing of large surfaces of buildings large size window glass are used, that produce as unpolished and polished, in order to avoid optical distortions.

Glass packs are applied for prevention from damping and freezing of glass, and also for the cutback of labor consumption and use of industrial methods at glazing of window cuts. They consist of two or few flat glasses, connected by welding or agglutination. Between them appears reserved space in thick 15...20 mm, filled with dry air (Figure 8.1). In the case of the glass packs use, heat-insulation and sound-proofing are improved.

If it is required to avoid transparency of the light openings or glass partitions, then the *figured glass* is applied. This glass has light-diffusing ability, is decorative and enables to create expressive interiors, especially in buildings of the cultural and medical purpose. It is produced colorless and colored.

A wire mesh is squeezed in glass for content of fragments at the damage of it. R *einforced glass* is used for finishing of lanterns, construction of partitions and arranging of protections of balconies, where enhanceable mechanical strength of material is required.

Reinforced glass as well as figured, is produced colored and colorless. Colorless glass is applied mainly for glazing of the light openings, and colored - in the protection of balconies, for arranging of internal partitions. The surface of reinforced glass can be smooth, corrugated, figured. To promote mechanical strength of glass, it is hardened.

Tempered sheet glass is used for production of glass doors, protections of balconies and steps, elevator shafts. Safe character of destruction is peculiar to this type of glass. The possible overfall of temperatures during exploitation reaches 270°C. Value of bending strength is in 5.5 and compressive strength in 1.35 times higher, than in ordinary window-pane. Tempered glass is not subjected to cutting, drilling and other types of tooling. Painted by the colored ceramic fusible paints tempered sheet glass is applied for revetment of buildings facades.

Unlike ordinary glass the special *uviol glass* passes biologically active ultraviolet rays. It is produced from particularly pure materials, containing no more than 0.03 % iron oxides. To achieve maximal effect, uviol glass with minimum possible thickness is applied. Ability of passing ultraviolet radiation recommences, when uviol glass is heated to 300...450°C. Uviol glass is applied for glazing of windows in child's and medical establishments, in buildings of the health resort purpose. It is desirable to apply glasses absorbing ultraviolet rays, in libraries, museums, art galleries, which influence on quality of paper, fabrics, paints, varnishes and others like that. Oxides of chrome, selenium, vanadium are added in particular to the complement of glass for this purpose, the painted glass with oxide-metallic coating is also used.

Glass with semi-transparent mirror coating is used for the external glazing. Colored and colorless mirror coatings get causing on glass of films of oxides of titan, iron, cobalt, copper. Glazing with the painted glasses is applied in buildings with the large areas of glazing to avoid overheat of apartments. Neutral glasses reduce transmission in all optical range of sun spectrum. Glass, painted by the oxides of iron, zinc, copper and other metals, or glass with superficial pellicle oxide-metallic coatings has an ability to heat absorption. They are set in the external row of glass packs and for intensive heat removal provide required ventilation. Elastic putties and rubber seals are applied for the removal of undesirable tensions.

Glass Products

Promoting of strength and longevity of light transparent protections, decreasing the amount of wood and metal, improving operating qualities of glazing is possible by the hollow and profile types of glass. These products are intended for filling of the various light openings, arranging of external and internal light filter protections in buildings and structures of different destination.

Glass blocks are produced rectangular (they can be also round and angular) from colorless or painted glass. The surface of facial walls of transparent glass blocks is smooth, light-diffusing - corrugated, what eliminates visibility. Glass blocks should have value of compressive strength not less than 1.5 MPa, resistance to the shock action – 0.8 MPa.

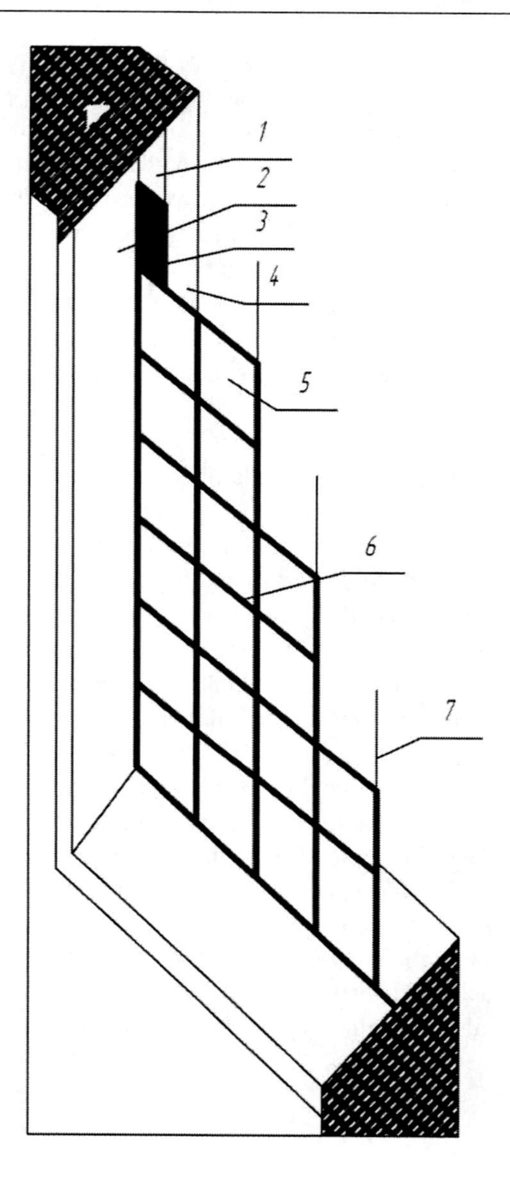

Figure 8.2. General view of light opening, filled by glass blocks:
1- bitumastic; 2- elastic gasket; 3- sealing gasket of lateral gaps; 4- concrete panel; 5- glass block; 6- cement seams; 7- reinforcement.

Structures from glass blocks form the soft diffused lighting and have enhanceable heat-insulation and sound-proofing.

Enclosures from glass blocks (Figure 8.2) are arranged directly in window opening (without establishment of frame) on cement-sand mortar or installed the glass reinforced concrete panels made in advance. Such panels due to structural peculiarities and production technology are similar to the combined reinforced concrete structures. They can be installed in window openings in place of the double glazing. Taking into account the considerable coefficient of thermal expansion, glass reinforced concrete panels are insulated from load-bearing panels with compensative joints, filled by elastic gaskets.

Structural glass is profiled glass with channel, box-like or other section (Figure 8.3). Due to large overall dimensions (length up to 8 m) at installation profile glass there is no necessity to apply intermediate frames.

For the improvement of light-diffusing ability and original appearance the surface of glass from the side of cavity is executed as corrugated. Structural glass is reinforced with a steel wire, as a result its fire-resistance rises, and at destruction fragments do not fly away. Value of bending strength for structural glass exceeds 10 MPa, optical transmission is 60...80%. For compression and sealing of joints between the elements of structural glass various polymeric mastics and porous gaskets are used.

Enclosing structures from structural glass, which are erected from the precast panels of factory-made, are the most perspective. Such panels consist of frame, made of wood, concrete or metal, and filling - structural glass.

Glass pipes are intended for construction of pipelines, which transport chemically aggressive liquids, gases, food products, friable and other materials. They perform in the conditions of temperature overfall 40...80°C and working pressure 0.2...0.7 MPa. The diameter of glass pipes achieves to 2000 mm, length – 1.5..3.0 m. They are not subjected to corrosion, their carrying capacity is higher, than cast-iron and steel pipes at an identical internal friction. They are used also for the gasket of electric wiring, making of the heated panels and others like that. Basic lack of glass pipes is limited impact resistance.

High chemical resistance of glass and mechanical strength of steel are combined in steel pipes, lined with glass. Considerably promoting strength of glass pipes is possible, covering them with a layer of *fiberglass plastic*.

Flags from *marbled glass* - opaque painted glass with the polished surface are used as facing material. Marbled glass has marble-alike structure. A reverse side is executed as corrugated or rough. Marbled glass is produced by rolling from wastes of glass, with additive of metallurgical slag. Sizes of flags are 250×140 and 500×500 mm, sheets-200×3000 mm, thickness -5...12 mm. Profiled elements are also made from marbled glass.

Similar to marbled glass due to structure and properties are *synthetic granite, glass-marble, glass crystallite* and other facing materials.

Glass crystallite - double-layer material, the facial surface of which is smooth, polished, coloured imitates ornamental stones, underlayer is rough, provides reliable adhesion with cement-sand mortar. For the decline of density of flags in the underlayer of glass crystallite gas-forming admixture is included.

Carpet-mosaic glass tiles are used for external revetment of walls (except for basements and cornices), internal finishing of walls and columns, making of decorative- artistic panels. It is obtained from opaque and semi- opaque glass by rolling. Tile is supplying in carpets, glued on a paper. After steam curing the facade side of panel is cleaned from a paper and glue.

The *enameled glass tiles* are cuted from sheet glass (as a rule, from wastes). One of surfaces of tile is covered by the coloured or white enamel. Tiles are applied for interior walls tiling of the buildings with enhanceable sanitary-hygienic requirements.

Facing glass tiles also can be manufactured by pressing from opaque glass mass or glass scrap. Reverse surface of tiles is made as corrugated for the best bond it with mortar.

Figure 8.3. Types of profiled glass:
a - channel section; b - box-like section.

Smalt - pieces of the stained opaque glass of various form, which are collected in carpets on a paper. They are used for the decorative finishing of buildings and structures, implementation of the mosaic works.

The basic component of large group of materials is *fibreglass*. Depending on length of filaments fiberglass is divided into continuous and staple (length to 30 cm). Due to diameter a fiberglass can be: ultrafine <1; superfine – 1...2; fine – 3…10; incrassate – 11…20; thick – over 20 μm. Ultra-, super and fine fiberglasses have so high flexibility that can be processed by textile technology. Their tensile strength is 200...500 MPa, it grows to the extent of diminishing of diameter. Electric insulating ribbons and fabrics, heat-insulated and sound-proof glass cotton wool, and also plait, nets and linen are produced from fibreglass.

Linen is applied for reinforcing of various waterproofing materials and for manufacturing of fiberglass plastic.

Glass-fibre materials are used effectively for protecting from the silting-up of the drainage systems. They differ by good filtration properties, longevity and resistance in aggressive subsoil water. Polyvinyl acetate emulsion serves for glass linen as binder. Filtration ability of glass linen in mineral soils is saved for long time. Glass cloth has considerably less water permeability, than glass linen and more subjected to the influence of mudding. They are recommended for application only in sandy soils.

It is important to execute the special safety measures at transporting and keeping glass and glass products. Glass is folded in wooden boxes or containers and interlaied with the wood shaving or corrugated paper.

8.3. Brief Technology of Glass Manufacture

Silicate glass, the basic components of which are SiO_2, Al_2O_3, CaO, MgO, Na_2O or K_2O, is applied in construction. Thus, there is about 72 % SiO_2 in composition of window glass, 10% CaO+MgO and 15% Na_2O. Every oxide has the certain destination. Silica creates spatial structure of glass and determines it's the most important properties. The oxide of sodium accelerates the process of glass formation, reduces the temperature of glass melting, facilitates degassing of glass mass, but promotes the coefficient of expansion and reduces thermal and chemical resistance of glass. The oxides of calcium and magnesium give chemical resistance to glass. Basic components are included in a glass batch with sand, feldspars, soda, limestone, dolomite and other materials. It is possible to enter in a batch up to 60...70 % ash and slags of heat power plants and to get glass of type of marbled glass. Glass on the basis of fuel ash and slags has comparatively low temperature coefficient of linear expansion ($54...65\times10^{-7}$ degree^{-1}), enhanceable strength (80...100 MPa) and water resistance.

The beaten glass, and also auxiliary materials - oxidants, restorers, clouding admixture and others like that are used for improvement glass mass properties and acceleration of melting. Raw materials, applicable to glass mass, are processed to previous preparation. Sands, which contain the enhanceable amount of iron oxides, are beneficated by washing in hydrocyclones or other methods. Dried up and the ground up components of batch are mixed up carefully. To prevent demixing, a batch is briquetted or granulated. Melting of glass can be done in the bath-type furnace of continuous action. It is possible to contain over 2000 tons of glass mass in such bath-type furnace. In bath-type furnace materials continuously move from the loading opening of furnace to the unloading opening; gets in the areas of different temperature and transform into prepared glass mass. The process of glass melting consists of five basic stages: silicates formation, glass formation, degassing, homogenization and cooling for required viscosity.

Glass mass can be produced in different ways depending on the type of products. Sheet glass is obtained by the vertical drawing out on special machines (Figure 8.4), and rolling on the layer of molten metal.

The last one, so-called *fluate-method* is the most progressive method of production of the polished glass. Its essence is contained in continuous merging of glass mass from the reservoir of glass-melting furnace on the surface of molten metal, spreadind on it like layer of even thickness and transforms into the ribbon of glass with the polished surface.

Glass blocks are produced by pressing of separate semi-blocks with the their following welding by the special welding automats. At the production of structural glass ribbon, which is made by the continuous horizontal rolling, is stretched out through a forming device for a making of required channel or box-type section. Glass pipes are produced on the production lines of the horizontal or vertical drawing.

Glass fiber is manufactured as filaments from molten glass mass, which stretch through special bushings and winded on a reel which is quickly revolved.

During cooling of glass as a result of overfall of temperature afterstrains are appeared between superficial and internal layers, which can be removed or weakened by special heat treatment of products.

8.4. Glass-ceramic Materials

Pyrocerams, slagcerams and stone casting belong to glass-ceramic materials. The general feature of these materials is a presence of both crystalline and glass phases in their structure, which provides high mechanical properties, thermal and anticorrosive resistance, low wearability etc. Glass-ceramic materials are applied for production of proof to the abrasion revetments, responsible parts of structures and others like that. Glass-ceramic pipes use for manufacture of heat exchangers. Due to ability of some glass-ceramic materials to take in slow neutrons and heat resistance they are applied in production of bars of nuclear reactors and for biological protection.

Pyrocerams are products of crystallization of glasses. The process of production of pyrocerams includes the manufacturing of glass of certain composition, forming products from it and special heat treatment. There are technical pyrocerams and slagcerams.

Slagcerams are widely used in construction. Slags of ferrous and non-ferrous metallurgy, and also ash from incineration of anthracite coal are the raw material materials for slagcerams. The production of slagcerams consists in slag glasses melting, forming products from them and their following crystallization. A batch for the making of glasses consists of slag, sand, alkaline and other admixtures. The most effective is the use of molten metallurgical slags that saves to 30…40% amounts of heat, which is outlaid on melting.

Slagcerams as sheets, pressed flags, pipes and other products are manufactured at the plants. Formed products are supplied to crystallizer, where the certain mode of thermal treatment is automatically supported.

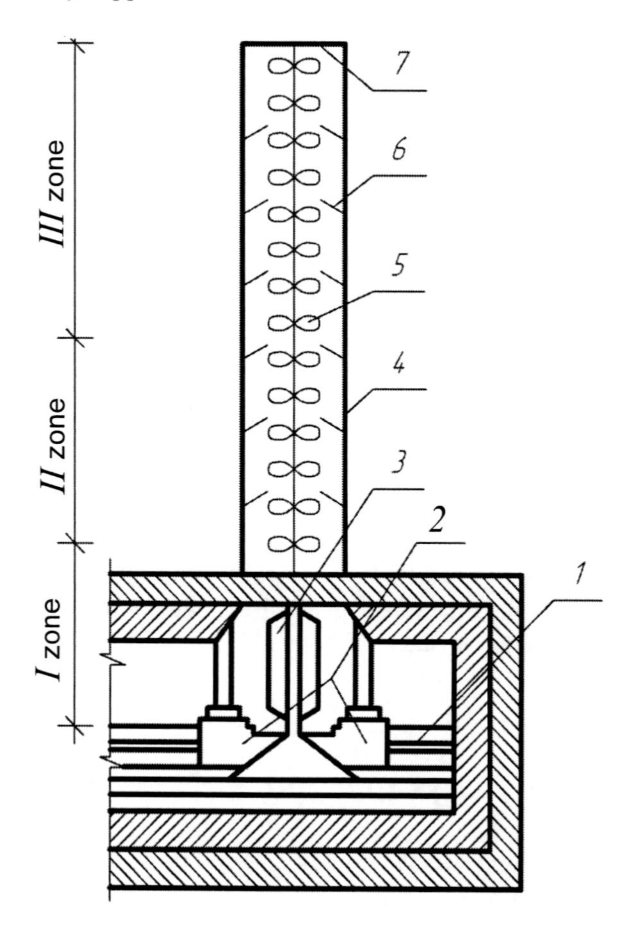

Figure 8.4. Vertical drawing machine for making glass:
1- glass mass; 2- gripper for glass mass; 3- refrigerators; 4- mine of machine; 5- rollers; 6- devices for the delete of cullet; 7- breakage area.

Slagcerams differs from the number of construction materials with higher physical and mechanical properties. Due to high strength they approach to cast-iron and steel, though in three times lighter than them (Table 8.4). The heat-resistance of slagcerams is 150...200°C. Indexes of chemical resistance and resistance to abrasion are especially high. Slagcerams can be processed by the different methods: grinding, polishing and cutting.

Basements and facades of buildings are faced by the flags of facial slagceram, also internal walls and partitions, balconies are decorated. Slagceram is an effective material for the stairs, window-sills and other structural elements of buildings. It is applied also for protection of building constructions and apparatus in chemical, mining industry and other industries.

Stone and slag casting is stone product, obtained by melting of igneous or sedimentary rocks and slags, pouring of fusion in forms and heat treatment of products with the purpose of passing of crystallization and destressing.

For the making of stone poured products fusible rocks are used as rule - basalt, diabase and others like that. In the batch admixtures which reduce the temperature of melting and increase ability to crystallization (fluor-spar, dolomite, chromite etc.) are entered.

Stone poured products are exceeded than natural stone and row of other materials due the density, firmness to the chemical action, resistance to wearing and other properties. Wearing of products from a stone casting in 3...5 times less than granite, basalt or diabase, and equals 0.016...0.1 g/cm^2, and ultimate compressive strength is 200...400 MPa.

The basic types of stone poured products are flags, pipes and other details, intended for work in severe climatic and other operating terms.

Table 8.4. Properties of pyrocerams and other construction materials

Index	Pyroceram	Slagceram	Rock glass-ceramic	Window-pane	Granite	Steel
Bulk density, kg/m^3	2600... 2900	2600... 2800	2900... 3000	2500... 2600	2600... 2800	7800
Ultimate strength, MPa:						
at a compression	800...1000	500...650	700...850	600...700	100...250	420...550
at a bend	100...225	90...130	130...170	60...70	10...15	400...1600
Modulus of elasticity, E 10^3, MPa	72...135	90...100	40...50	60...70	8...25	210...220
Acid resistance, %	97,8...98,9	98,8...99,8	99,8	55...58	95...95,5	–
Alkali-resistance, %	–	90...94,7	94...96	98,9	–	–
Wearability, g/cm^2	–	0,01	–	0,5...0,6	–	–

Various products are poured from molten metallurgical slags: stone for roads and floors of industrial buildings, tubing, kerb-stones, anticorroding tiles, pipes etc. The poured products from slag fusion are more economic advantageous, than stone casting, approaching them by mechanical properties. The bulk density of the poured products from slag equals 3000 kg/m^3, and ultimate compressive strength is 500 MPa.

Due to wear resistance, heat-resistance and a series of other properties the slag casting exceeds the reinforced concrete and steel. The poured products from a slag are more effective than steel in different linings, for example hoppers for transporting of abrasive materials (ores, agglomerates, crushed stones, sand and etc). Their service life is in 5...6 times longer than steel linings.

The poured paving stones for roads and floors of industrial buildings are not less effective. Service life of highways from a slag poured paving stones between major repairs is twice greater, and performance is cheaper, than asphalt one. Tubings, heat-resistant blocks, chemically proof products are poured from a slag.

Slag products for the decline of internal tensions are reinforced with steel reinforcement. Steel reinforcement prevents cracks formation. Reinforced slag products can be used in place of the precast reinforced concrete, because they have higher strength. However the lack of these products is some decline of strength of steel at the high temperature of slag fusion, and also comparatively high labour intensiveness of production.

For production of the porous slag casting pore-formation of slag fusion is executed. The products of various configurations are formed from porous slag fusion. Depending on the degree of pore-formation the bulk density of poured products is 350...1500 kg/m^3, ultimate compressive strength is equal to 1.5...30 MPa.

Self-Assessment Questions

1. Explain the significance of glass in modern construction.
2. Tell about the essence of vitreous state.
3. Tell about the types and compositions of glass.
4. Describe the basic varieties of a sheet glass.
5. Describe the main features of glass blocks, structural glass, glass pipes and facing products from glass.
6. Tell about the technological process of glass products manufacture.
7. Tell about the properties of glass.
8. Describe the features of structure and properties of glass-ceramic materials.

Chapter 9

Mineral Binders

Materials for bonding of heterogeneous components into artificial conglomerates of building destination (concrete, mortars, mastics, etc.) are called *constructional binders*. Depending on the composition they can be divided into two groups: mineral (inorganic) binders and organic binders.

Mineral binders are the powdery materials, which form a plastic mixture when mixed with water or water solutions of salts, and gradually hardens into a stony state. Depending on the conditions of hardening, the mineral binders may be categorized as non-hydraulic (air-hardening) or hydraulic.

Non-hydraulic binders can harden and gain strength only in the air conditions. Gypsum, magnesia binders, air-hardening lime and liquid glass are categorized as non-hydraulic binders.

Hydraulic binders are characterized by their ability to harden in air and in water. Portland cement, alumina cement, hydraulic lime, and alkaline-slag are some of the well-known hydraulic binders. Some binders, such as lime-silica binders, harden intensively only in autoclaving conditions.

The hydration and hydrolytic dissociation are characteristic for many mineral binders at the water or at water solutions of salts tempering. On the contrary, organic binders harden due to the coagulation structurization, condensation and polymerization reactions. For example, clays are known to harden by the coagulation mechanism.

Gypsum, limestone, clay, marl, or waste materials and by-products from different industries - slags, ashes, sludges, etc serve as the raw materials for production of mineral binders. The burning of the raw meal to pass the necessary physical-chemical processes is common working operation for most of mineral binders. They are usually ground to a powdery state to improve chemical reactivity. Different additives such as: plasticizers, hardening accelerators, micro-fillers and others are added to the mix for regulation of binding properties.

Usually, binders are used together with aggregates and dispersed or fibrous fillers are incorporated to reduce cost and to improve the technical properties of products (shrinkage reducing, increasing of the crack growth resistance, heat-resistance and others).

Most constructions works use hydraulic binders, such as, Portland and alumina cement and their modifications to produce the most widespread construction materials - cement concretes and mortars.

9.1. Gypsum Binders

Materials, which consist of semi-hydrate gypsum $CaSO_4 \cdot 0.5H_2O$ or anhydrite $CaSO_4$ are called gypsum binders. They are produced as a result of the thermal treatment and grinding of raw materials such as, dehydrate or anhydrite calcium sulfate. Dehydrate - $CaSO_4 \cdot 2H_2O$ is a mineral constituent of various rocks - gypsum stone, clay-containing gypsum and others, and also constituent of artificial products - industrial wastes (phosphogypsum, borogypsum etc). On heating a dehydrate, a series of chemical processes takes place, leading to the formation of substances with cementing properties. The main one being the modification of the semi-hydrate gypsum (semi-hydrate) - the product of partial dehydration of dehydrate. The basic representatives of gypsum binders are *gypsum plaster and alpha gypsum(*–high-strength gypsum). Both binding materials are manufactured at low-temperature burning of gypsum (120-180 ° C).

When the gypsum heating is realized in an open apparatuses - gypsum boilers or rotary kiln (Figure 9.1) and crystallization water is selected in the form of water vapor, small crystals β - $CaSO_4 \cdot 0.5\ H_2O$ are formed. For water of crystallization derived in the liquid state, comparatively large crystals of α-semi-hydrate are formed, that is the main phase of the high strength gypsum.

The differences between the binding properties of β- and α-semi-hydrates can be explained by their different water demand due to the peculiarities of crystal structure. The α-semi-hydrate has higher strength due to lower water demand and consequently less porosity in the hardened state.

The gypsum binders industry produces mainly gypsum plaster. The production process involves grinding and heat-treating of the gypsum stone, and fine milling of the finished product. Spread apparatus for heat treatment of gypsum plaster are gypsum frying boilers. Alpha gypsum is produced in the autoclaves. Other thermal units are also used for gypsum binders burning.

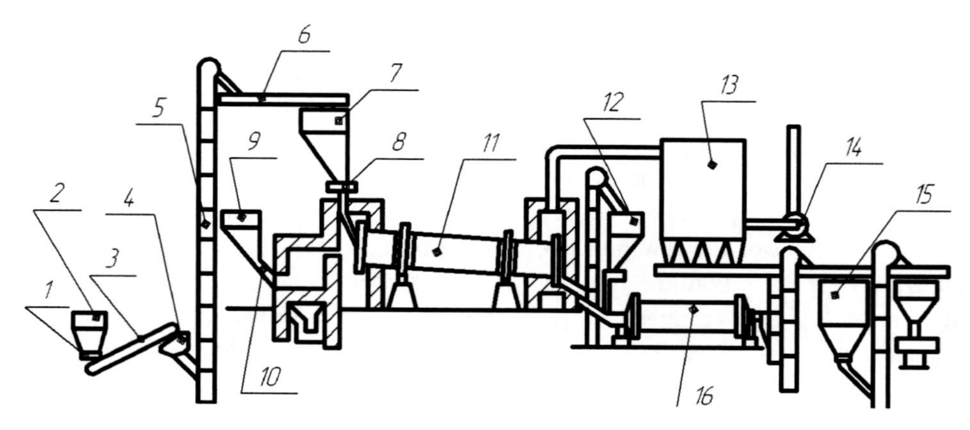

Figure 9.1. Chart of gypsum plaster production in rotary kiln:
1- tray feeder; 2- bin of the gypsum rocks; 3- conveyor belt; 4- hammer crusher; 5- elevator; 6- auger; 7- bin of the crushed gypsum rocks; 8- feeder; 9- coal bin; 10- firebox; 11- rotary kiln; 12- bin of the fired crushed rock; 13- dust precipitation chamber; 14- aspirator; 15- gypsum plaster bin; 16- ball-race mill.

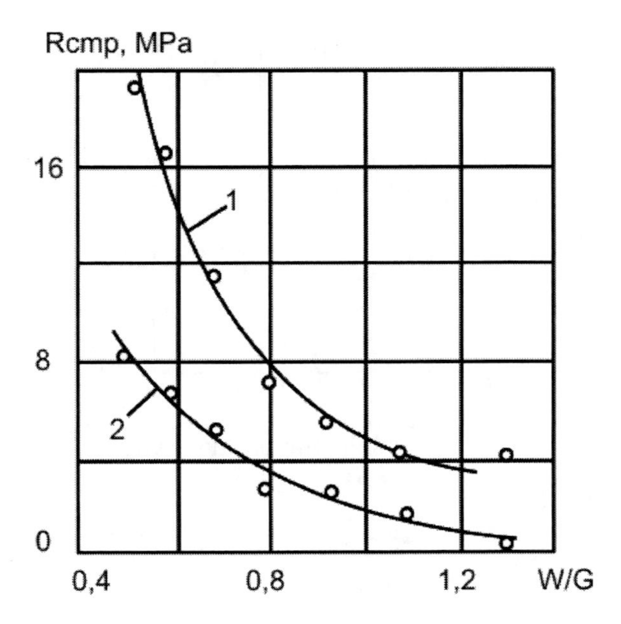

Figure 9.2. Relationship between water-gypsum ratio and ultimate compressive strength of gypsum binder:
1 – strength of dry specimens;
2 – strength of wet specimens.

The setting and hardening of gypsum binders is conditioned by the transition of the semi-hydrate into crystalline dehydrate gypsum. The hydration and crystallization of the dehydrate gypsum runs quickly and finishes in a few minutes after tempering. The strength of gypsum increases as it dries. The main features of a gypsum plaster are: rapid setting and hardening, enhanced water demand and porosity, the ability to grow in volume up to 0.5-1% in the early-hardening period, tendency to suffer creep strain, and low water-resistance.

Quality of the gypsum binders is defined by the time of setting, fineness of grinding, water demand (Figure 9.2.), compressive and bending strength. Depending on the time of setting, gypsum binders may be divided into such groups: rapid, normal and slow setting. The initial setting time occur not earlier than 2 and 6 minutes, and the final set, not later than 15 and 30 minutes respectively for the first two groups. An initial setting time of 20 minutes or more is typical for gypsum binders of the third group.

Fineness of gypsum binder grinding is evaluated by considering the amount retained on a sieve size of opening 0.2 mm.

Gypsum plaster needs up to 50-70% alpha gypsum and up to 30-40% of water for preparing a paste with normal consistency. Theoretically, 18.6% of water from the weight of binder is needed for the hydration of the semi-hydrate gypsum and all the excess water, which is removed during drying forms the voids that reduce the strength of binder. Set retarding admixtures, allow a reduction in normal consistency of up to 10-15%, and facilitate a reduction in the water demand. Adding of the superplasticizers is also effective.

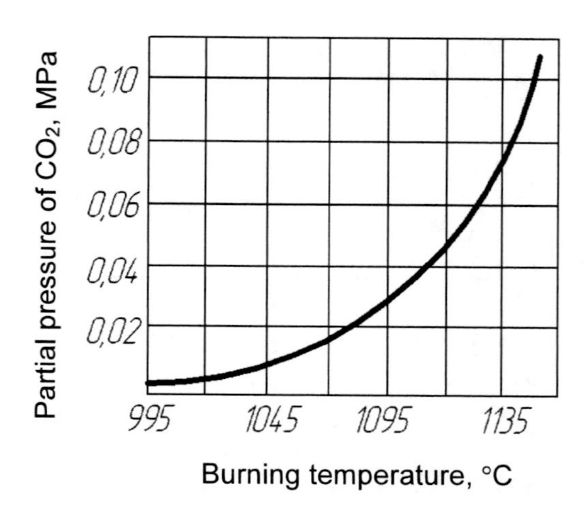

Figure 9.3. Relationship between the burning temperature of $CaCO_3$ and partial pressure of CO_2.

It is possible to use the additives that reduce solubility of semi-hydrate at the certain concentration (ammonia, ethyl alcohol, etc.), substances that reduce the rate of crystals formation at the gypsum hydration (lignosulfonates, keratin retarder, etc.) for increasing of setting time of gypsum binders.

All gypsum binders are graded depending on the strength. The strength of the gypsum binders is determined by compression and bending test of a $40 \times 40 \times 160$ mm beams at the age of 2 hours after tempering with water.

The gypsum binder must be protected from moisture during the transportation and storage.

Water stable *gypsum-pozzolan cements* is successfully used in building construction. These binders are produced by carefully mixing of 50-70% gypsum, 15-25 Portland cement and 10%-25% hydraulic admixtures. They have some advantages in comparison with the gypsum binders and Portland cement. These include: rapid hardening and strengthening, enhanced water and sulfate-resistance, plus low creep.

With respect to the scale of use, the Gypsum binder is only surpassed by cement and lime in the modern building industry. However, due to the simplicity of the technology, relatively low heat and power inputs and other advantages, their value and use is growing. Fuel requirement for production of 1 kg of the gypsum binder is 4 times lower than that used for production of 1 kg of Portland cement, and specific investments are half that for 1 kg of Portland cement.

9.2. Air-Hardening Lime

Product of carbonate rocks burning, which consists mainly of calcium oxide is called air-hardening lime.

The raw materials - chalk, limestone and dolomite should contain no more than 6-8% of clay impurities for an air-hardening lime. The hydraulic lime is manufactured with a greater content of impurities. Depending on the content of active magnesium oxide, lime may be categorized as: high-calcium (MgO <5%), magnesian (MgO = 5-20%) and dolomite (MgO = 20-40%) and according to the fractional composition - as lump or powdery.

Types of lime include: lump quicklime, unslaked milled, hydralime and lime paste.

The *lump quicklime* is a semi-finished product for manufacturing of other types of lime, and it forms directly as a result of raw material burning.

The reaction of calcium carbonate dissociation $CaCO_3 = CaO + CO_2$ is a base of the process of lime manufacture. Heat of reaction is 178.16 J per 1 mole. The degree and speed of completion of the dissociation of calcium carbonate depends on the partial pressure of CO_2, temperature, content of impurities, etc. Partial pressure of CO_2 reaches atmospheric approximately at 900 °C (Figure 9.3). The temperature in production terms can reach 1000-1200 ° C. The shaft and rotary furnaces mainly are used for burning of lime.

Shaft furnaces are used the most due to the simplicity of their design and operation, high thermal efficiency, relatively small capital cost of construction and high volumes of products. Shaft furnaces work continuously: through certain intervals in the upper part, limestone, anthracite or cokes are loaded, and from the lower part ready lime is unloaded. The air required for combustion comes from below and it cools the lime (Figure 9.4).

The rotary furnaces are also used for burning. They have high productivity and allow the use of small fraction of carbonate rocks to manufacture a product which has more stable properties. The furnaces for speed burning of lime in the boiling bed are also effective.

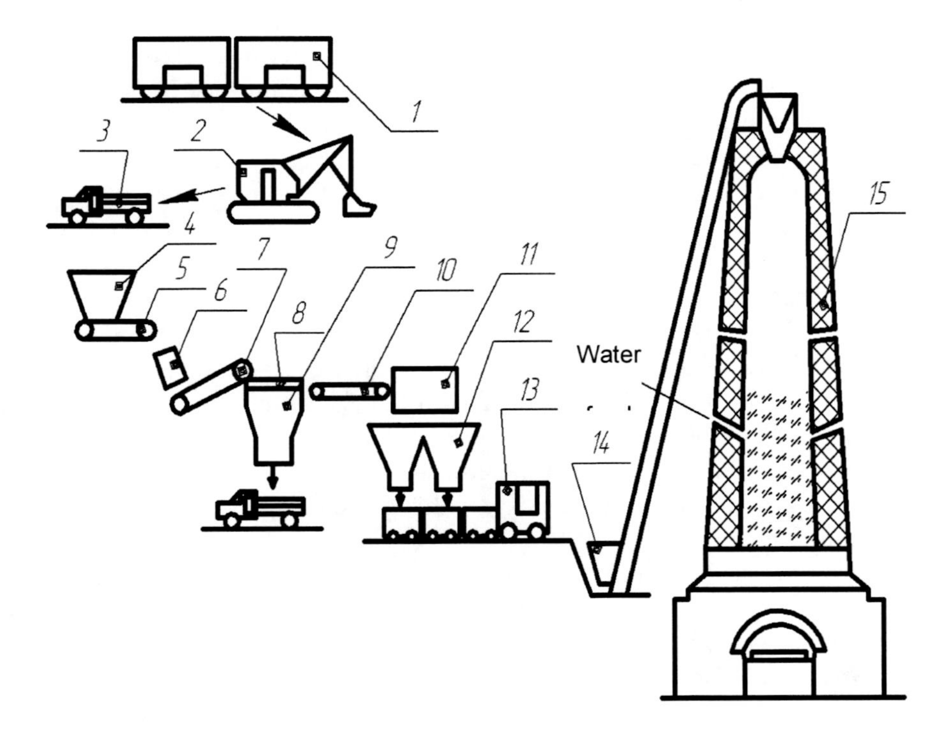

Figure 9.4. Chart of the lump quicklime production:
1- cars with a lime; 2- excavator; 3- dumper truck; 4, 9, 12- bins; 5- feeder; 6- crusher; 7, 10- conveyors; 8- sieve; 11- cylinder sieve; 13- trolleys; 14- burden-charging carriage; 15- shaft furnace.

Table 9.1. Requirements for high-calcium lime

Indexes	Quicklime			Hydrated lime	
Content of active CaO+MgO, by a dry matter, (minimum %):					
without additives	90	80	70	67	60
with additives	65	55	–	50	40
Content of CO_2 (maximum %):					
without additives	3	5	7	3	5
with additives	4	6	–	2	4
Content of the unslaked grains (maximum %)	7	11	14	–	–

Quality of quicklime depends on the content of free active calcium oxide and magnesium oxide (activity) in it and its structure. The temperature of burning influences the reactivity of CaO and MgO. With increased temperature of burning-of limestone, the size of CaO and MgO crystals is increased. As a result of recrystallization and sintering of CaO and MgO, the lime is compacted and its reactivity decreases. The presence of particles which are slowly slaked in the lime causes internal tensions in hardenings mortars and products and results in lower strength and frost-resistance, and may lead to the appearance of cracks.

In Table 9.1 shows the requirements for high-calcium quicklime and hydrated lime. The important indicator of quality is the degree of dispersion for milled and hydrated lime, which is determined by the screenings on sieves № 02 and № 008 (opening sizes of 200 and 80 μm), which must be less than 1.5 and 15%.

The active mineral additives: ash, slags and others can be added during grinding of lime to increase the water resistance and other properties of lime mortars. It is also possible to add quartz sand and gypsum.

The important feature of quicklime is the slaking ability with transition of calcium and magnesium oxides into their hydrates. This process proceeds according to the reaction: $CaO+H_2O = Ca(OH)_2+64.9$ kJ and is accompanied with the disintegration of lime pieces into fine particles. Duration of this process for a lime, which slakes quickly, lasts not more than 8 minutes, with moderate slaking rate - not more than 25 minutes and for slow rate - over 25 minutes. The process of slaking occurs quickly with high content of CaO and diminishing of burning temperature.

Depending on the amount of water, taken for slaking, - hydrated lime or lime paste is produced. In the first case, as a rule 0.6-0.8 parts of water are added on one part of lime by weight, and in the second case – 2-3 parts.

Hydrated lime is a finished product, which after water addition turns into lime paste. It is easier to transport and store in the packed state than quicklime. The lump quicklime and the milled lime are often slaked before the paste forming.

The lime is slaked using special devices. For small building works, the slaking of lime into the paste is made in temper boxes with plenty of water. The lime milk, formed in them is poured off to the temper pit, where the particles are fully slaked; lime is dehydrated to the consistency of the paste due to evaporation and water draining through the wooden walls of the pit in to the ground. In the temper pit lime comes through the storage at least 15 days to complete slaking and produces a plastic fine mixture. There is almost 50% of water in the lime paste, as a rule.

There are three types of lime hardening: carbonate, hydrated and hydrosilicate.

Carbonate hardening takes place in the mortars based on the slaked lime at normal temperature and pressure. Two process runs at the same time – crystallization of $Ca(OH)_2$ from the saturated water solution and the formation of $CaCO_3$ according to the reaction:

$$Ca(OH)_2 + CO_2 + H_2O = CaCO_3 + 2H_2O.$$

Crystals of the formed $CaCO_3$ accrete with each other, with the particles of $Ca(OH)_2$ and sand and form an artificial stone. Carbonate hardening process flows very slowly.

Hydrated hardening of lime is effected at the water tempering of the milled quicklime and takes place under certain hydration conditions (water content, heat-sink cooling). Milled quicklime dissolves in water with the formation of oversaturated solution. The effect of hardening is caused by accreting of the formed particles of calcium hydroxide.

Hydrosilicate hardening of lime occurs in lime-sand and other silicate products in conditions of high temperature and pressure of steaming, i.e., in autoclaves. The essence of it is the interaction of calcium hydroxide, silica and water and formation of new compounds - hydrosilicates that cements the sand grains. Production of lime-sand brick and silica concrete is based on the hydrosilicate hardening.

The main advantages of lime are high plasticity which gives good placeability and water-retaining ability to lime-based mortars and concretes and prevents the stratification of mixtures.

The average density of the lump quicklime is 1600-1700 kg/m^3 at normal temperature of burning. It grows to 2900 kg/m^3 as the temperature and duration of burning is increased. The average density of milled quicklime in the bulk state is 900-1100 kg/m^3, hydrated lime is 400 -500 kg/m^3, and lime paste is 1300 -1400 kg/m^3.

A lime paste hardens very slowly. It is possible to strike specimens of mortars after 5-7 days. The process of setting of mortars for the milled quicklime is completed between 15-60 minutes after tempering. The lime mortars made with quicklime have large shrinkage when hardening occurs in air conditions.

Strength of materials and products based on lime and its resistance to water and to freezing and thawing depends on the type of the hardening. The highest values of these properties are exhibited by hydrosilicate and the least by the carbonate hardening. After one month hardening at normal temperature (10-20 °C), mortars based on lime paste increase their strength to 0.5-1.5 MPa and those based on the milled quicklime to 2-3 MPa. It is possible to produce lime – sand (silica) concretes with a compressive strength up to 30-40 MPa and more for the hydrosilicate hardening.

Lump quicklime is transported in boxcars in bulk or in containers. Transportation of milled quicklime and hydrated lime is carried out in paper bituminizated bags, in air-tight containers or cement trucks. It is necessary to protect the lime from moistening in the period of transportation and storage. Duration of the milled quicklime storage should not be more than 20 days, because its chemical activity quickly decreases due to interaction with atmospheric moisture.

The air-hardening lime is used for mortars which work in air dry conditions; dense and porous silicate products of the autoclave hardening; mixed hydraulic binders and lime paints.\

9.3. Soluble and Liquid Glass, Magnesia Cements

Soluble glass is a glassy material, consisting of alkaline silicates. Its general formula is $R_2O \cdot nSiO_2$, where, R_2O - alkaline oxides (Na_2O, K_2O), n – silica modulus.

The silica modulus value can vary from 1 to 6.5. The sodium soluble glass with silica modulus 1.5-3 is usually used in construction, potassium glass is used basically to produce silicate paints. The soluble glass is produced in the form of the solid monolith, which is crushed into pieces or into fine product. The raw meal that includes silica and alkaline components, such as sand and soda, or sodium sulfate, is used for its production. The batch is fused in the glass-making furnaces at 1100-1400 ° C.

Silicate- boulder is formed with slow cooling of the glass bath in the air, and silicate finely crushed soluble glass by cooling in running water. Dissolution of the silicate- boulder in water is carried out in autoclave conditions under a pressure of 0.3-0.8 MPa. The finely crushed soluble glass can dissolve under atmospheric pressure and temperature of 90 - 100 ° C.

Water solution of soluble glass is called also a *liquid glass*. It can be produced from silicate- boulder or silicate finely crushed soluble glass by dissolution, or directly from the autoclave treatment of amorphous silicate products in solutions of caustic alkalis. Liquid glass is a colloidal solution with density of 1.4-1.5 g/cm^3. In closed vessels it can be stored for a long time but in the air conditions it slowly hardens. The essence of the hardening process of the liquid glass is the evaporation of the liquid phase, increasing the concentration of free colloidal silica, and the following coagulation and sealing. Air carbon dioxide neutralizes the alkalis, which are contained in water solution and promotes the silica coagulation. The fluorosilicate sodium additive and some other matters significantly speed up the liquid glass solidification. The liquid glass is characterized by high adhesive ability, which is 3-5 times higher than that of cement and other mineral binders. It is widely used for gluing of cardboard, paper, wood and silicate products, etc.

Silication – involves pressurization of liquid glass in soil together with solutions of additives through a system of the perforated pipes. It is used for soil fixing under foundations and as protection from ground-waters at passing of mines and tunnels.

Liquid glass serves as a sealer compound for quartz fluorosilicate in *acid-proof cements(-* products of the combined grinding or mixing of quartz sand and fluorosilicate sodium). There are two types of such cements: I - for acid-proof putties, II - for concretes and mortars. They differ with respect to the content of fluorosilicate sodium and time of setting. Acid-proof cements are used for agglutination of chemical-resistant materials, casing of different apparatus and protective coverings of building constructions. Acid-proof cements cannot be used in conditions of alkali action, fluorohydrogen and fluorosilicic acids, boiling water and stream.

The high heat resistance is also an excellent feature of mortars and concretes on the basis of liquid glass next to an acid resistance.

Magnesia cement is a product of the moderate burning of magnesite (caustic magnesite) or dolomite (caustic dolomite) tempered with solutions of electrolytes: magnesium chloride and sulfate, iron sulfate and others. Caustic magnesia is produced at 800-850 ° C, and caustic dolomite - at 650-750 °C.

Temperature rise, above the optimum, leads to compaction of MgO crystals, and during the burning of dolomite – to $CaCO_3$ dissociation, which impairs the quality of magnesia binders.

The highest strength of magnesia cement is obtained with use of magnesium chloride as sealer compound. So, caustic magnesite, tempered with the aqueous solution of magnesium chloride with a density of 1.2 g/cm^3, shows a compressive strength of 50-60 MPa and higher at 28th days of hardening. Magnesia binders are the best binders for materials using wood sawdust (xylolite) and chips (fibrolite). That differs with respect to high impact strength, heat - and sound insulation properties. The strength of products based on magnesia binders become reduced in water and in humid atmosphere. Phosphate additives can be used to increase their water resistance. Using of the salts solutions at tempering of the magnesia binders is sharply limits their application in construction.

9.4. Hydraulic Lime and Lime Containing Binders

Hydraulic lime is the product, produced by the moderate burning of limestone, which contains 6-25% fine-grained clay and sand impurity substances. The silicates and calcium aluminoferrites, which provide lime the hydraulic properties, are formed at the burning of such fine-grained marl limestone along with calcium oxide.

There are weakly hydraulic and strongly hydraulic limes. The content of active calcium oxide and magnesium in the first one are 40-65%, in the second one 5-40%. The main characteristic of the raw materials for hydraulic lime is the hydraulic index (– ratio of the calcium oxide percentage to the total content of alumina, silicon and ferric oxides). It is 9-4.5 for the weakly hydraulic lime, and 4.5-1.7, for strongly hydraulic limes. Weakly hydraulic lime on interaction with water intensively slakes and scatters in the powder, whereas the strongly hydraulic limes slowly slakes or does not slake completely. The initial setting time of hydraulic lime occurs after 0.5-2 hours, the final set takes from 2 to 16 hours. With increasing free calcium oxide content the hardening accelerates.

The ultimate compressive strength for samples of lime-sand mortars of the proportion 1:3 (one part by weight of lime, 3 parts of normal sand) after 28 days for weakly-hydraulic lime must be at least 1.7 MPa, strongly hydraulic - 5 MPa. Hydraulic lime is used to manufacture building mortars for masonry and wall plastering in dry or wet environments. The mortars, based on it, are more durable than on air lime, but less plastic. Hydraulic lime can be used also to produce mixed lime containing binders and concrete of low strength classes.

Hydraulic lime containing binders include the binding materials, produced by grinding or mixing of air-hardening or hydraulic lime with *active mineral additives*(- natural and artificial substances that when mixed with lime and water gives a paste, which after hardening in the air conditions can continue to harden in water). Examples of natural active mineral additives are tuffs, trasses, diatomites, etc. The blast-furnace and fuel slags, ashes and others can be attributed to artificial additives.

The ability to bind $Ca(OH)_2$ at normal temperature (activity of mineral additives) is characterized by the amount of absorbed CaO from the lime mortars within 30 days. In binders, which include lime, the content of active calcium and magnesium oxides is 10-30% by the weight.

Hardening of these binders at normal conditions is slow. It is stipulated mainly by the interaction of calcium hydroxide with active silica and formation of hydrosilicates. Compressive 28 days strength of these binders is in the range of 5-20 MPa. The hardening rate at temperature 80-100 °C and at enhanced humidity is substantially accelerated.

A significant deficiency of lime-containing binders, is their low stability in air conditions. Stability may be improved by addition of gypsum additives, calcium or sodium chlorides, and by the replacement of air-hardening lime with hydraulic lime. They are also characterized by comparatively low freeze-thaw resistance. It is expedient to use them for mortars and concretes of low strength in underground or submerged structures and to manufacture products using thermal treatment, especially autoclaving.

9.5. Portland Cement. Technology Bases

Portland cement is a hydraulic binder, which hardens in water and in the air conditions. It is produced by compatible milling of clinker, necessary amount of gypsum and other additives. Portland clinker is a product of sintering of raw material mixture of certain chemical composition which provides after the burning prevailing content of minerals-silicates.

Cement industry produces a large amount of cements types based on Portland clinker, which can be conditionally divided into cements of the common and special designation.

Cements of the common designation are divided into five types according to composition and compressive strength at 28-days (EN 197-1:2000):

- Cem I – Portland cement (may contain 0 to 5% mineral additives),
- Cem II – Portland-composite cement (6 to 35% blast furnace, granular slag or 6 to 20% mineral additives);
- Cem III – Blast furnace cement (36 to 80% blast furnace granular slag);
- Cem IV - Pozzolanic cement (21 to 55% mineral additives);
- Cem V - Composite cement (36 to 80% mineral additives).

Typical compositions of the different types cement are given in the Table 9.2.

According to the early age strength(after two or seven days of hardening) cements are divided into two kinds: cement with normal strength in early age and high-early-strength cement.

The type, class, and special signs (high strength in the early age – R; rationed mineralogy - N) are indicated at the symbolic notation of cement.

The limestone and clay rocks or their natural mixtures – marls are the basic raw materials for obtaining Portland cement. The various waste products are also used: slags, ashes, nepheline sludge and etc.

The Portland cement production consists of two stages: production of clinker and its fine milling with additives.

The chemical composition of *Portland clinker*, as a rule, falls between the following percentage limits:

CaO – 63-66	MgO – 0.5-5
SiO_2 – 21-24	SO_3 – 0.3-1
Al_2O_3 – 4-8	Na_2O+K_2O – 0.4-1
Fe_2O_3 – 2-4	$TiO_2+Cr_2O_3$ – 0.2-0.5

The basic technological operations of cement clinker production (Figure 9.5) are:

- excavation and preparation of raw materials, which include mining, grinding and drying as a necessity;
- production of a homogeneous raw material mixture by milling and mixing of components;
- firing of raw material mixture, which provides passing of physical and chemical processes of clinker formation;
- cooling of the product.

There are two basic methods of clinker production - wet and dry. For the wet method, the raw material mixture is prepared and burned as a water suspension – sludge with humidity 35-42%. A dry method includes drying of materials to its compatible grinding. Raw material mixture at a dry method goes on burning in the powdery or granular state.

In many countries, a wet method of production of clinker has been the most widespread. This is mostly due to the need to achieve high homogeneity of raw material mixture, which favorably affects quality of clinker. However, the dry method of production involves 1.5 - 2 times less heat consumption than the wet method. Presently, due to creation of effective constructions of clinker burning furnaces the dry method has the wider distribution.

Basic raw materials, required for the production of cement, are mined in careers near by plants by the opened method. Grinding down of raw materials is a two-stage process involving (a): crushing in crushing machines (jaw, hammer crusher, roll crusher, gyratory breaker) and (b): milling i.e., fine grinding in ball or other mills.

Grinding of the raw materials to a high fineness is characteristic for modern technology of Portland cement, because increased fineness and homogeneity of mixture accelerates the physical and chemical processes.

Firing of the homogeneous raw material mixture is a main technological operation in the production of Portland cement. Firing is carried out, as a rule, in rotary furnaces. Modern furnaces used for the wet method are typically 185-230 m in length, and 5-7m in diameter. The productivity of a 5×185 m furnace is 1800 t/day, and a 7×230m rotary furnace produces about 3000 t/day. For the dry method, furnace sizes diminishes considerably without a corresponding decline in the productivity. So, rotary furnaces with cyclone heat-transfer apparatus have sizes of 5×75 and 7×95 m, and a daily productivity of 1600 and 3000 t respectively.

Table 9.2. Typical composition of common cements (percentage by mass)

Type of cement	Clinker	Mineral additive	Type of cement	Clinker	Mineral additive
CEM I	95-100	-	CEM II/A-L	80-94	Limestone (L) 6-20
CEM II/A-S	80-94	Blast furnace slag (S) 6-20	CEM II/B-L	65-79	21-35
CEM II/B-S	65-79	21-35	CEM II/A-LL	80-94	Limestone (LL) 6-20
CEM II/A-D	90-94	Silica fume (D) 6-10	CEM II/B-LL	65-79	21-35
CEM II/A-P	80-94	Natural Pozzolan (P) 6-20	CEM II/A-M	80-94	Composition of additives (M) 6-20
CEM II/B-P	65-79	21-35	CEM II/B-M	65-79	21-35
CEM II/A-Q	80-94	Natural Pozzolan calcine (Q) 6-20	CEM III/A	35-64	Blast furnace slag (S) 36-65
CEM II/B-Q	65-79	21-35	CEM III/B	20-34	66-80
CEM II/A-V	80-94	Fly ash siliceous (V) 6-20	CEM III/C	5-19	81-95
CEM II/B-V	65-79	21-35	CEM IV/A	65-89	Pozzolan, silica fume, fly ash 11-35
CEM II/A-W	80-94	Fly ash calcareous (W) 6-20	CEM IV/B	45-64	36-55
CEM II/B-W	65-79	21-35	CEM V/A	40-64	Blast furnace slag (S) 18-30 Pozzolan additives (P,D,V,W) 18-30
CEM II/A-T	80-94	Burnt shale (T) 6-20	CEM V/B	20-38	Blast furnace slag (S) 31-50 Pozzolan additives (P,D,V,W) 31-50
CEM II/B-T	65-79	21-35			

Notes:

1. Up to 0-5% minor constituents may be added to all types of cements.
2. In Portland-composite cements CEM II/A-M and CEM II/B-M, in Pozzolanic cements CEM IV/A and CEM IV/B and in composite cements CEM V/A and CEM V/B, all main constituents other than clinker shall be declared by designation of the cement.

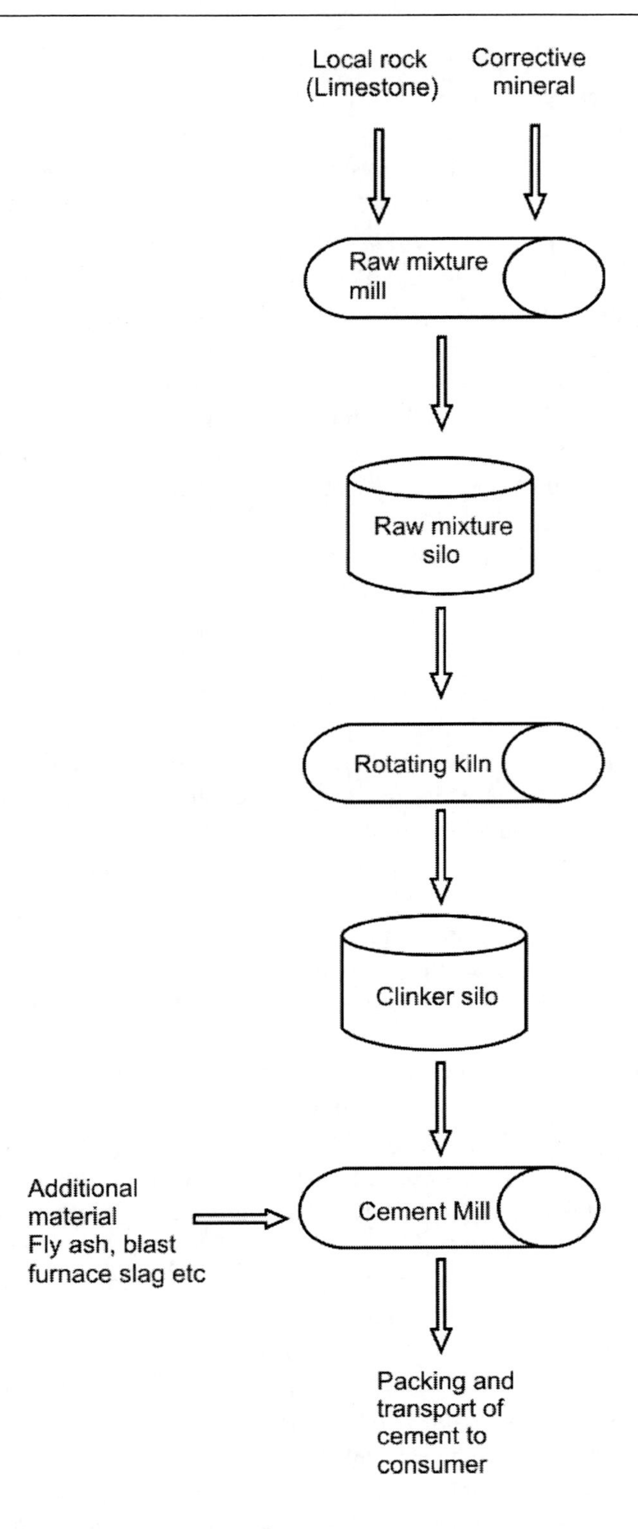

Figure 9.5. Scheme of Portland cement manufacture.

The material passes through the sloping drum of the furnace, and on coming into contact with the hot gases, experiences a series of complex physical and chemical transformations. Firstly, physically bound water is removed from the raw material mixture at about 110°C, and then the chemically bound or constitution water is removed from the clay component at 600-700 °C. The dissociation of $CaCO_3$ occurs at 750-1000 °C. At the temperature about 1300 °C exothermal reactions results in the formation of basic compounds - clinker minerals: dicalcium silicate $2CaO{\cdot}SiO_2$ or shortly C_2S, tricalcium aluminate $3CaO{\cdot}Al_2O_3$ (C_3A), tetra-calcium aluminoferrite $4CaO{\cdot}A_2O_3{\cdot}Fe_2O_3$ (C_4AF). At a temperature of about 1300°C partial melting of raw material begins. CaO and $2CaO{\cdot}SiO_2$ dissolve in the melt and leads to the main reaction of clinker formation (- reaction of two-calcium silicate satiation CaO to the tricalcium silicate - $3CaO{\cdot}SiO_2$ (C_3S)). The formation of C_3S takes place in the sintering zone of the furnace at a temperature range of 1300-1450°C. This ends the formation of conglomerates of clinker grains-granules, most of which have the sizes 10-60 mm depending on the construction and mode of operations of furnace.

The cement clinker is then cooled and stored. After cooling and storage comes the milling in the ball mills with a capacity of 50-100 t/hour. The required amount of gypsum stone is added into to the mills, which provides content in cement to 3.5% SO_3 to regulate the time of setting. Other active mineral additives can also be added into Portland composition at the milling, to improve water resistance and some other properties. The type and content of additives depends on the type of cement. Specific designation is given to identify Portland cements containing mineral additives. Portland cement (type II), is divided into cement with additive of granulated blast furnace slag, pozzolan, flue ash and limestone and other additives. Portland cement of group A includes 6-20% additives; group B - 21-35% (Table 9.2). The blast furnace cement (type III) group A includes 36-65% of granulated blast-furnace slag, group B - 66-80%, in pozzolanic cement (cement type IV) – respectively 11-35 and 36-55% of flue ash or its mix with the pozzolanic component. It is allowed to enter 18-30% blast-furnace granular slag and 18-30% of flue ash and pozzolans into composite cement group A, and group B - respectively 31-50 % blast-furnace slag and 31%-50% flue-ash and pozzolanic additives.

A clinker milling is the most power-consuming technological operation in cement production. More than 60% of total electric power, which is consumed on the production of cement, has been known to be spent on milling. At a milling plant, special surface-active substances can be entered for intensification of grinding down or providing cement with the special properties (technical lignosulphonate (LST), fatty acid and others), in an amount not more than 0.3% (by mass) of cement content.

After milling, the cement is transferred to the silo section, equipped with the pneumatic devices for aeration, homogenizing and unloading.

Portland cement is shipped in valvular paper bags or in bulk in special transport vehicles, cement trucks or cement carriers, covered carriages, etc. Cements should be saved separately according to the types and classes in the stationary and movable automated storages (silos) or in the special containers and bunkers which have a device for a reception and delivery of cement. The serial number, complete designation of cement and its type, kind and amount of additives, normal consistence of cement paste and other information are indicated in a document of cement quality.

9.6. PROPERTIES OF PORTLAND CEMENT. CORROSION OF CEMENT STONE

Composition and Properties of Portland Cement

Properties of Portland cement depend on the composition and structure of the clinker. An enhanced calcium oxide content in the clinker, bound in minerals, results in the production of cement with high activity and speed of strength gain with age. The content of free CaO in a clinker ranges from 0 to 2%. As a rule, effort is made to keep the CaO content to the minimum during the reactions of clinker formation. This is because free calcium oxide remaining in the clinker may cause volume instability and may lead to reduction in the strength of a cement stone.

A high Magnesium oxide content may have a negative influence on the properties of cement. Therefore, the MgO content in Portland cement must not exceed 5%.

Harmful influence of free calcium oxide and magnesium oxide are characterised by their capacity for the slow slaking and development of internal tensions in hardened concretes and mortars.

The most important mineral compounds in the cement clinker are called alite and belite. An *alite* - is a solid solution of tricalcium silicate (C_3S) and little amount of Al_2O_3, MgO and others. Solid solution in this case is the result of interstitial of the indicated oxides into the crystalline lattice of tricalcium silicate. Alite determines the basic properties, of the Portland cement, the early strength and rate of strength gain.(Figure 9.6).

Belite β is the second most important of the clinker minerals. It is the solid solution of dicalcium silicate (C_2S) and tenuous amount of Na_2O, K_2O, Fe_2O_3 and others. It hardens slowly but steadily increases its strength over a long time.

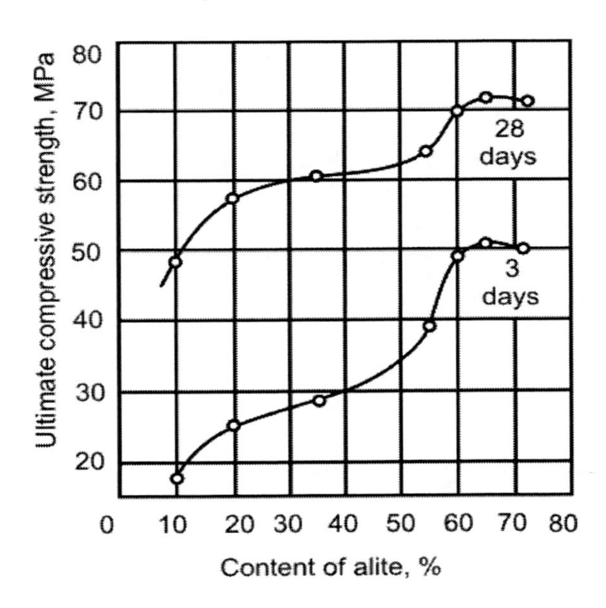

Figure 9.6. Influence of content of alite on cement strength.

Microscopic examination of cement clinker reveals clearly discernable prismatic crystals of alite and round-belit. The aluminate and aluminoferrite phases are the complement of intermediate matter which lies between them.

The aluminate phase is present in the form of tricalcium aluminate $3CaO{\cdot}Al_2O_3$ (C_3A). It is a high-early-strength mineral which is characterized by fast and highly exothermic reaction.

The aluminoferrite phase is the solid solution of different aluminoferrites and for many clinkers near on composition to tetracalcium aluminoferrite (C_4AF). It contributes very little to the strength gain.

Usually, the mineralogical composition of Portland clinker varies within the following percentage range: C_3S – 45-60; C_3A – 5-15; C_2S – 15-30; C_4AF – 10-20. However, the mineralogical composition for some special types of Portland cement may not correspond to these limits. Increased content of mineral-silicates (especially alite) improves strength and other properties of cement, but complicates the burning of the clinker. The rational compositions of clinker are selected at the production of cement, which provide both high quality of cement and optimum conditions of rotary furnace working.

Grinding cement clinker to a high fineness is the necessary condition to ensure high reactivity and binding properties. To ensure adequate fines, not less than 85% by weight of cement must pass through when sifted through a №008 sieve. Increasing the fines of cement also increases the surface area of the cement particles which enhances chemical reactivity and strength gain in proportion to the specific surface area. The specific surface area for Portland cement ranges from 2500-3500 cm^2/g and is determined based on the speed of air passing through the layer of cement powder.

The *absolute density* of Portland cement varies between 3 – 3.2 g/cm^3. The *bulk density* of cement depends on the degree of powder compression: in the loose state, it is 960-1300, but varies between 1600-1840 kg/m^3 in the compacted state.

Cement, mixed with water, forms a plastic paste. The water demand of cements is estimated by the amount of water (in percents by weight of cement), which is necessary for forming the paste with *normal consistency.* The concept of normal consistency is conditional and determined by the standardized penetration of pestle of Vicat apparatus in cement paste. Portland cement is characterized by the comparatively small water demand. Its normal consistency ranges between 24 to 29%. High content of aluminates and mineral additives of sedimentary origin (tripoli powder, diatomite and others) and increased milling fineness increases the water demand of cement. Introduction of surface-active additives, such as so-called superplasticizers, can reduce the water demand. Increasing of water content affects the properties of cement such as: strength, shrinkage deformations, freeze-resistance and others, unfavorably. This is because increasing the water content beyond the theoretical necessity for good consistency increases the total porosity of cement stone.

Setting is the first stage of hardening of the cement paste. All the period of setting is divided into initial and final. Initial setting time of cement paste is considered to be the time period from the moment of mixing till the moment, when the needle of Vicat apparatus will not reach the plate on which a ring is set on 1-2 mm. Final setting time is considered to be the time from the beginning of mixing till that moment, when penetration of the needle into the paste will not be more than 1 mm. Initial and final setting of cements is rationed within the limits, comfortable for mortars and concretes preparation. Initial setting time does occur before, a period of between 45-60 minutes for cement. As a rule, it is observed within 2-4 hours from the moment of mixing. The final setting time for cement should not occur later than 10 hours. The indicated requirements are provided as a result of adding of gypsum additive to Portland cement. Gypsum – dehydrate slows the setting of Portland cement. The slowing action of gypsum is caused by the formation on the surface of grains of C_3A (the most fast-hardening phase of cement), protective coats of a new compound – hydrosulphoaluminate. This compound is the product of interaction of gypsum, tricalcium aluminate and water.

Strength of cement is determined after 28 days of hardening of specimens. The standard prism(beam) specimen, 40×40×160 mm in size, is made from cement-sand mortar 1:3 (1 part of cement, 3 parts of normal sand) using standard conditions. The cement strength increases intensively at early age, but the rate of strength gain slows considerably at later ages, The strength of high-early-strength cements is determined also after 2 days of hardening. Mechanical and physical requirements of cement accordingly to EN 197-1:2000 are shown in the Table 9.3.

Cement plants should determine strength of cement also at steaming in the age of one day and to specify it in the certificate of cement quality.

Also another accelerated methods can be used for approximate determination of cement strength.

The cement strength is dependent on the multiplicity of complex of factors. Cement composition is one of the basic factors. Not only content of separate minerals, but also their microstructure influences the indexes of the cement strength. In the last years the additives which promote the cement strength has received great attention.

The compressive strength of cement, especially at early ages, increases as the finnes and specific surface area increases. Figure 9.7 shows that in ultra-high-early-strength cements, made with very fine particles below 10 μm, 82% of the ultimate strength was reached after 7days whilst for the mixture made with particles between 25-50 μm, only about 30% of the ultimate strength was developed (Figure 9.7).

Conditions of storage, using and hardening of cement substantially influence on the strength of cement concretes and mortars. It is not desirable to hold in stock Portland cement for a long time, as its activity goes down under the action of humiduty and carbonic acid of air.

Table 9.3. Mechanical and physical requirements for cement

Strength class	Compressive strength, MPa			Initial setting time, min	Soundness (expansion), mm
	Early strength		Standard strength		
	2 days	7 days	28 days		
32,5 N	-	≥16.0	≥32.5	≤52.5	≤10
32,5 R	≥10.0	-			
42,5 N	≥10.0	-	≥42.5	≤62.5	
42,5 R	≥20.0	-			
52,5 N	≥20.0	-	≥52.5	-	
52,5 R	≥30.0	-			

Note: N - normal early strength, R - high early strength.

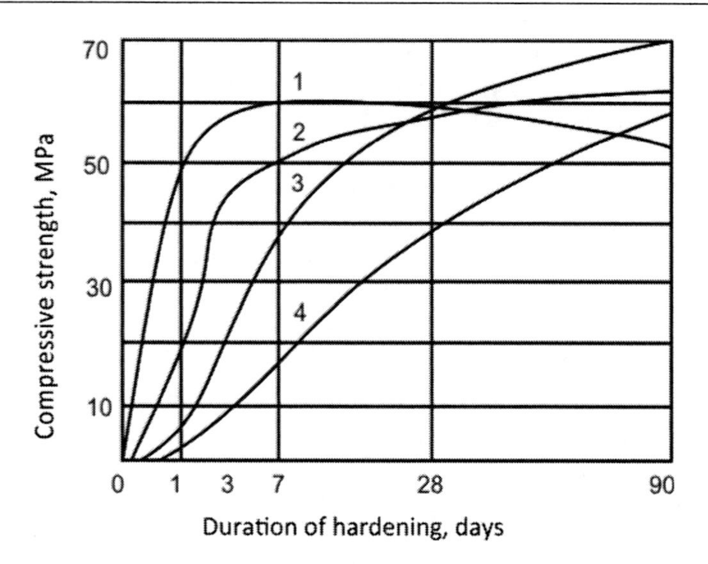

Figure 9.7. Influence of cement particles size on kinetics of cement hardening:
1 - cement particles size <3 μm; 2 – 3-9 μm; 3 – 9-25 μm; 4 – 25-50 μm.

Hydration and Hardening of Cement

New chemical compounds - hydrates appear as a result of interaction of minerals, which are contained in cement, with water. *Hydrosilicates* are basic compounds, which are formed during the hydration of minerals – silicates. It is possible at certain conditions to describe the process of hydration of C_3S by equation:

$$2(3CaO{\cdot}SiO_2)+6H_2O=3CaO{\cdot}2SiO_2{\cdot}3H_2O+3Ca(OH)_2$$

Composition of hydrosilicates depends on temperature and on the concentration of calcium hydroxide.

Tricalcium aluminate in the presence of gypsum, which is contained in cement, and water, forms *calcium hydrosulphoaluminate (ettringite),* which slows the process of setting of cement paste:

$$3CaO{\cdot}Al_2O_3+3(CaSO_4{\cdot}2H_2O)+25H_2O=3CaO{\cdot}Al_2O_3{\cdot}3CaSO_4{\cdot}31H_2O$$

After the formation of hydrosulphoaluminate, another hydrate, *calcium hydrosulphoferrite* or solid mortar of these two compounds appears. Chemical reactions begin just after mixing of cement with water. Ettringite and calcium hydroxide are the first new hydrate formed.

The mechanism of cement hydration is very complex. Accordance to modern views, new formations crystallize from oversaturated solution in two stages. There is the formation of framework with the origin of contacts of accretion between the crystals of new formations during the first stage. Growing of crystals, which are joined between themselves, is also possible. During the second stage, new contacts does not develop, but there is only the growing of the framework which has appeared already. It results in the increase of the strength of the cement stone, but internal stretching tensions can develop. A decisive role is played by oversaturation of solution. At low oversaturation the amount of crystals is small and they are not accreted. For the maximum strength of artificial stone, the optimum conditions of hydration are needed, which provide the origin of new hydrate compounds of sufficient size at minimal tensions.

Hardening of Portland cement largely depends on the temperature and moisture conditions. So, the decline of temperature from 20 to 5°C slows the rate of hardening by 2-3 times. Increasing the temperature to 80°C increases the speed of hydration by upto 6 times. At a temperature below -10°C, the hydration of cement become practically halted.

The normal progress of the processes of hardening is possible only in conditions of sufficient environmental humidity and temperature, but temperature increase must not be accompanied with drying. The acceleration of physical and chemical processes of hardening of Portland cement by means of thermal treatment (steaming-out, electro-warming and others like that) allows obtaining concrete and reinforced concrete products with the required strength in the short terms.

Cement stone corrosion. A cement stone is the basic component of concrete, and it corrodes under act of aggressive environment.

Natural and industrial water solutions, which contain the different amounts of dissolved matters (acids, salts, alkalis), and also some organic liquids are the most widespread liquid aggressive environments.

Atmospheric water can contain the enhanced amount of salts in seaside areas and saline lands. Chemistry of river water depends on the type of sources and river bed rocks. By the degree of mineralization, that means content of salts, river water may be devided on four groups: I - small mineralization (to 200 mg/l), II - middle (200-500 mg/l), III - enhanced (500-1000 mg/l), IV - high mineralization (over 1000 mg/l). Natural and ground waters also substantially differ by composition. The soft ground water, which formed as a result of snow melting and rain falls, is characteristic for the north and mountain areas. The high mineralized ground waters are often meet in the south areas.

Aggressive to cement concretes gases contain vapors and aerosols of different acids and salts. The solid corrosive environments are dry mineralized soils and different granular chemical matters: fertilizers, paints, insecticides, fungicides, herbicides, etc. Corrosive processes in gaseous and solid environments occur only in the presence of liquid phase.

Corrosive processes under the action of different environments, which operate on concrete and reinforced concrete structures, may be divided into three types.

Corrosion of the first type is conditioned by the dissolution of some components of cement stone and mainly the hydroxide of calcium - the product of hydrolysis of tricalcium silicate (corrosion of leaching). It goes intensively in soft waters, especially during filtration of water through a concrete. At leaching of $Ca(OH)_2$ disintegration of other hydrates, steady only at the certain concentration of $Ca(OH)_2$ begins and take place reducing of density and violation of structure of cement stone.

Corrosion of the second type is based on the exchange chemical reactions of interaction between components of cement stone and aggressive water solution with formation of easily soluble salts, which are eluted, or amorphous products without binder characteristics. Acid (i.e. carbonic acid) and magnesia corrosions are the kinds of this type of corrosion.

At carbonic acid corrosion hydroxide of calcium, which is contained in a cement stone, at first, interacts with CO_2 with formation of sparingly soluble $CaCO_3$, than the soluble hydrocarbonate appears at the excess of carbonic acid, which is eluted from concrete. For the concrete in sandy or gravelly soils which filter water at a speed of 0.04-0.1 m/sec, corrosion develops at content of CO_2 in water over 15-20 ml/l. At the increase of carbonate water hardness, that means the content of soluble calcium carbonates and magnesium carbonates, the amount of free CO_2 diminishes and carbonic acid corrosion becomes less dangerous.

The other acids: hydrochloric, sulfuric, nitric, acetic and other have also a destructive influence on a cement stone. Acids interact mainly with hydroxide, and then with the calcium hydrosilicates. The more soluble are salts, the more quickly cement stone and accordingly concrete are destroyed. So at the same concentration of hydrogen ions, the speed of corrosion under the action of solutions of sulfuric acid is lower than hydrochloric acid that is explained with the greater solubility of calcium chloride comparatively with a sulfate.

All natural waters contain some amount of the carbonic acid. The sulfur and some organic acids can be met in the peat waters. The large amount of free acids can be contained in sewage of industrial enterprises, and also in underground waters, fouled with the products and waste of different productions.

The magnesium salts (mainly magnesium chloride and magnesium sulfate) are contained in sea water, and also often are in underground waters. At interacting of them with calcium hydroxide besides to soluble calcium salts, the amorphous magnesium hydrate appears and mechanical properties of concrete are decreased.

Corrosion of the third type is also caused by the reactions of exchange of matters, dissolved in water with the components of cement stone, however their products are poorly soluble salts, crystallization of which takes place in pores and capillaries with the volume increase. The typical example of corrosion of this type is sulfate corrosion. It is caused by the ions of SO_4^{2-} the source of which are calcium, magnesium and sodium sulfates. The varieties of sulfate corrosion are sulphoaluminate and gypsum corrosion. At sulfo-aluminate corrosion take place a reaction between the calcium sulfate, which is contained in water, and hydroaluminate in a cement stone with formation of ettringite, that is accompanied with the considerable increase of volume and origin of destructive tensions. At the large concentration of sulfates (over 1000 mg/l SO_4^{2-}) gypsum is crystallized in capillaries of a cement stone (gypsum corrosion).

Corrosion of cement stone can be also caused by interacting of alkalis of cement with active silica which are contained in such minerals, as an opal, chalcedony and other, which are met in the concrete aggregates. Harmful tensions are the results of jelatinous alkaline silicates formation.

Finding-out of reasons and mechanism of corrosion of cement stone allows to choose the method of increase of its resistance. An increasing of concrete density due to diminishing of water content and water-cement ratio of concrete mix, and also adding of plasticizers, polymeric and other additives is positive in all of cases. More resistant to the corrosion of leaching and sulfate corrosion are sulfate-resistant cements with specified amount of tricalcium aluminate and also cements, that contain active mineral additives, which bound $Ca(OH)_2$ in slightly soluble compounds.

9.7. Varieties of Cements Based on Portland Clinker

The different exposure conditions to which concretes and mortars are subjected in service in various environmental and constructional situations has necessitated the production of a wide assortment of cement types based on Portland cement clinker. Greater parts of the volume output of Portland cement in the world contain mineral additives. The use of different mineral additives results in the economy of the most expensive and power-consuming intermediate product – Portland clinker and utilization of different wastes. These cements are more water and corrosion-resistant, than the cement without the additives. Portland cement without additives is used for the production of ultra-strength, frost-resisting concretes and in a number of other cement types.

Rapid Hardening Cement and High-strength Cements

The clinker of rapid-hardening cement contains an enhanced amount (60-65%) of the most active minerals - the tricalcium silicate (C_3S) and tricalcium aluminate (C_3A). This type of cement is milled to a much higher fineness, and higher specific surface area (3500 cm^2/g). Increasing the C_3S content of the cement clinker to 60-65% and specific surface area to 4000 cm^2/g or more produces *cements* with *high early*-compressive strength of 25-30 MPa at 1 days and 80-90 MPa at 28 days. However, their production involves a sharp growth of the power inputs, whilst the productivity of equipment become reduced.

It is rational to use high-early-strength cement and high strength cements for the production of precast concrete, which provides rapid growth of the strength of products and declining of binder content per 1 m^3 of concrete.

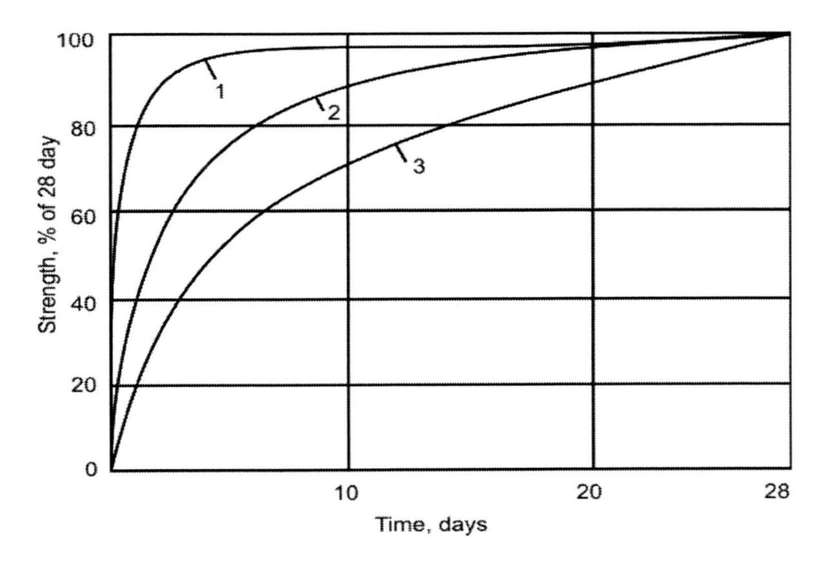

Figure 9.8. Curves of cement hardening:
1 – alumina cement; 2 – rapid-hardening cement; 3 – ordinary cement.

For rapid-hardening Portland cement (Figure 9.8.) an additional requirement is achievement in 2 days of hardening of the compressive strength not less than 15-25 MPa.

The number of new methods of hardening acceleration and increase of cement activity has been developed within the last few years.

The technology has also been developed for *extra-rapid hardening cement* that contains fluorine. Limestone, slags of the meltback of aluminium, calcium fluoride and special additive for retaining fluorine in a clinker are used as raw materials. Cement, modified by fluorine, achieves compressive strength of 5-8 MPa after 6 hours of normal hardening. The use of fluoride cement allows the production of concrete products having the required strength after about 1 hour of thermal treatment.

The production of sulphoaluminate clinkers by adding over 5% of sulfates to the raw meals is the future direction for the production of rapid-hardening binders. The calcium sulphoaluminates produced, gives the clinker a capacity for rapid hardening, and high hydraulicity. The phosphogypsum and other gypsum wastes can be the source of sulfates for the production of sulphoaluminate clinker.

Extra-rapid-hardening cement is produced by the milling of a clinker with special composition (sulphoaluminate belite) and gypsum. A strength of 7-15 MPa may be reached in about 6 hours after tempering at a temperature of 20°C. This enables its use for producing precast concretes and structures without thermal- moisture treatment. This type of cement is characterized by a short setting time(– initial set is not earlier than 5 minutes, and final set of not later than 1 hour. Concrete made with this type of cement needs to be applied immediately after mixing.

Technology of binders of low water demand is based on fine milling of Portland cement with addition of mineral additives or without them and enhanced dosages of dry superplasticizer. Depending on the content of mineral additives these binders are divided into different types. The use of these binders allow to produce concretes with the strength to 100 MPa and higher. At the same time the production of binders of low water demand is related with certain difficulties, including the necessity of the dry a superplasticizer using, etc. Method, based on making of fine multi-component cements with different mineral additives and specific surface area 4500-5000 cm2/g and adding of liquid superplasticizers at production of concrete mixes, is more simple.

High-early-strength and sulfate-resistant blastfurnace cements are also produced along with the ordinary one. Clinker, which contains not more than 8% C_3A and slag which contains not more than 8% Al_2O_3 is used for its manufacture.

The quality of blastfurnace slag is characterized, mainly by the value of the coefficient of quality which represents the ratio between the total percentage of oxides CaO, MgO, Al_2O_3 and the content of SiO_2 and Fe_2O_3.

Blast furnace cements, has a higher firmness in soft and mineralized waters, and enhanced heat-resistance compared to Portland cement. This can be explained by the denser microstructure of the former due to reactions involving the silica (and/or alumina) and calcium hydroxide to form extra hydration products; and through thermo-gravimetric analysis, by the insignificant content of free hydroxide of calcium in the cement stone. It hardens more intensively at thermal-moisture treatment, more slowly at a reduced temperature and has a lower freeze-thaw resistance. The compressive strength of blast-furnace cement is usually 30-50 MPa at 28 days.

Blast furnace cements are used for underwater structures which are subjected to the action of fresh waters, and also for making of precast elements with the use of thermal-moisture treatment, and for preparation of building mortars. It is not allowed to use these cements for structures, exploitation of which needs enhanced freeze-thaw resistance and for building works at subzero temperature without artificial heating, and also in a hot and dry weather without careful keeping of the moisture conditions during hardening.

Blast furnace cements are economically effective types of cements. The Power consumption at blast-furnace cement production on average to 25% is lower than energy consumption at production of Portland cement.

Sulfate-resistant Cements

The sulfate-resistant Portland cement and Portland cement with mineral additives, pozzolanic Portland cement and sulfate-resistant blastfurnace cement belong to this group of cements.

Sulfate-resistant Portland cement differs from the ordinary Portland cement because it has an enhanced resistance to sulfate aggression in the conditions of systematic alternate freezing and thawing or moistening and drying. The main feature of sulfate-resistant Portland cement is that it has a mineralogical composition with low content of tricalcium aluminate (C_3A). The mineralogical composition of such clinker must comply with three conditions: 1) content of C_3A is not more than 5%; 2) content of C_3S is not more than 50%; 3) sum of C_3A+C_4AF is not more than 22%.

Next to sulfate-resistant Portland cement, to which it is not allowed to add mineral additives, industry produces *sulfate-resistant Portland cement with mineral additives. These may* be granulated blast furnace and electrothermophosphoric slags at content of Al_2O_3 not more than 8 %.

The content of C_3S in the clinker of sulfate-resistant Portland cement with mineral additives is not specified. The compressive strength of this cement at 28 days is 40-50 MPa. The sulfate-resistance of cements, as well as a range of other properties, are improved by addition of surface active admixtures.

Pozzolanic Portland cement is a hydraulic binding agent, which is produced by inter-grinding cement clinker with acid active mineral (hydraulic) additive and gypsum.

The presence of active mineral additives in pozzolanic cement as well as in Portland blast-furnace cement promotes water- and sulfate-resistance. The limitation of tricalcium aluminate ($C_3A < 8\%$) is also desirable in pozzolanic Portland cement. The low density of pozzolanic Portland cement (2.7-2.9 g/cm^3) results in an enhanced volume of cement paste from it. This provides a more compact microstructure of the cement stone and ensures higher water impermeability of concretes, based on pozzolanic Portland cement.

Pozzolanic Portland cement at ordinary and especially at low temperatures hardens slowly. The strength development of this cement is especially slowed during hardening in air conditions. Adding some mineral additives (diatomites, tripoli powders and others like that) with high specific surface area and fineness to cement, increases the water demand and accordingly the cementitious content of concrete mixes. Some concretes and mortars, based on pozzolanic Portland cement, are not frost-resistant. Large shrinkage and low resistance to alternate moistening and drying are possible imperfections also observed in such cements.

It is most rational to use pozzolanic Portland cement for underground and underwater structures on account of their enhanced water impermeability and resistance both in soft and in mineralized waters. It is possible to use this type of cement for ordinary structures, which are exposed to conditions of high humidity, but not subject to freezing and thawing.

Masonry cements are made by inter-grinding Portland clinker, active mineral additives and filling agents. The content of clinker must be not less than 20% of the weight of the binder. Ash, quartz sand, limestone and others like that are used as fine milled additives. Such cements have a compressive strength of 5-20 MPa at 28 days. Due to the slow hardening characteristics, these cements are used at air temperatures not below 10°C for making of masonry mortars and plasters, and also for concrete of low strength, to which frost-resistance is not a requirement.

Cements with AdMixtures of Surface-active Agents

Surface active agents can be added in an amount not more than 0.3% by weight in different types of Portland cements at grinding. During the process of cement grinding, surface active agents are adsorbed on its grains and give to it a number of new properties depending on whether the surface agent added is a hydrophilic or hydrophobic agents.

Hydrophilic agents are absorbed on the surface of cement grains and form hydrophilic films which are instrumental in the better moistening of the particles with water, diminishing of their tripping and increasing the plasticity of cement paste. *Plasticized Portland cement* is produced by adding hydrophilic agents, in an amount 0.15-0.3% during the process of grinding.

Plasticity of cements is determined by the degree of cone spread of a cement-sandy mortar on a shaking table. The mortar from a mix, based on plastificised portland cement with standard sand composition of 1:3 (by weight) and a water-cement ratio (W/C) of 0.4 after a 30 shakings on a shaking table has a cone spread of not less than 135 mm. This index is 106-115 mm for not plasticized Portland cement

The use of plasticized Portland cement allows the production of concrete mix of better plasticity and facilitates its compaction and placing without a corresponding increase of water content. It can also be used to reduce the cement content by 8-12%, with unchanging plasticity and water-cement ratio. By keeping the cement content and required plasticity constant, the water-to-cement ratio of concrete mix diminishes, which results in the increase of strength, frost-resistance and water impermeability of concrete.

Hydrophobic Portland cement is produced by adding hydrophbic agents to the cement composition during the process of grinding. These agents form the thinnest adsorption layers - unwettable membranes on grains of cement such that a drop of water on the surface of hydrophobic cements must not be sucked for 5 min.

Hydrophobic property of cement stipulates has as its main advantage – a capacity to protracted storage. Hydrophobic cements do not reduce activity during 1-2 years in unfavorable conditions; compared to ordinary cements which may lose up to 30% of their initial strength within 3-6 months. The hydrophobic films are destroyed during the mixing with water of these cements that provides the normal flowing of processes of hydration and hardening.

The cement hydrophobization is instrumental in the increase of density and improvement of cement stone structure, which allows attainment of greater frost-resistance and water impermeability of concretes, their firmness in corrosive environments, to decrease the leaching in the plasters and finishing mortars and others like that. The adsorption action of hydrophobic admixtures is instrumental in the increase of plasticity of concretes and mortars.

Surface active agents can slightly slow down the strength growth in the first periods of hardening.

Extended and Straining Cements

The shrinkage which causes stretching tensions in a concrete even to formation of cracks is characteristic of ordinary cements during hardening in air conditions. Special cements which increase their volume during hardening are called *extended cements*. Expansion of cements during early age hardening, compensates the negative influence of shrinking deformations in the future. Extended cements as a rule are composite and consist of basic binding agent and additive which cause expansion. Expansion of cement stone takes place, as a rule, due to the chemical interaction of components of additive and formation of calcium hydrosulphoaluminates and also $Ca(OH)_2$ and $Mg(OH)_2$.

Straining cements are the variety of extended cements. They have energy of expansion, sufficient for armature tensioning in concrete constructions. They are divided into cements with small (1-2 MPa), middle (4 MPa) and high (6 MPa) energy of expansion. Cements which are strained are produced both for the conditions of thermal treatment and for the normal hardening. Application of these cements allows to produce effective prestressed structures. Their use allows lower expenses on reinforcement by up to 30-50%.

Self-stressing Portland cement is produced by fine milling of 65-70% Portland clinker, 16-20% high-aluminous slag and 14-16% of the gypsum. Relative linear expansion equals 3-4%.

Technology of straining cement is developed on the basis of sulphoaluminate clinker which contains the sulphoaluminate calcium as a basic mineral. Such clinker is produced by the burning of kaolin or ash mixed with limestone and gypsum.

White Portland cement is produced by compatible milling of white low ferrous clinker, mineral additives and gypsum.

Coloured Portland cements are produced by the adding of coloring agents in the white cements.

Apart from the cements considered above on the basis of Portland clinker, there are a range of other varieties: oil-well, air-entraining cements etc, not covered here.

9.8. Alumina Cement

Alumina cement is a high-early-strength binding agent, which is produced by the fine milling of clinker which consists mainly of low basic calcium aluminate.

The bases of alumina cement, as well as Portland cement, are four main oxides - CaO, Al_2O_3, SiO_2, Fe_2O_3. However, numerical correlations of these oxides are principally different.

Alumina cements contain mostly Al_2O_3 (30-50%) and CaO (30-45%). These oxides are provided by the principal raw materials - bauxite and limestone.

The clinker of alumina cement is produced by two methods: melting of raw meal, or its burning to sintering.

The degree of fineness of the cement must be such that not less than 90% by weight of the test material passes when sifting through a sieve № 008. The quality of alumina cement is determined mainly by the amount of calcium aluminate. The water demand of alumina cement is the same as Portland cement (23-28%), setting. but it is more sensible to control the water content during mixing to avoid premature

Initial setting time must begin not early than 30 minutes, and the final set - not later than 10 hour

After 5-6 hours of hardening, the strength of alumina cement reaches about 30% of ultimate strength. After 3-days hardening alumina cement usually achieve a compressive strength of 40-60 MPa.

Alumina cement is considerably more expensive than Portland cement in connection with the deficit of bauxite as raw material. It is practical to apply it, if its advantages are rationally used: at emergency works, at the speed of concrete construction building, especially at a temperature below zero, for heat-resistant concretes.

Gypsum-alumina cement and waterproof expanded cements are produced on the basis of alumina cements. *Gypsum-alumina cement* is produced by the compatible grinding of clinker of alumina cement and gypsum in the ratio 70: 30 (by weight), and *waterproof expanded cement* - by a compatible grinding or mixing of 73-76% alumina cement, 20-22% gypsum and 10-11% of the high-basic calcium hydroaluminates. Hydroaluminates are produced by the treatment of mixture of alumina cement with slaked lime. Gypsum-alumina cement expand during hardening in a moist environment, and waterproof expanded cement – also in the air conditions. The feature of this cements are short setting time, high watertightness of mortars and concretes even at early age.

Self-Assessment Questions

1. Discuss the nature and properties of mineral binders.
2. Explain the nature of mineral non-hydraulic binders, their classification and value in building.
3. Distinguish between the features of a building and alpha - gypsum?
4. Describe the technical properties of gypsum binders.
5. What types of lime are used in building construction?
6. What are the features and properties of building lime?
7. Explain the manufacture and properties of liquid glass?
8. Provide a classification of hydraulic binding agents.
9. Discuss the nature and properties of hydraulic lime.
10. Describe the process of manufacture of Portland cement clinker.
11. Describe the chemical-mineralogical composition of Portland cement clinker and its influence on the properties of Portland cement.
12. What are the basic properties of Portland cement ?
13. Explain the processes of the cement hydration?

14. Describe the corrosive environments which destroy a cement stone.
15. Describe the chemical reactions characteristic of the process of cement stone corrosion? What are the methods of preventing chemical corrosion?
16. Outline the varieties of cements based on Portland clinker.
17. What are the features of alumina cement and binders based on it?

Chapter 10

CEMENT-BASED CONCRETE

Concrete is an artificial composite material, obtained from the reaction and subsequent hardening of a rationally proportioned mixture of a binding material, aggregates, water and other substances.

Concrete may be produced from cement, lime, gypsum and other special binders. Concrete with special binders are those based on polymeric, magnesia, phosphate binders, liquid glass and other such materials.

Concrete produced from Portland cement binder is the most widely used in all areas of construction industry. The dominant use of concrete in construction is explained by the availability and large supplies of non-metallic materials - sand, gravel, and crushed stone. Concrete is characterized by the possibility to give different properties and to assume different shape of the mould in which it is cast and exhibits good durability and reliability performance.

Concrete is classified by the followings features: basic assignment, binder type and the type of aggregates, structure and average density.

Concrete may be categorized as structural or special (hydraulic, heat-resistant, radiation protective, chemically proof, heat-insulating etc.) depending on the basic assignment.

Concrete used for load-bearing and inclosing structures are categorized as structural concrete. They are made to satisfy specific mechanical strength requirements, and/or some other properties. Concrete which meets particular performance requirements of structures and elements are categorized as special concrete.

Dense, porous and special aggregates may be used for making concrete. Aggregates from, ore containing rocks, scrapped iron, chamotte and others, give special properties to the concrete.

The micro-structure of concrete can be dense, porous, macroporous or cellular. In concretes with dense micro-structure, all the space between grains of the aggregates is filled with hardened binder, including the pores of the entrapped air. Porous and macroporous structures is produced by means of foam- or gas-forming additives, such that the spaces between grains of coarse aggregate are not fully filled. Concrete with cellular structure consists of hardening mixture of binder, silica component and uniformly distributed pores as cells, formed by gas or foaming agents.

Concrete with density of 2000...2500 kg/m^3 is categorized as heavy-weight, 500...2000 kg/m3 – light-weight, over 2500 kg/m3 - extra heavy-weight and less than 500 kg/m3 – extra light-weight.

The increasing efficiency of concrete application in construction is influenced by the decline of materials content in the structures and labor intensiveness of their production.

10.1. REQUIREMENTS FOR THE INITIAL COMPONENTS OF CONCRETE

The technical properties of concrete in construction depend on the quality of the initial components and their proportion in the concrete mixture.

Cements. During the selection of cement, consideration should be given to the required concrete strength; intensity of its growth, aggressive influence of the environment, structural features of the elements and the conditions of concrete works conduction. The recommended ratio of compressive strength of cement and concrete changes from 3 for concrete with compressive strength between 10...15 MPa to 1.2...2 for concrete with strength between 20...50 MPa for a heavy-weight concrete. On reduction of these ratios, the cement contents increase, shrinkage deformations develop and crack resistance of concrete reduces. On increasing the ratio – due to insufficient cement content (if there are no additives of the filling agents and plasticizers) there could be a possibility of segregation of the concrete mixture, and reduction of the concrete density.

The chemical-mineralogical and material composition of cement influences the hydration rate, and the performance. In concrete subject to alternate freezing and thawing, moistening and drying, the application of cements with diminished content of mineral additives and tricalcium aluminate($C_3A < 8$ %) is desired. The low-heat cements are required for hydraulic concrete, which is placed in dams.

It is effective to apply high-early-strength Portland cement and blastfurnace Portland cement for the production of precast reinforced-concrete structures hardening at thermal and moisture treatment.

Mixing water. Portable-water is used for mixing concrete. It is also possible to use recycled and natural mineralized waters which contain the allowable amount of admixtures (Table 10.1).

Table 10.1 Allowable admixture contents in mixing water, mg/l

Type of the concrete	Soluble salts	Ions SO_4^{2-}	Ions Cl^-	Weighted particles
For pre-stressed reinforced-concrete structures	2000	600	350	200
For structures with ordinary reinforcement, including water spillway structures, zones of alternate water level of massive structures	5000	2700	1200	200
For the non-reinforced structures to which requirements on limitation of walls salinity are not specified	10000	2700	3500	300

Note. The pH-value of water should not be less than 4 and more than 12.5.

Dense aggregates. Dense aggregates have a density of more than 2 g/cm^3 The required concrete properties at the minimal possible cement content and water-cement ratio are provided by the selection of appropriate combination of fine and coarse aggregate.

Sand – is a natural or artificial mineral mixture of grains from 0.16 to 5 mm in size, serving as a fine aggregate. Quartz sands are the most widespread dense fine aggregates. Due to the high strength of the grains they can be utilized practically for concrete of all classes. Sands, which consist of loose grains of igneous rocks, limestone, dolomites and others are utilized for laboratory tests and technical and economic substantiation.

The sands made of industrial wastes can be used in many ways after testing. Depending on the strength of the initial rock, manufactured sands are divided into grades. Igneous and metamorphic rocks, used for the production of the manufactured sands should have an ultimate compressive strength of not less than 60 MPa.

The basic properties of concrete depend on the bond between the cement stone and the aggregates, which is influenced by the shape and character of their surface, the presence of clay, dust, other harmful admixtures and the chemical- mineralogical composition of the aggregates. Bond increases with acute-angled shape and rough surface of the grains. Smooth surface grains achieve less effective bonding at the cement-aggregate interface.

Clay and dust-like particles due to highly developed surface substantially increase the water requirement of concrete mixtures; they envelop grains of aggregates and diminish their adhesion to cement stone. Besides that, the very fine dust-like particles (< 0.08 mm) reduce frost resistance of concrete.

The permissible content of dust-like and clay particles is specified depending on the aggregate type and concrete works. For ordinary natural sand it should not exceed 3 %, ground up - 4 %, for fractional natural and manufactured sands - 2 % and 3% correspondingly. Special limitations for dust-like and clay admixtures contents are specified for sand, used in the production concrete for hydraulic structures, culvert aqueducts, transport structures. For example, content of dust-like and clay admixtures in sand should not exceed 2 % for concrete in the variable water level zone of hydraulic structures, above-water concrete - 3 %, underwater concrete and concrete in the internal zone - 5 %.

The inclusion of mica, sulfurous and sulfate compounds, iron oxides and hydroxides, minerals which contain the amorphous types of silica, organic admixtures and others like that are not recommended as they are harmful aggregate admixtures. Research findings have established that they impair concrete strength and influence negatively on the hardening process of cement stone.

Sand with sulfurous and sulfate admixtures content should be used for the concrete works of hydraulic constructions in amounts not exceeding 1% by mass. The mica content should not exceed 1 % for the concrete situated in the variable water level zone, 2% for the concrete situated above-water zone and 3 % for concrete in the underwater and internal zones.

Cement paste is used for the filling of voids in the mixture with aggregates and the formation of envelop, which greases and eventually cements the separate grains into a strong conglomerate. That is why, the lower the volume of voids and surface area of the grain, the more saving cement paste application is in concrete.

Sand fineness is characterized by the fineness modulus M_{fin}, which means the sum of complete residues (A_i, %), after screening the sand on standard sieves, divided by 100:

$$M_{fin} = \frac{A_{2.5} + A_{1.25} + A_{0.63} + A_{0.315} + A_{0.16}}{100} \quad (10.1)$$

A complete residue is the sum of partial residues on the current sieve and on larger sieves, which are included in the screening set. A partial residue is the ratio of mass on this sieve to the mass of the screened sample.

Description of sands according to fineness is represented in Table 10.2.

Table 10.2. Sands description according to fineness

Sand groups	Fineness modulus, %	A complete residue is on sieve No. 0.63 (by mass), %
Raised fineness	Over 3 to 3.5	Over 65 to 75
Coarse	Over 2.5 to 3	Over 45 to 65
Middle	Over 2 to 2.5	Over 30 to 50
Fine	Over 1.5 to 2	Over 10 to 30
Ultrafine	Over 1 to 1.5	To 10

Coarse, middle and fine sands are used as the fine aggregate for concrete. Fine sands cause the excess of cement content, especially for the high strength concrete. Using the sands with M_{fin} = 1.5...2 in concrete with ultimate compressive strength 20 MPa and higher is accepted only at the proper technical and economic substantiation.

Along with sand fineness, its voidage or intergrain space volume (P) has an important value defined as follows:

$$P = \left(1 - \frac{\rho_0}{\rho}\right) \cdot 100\ \%, \quad (10.2)$$

where ρ_0 - bulk density, ρ - absolute density of sand.

Voidage depends on the ratio of different grain fineness and on the grain distribution of sand. It can increase in the sands of unsatisfactory grain distribution up to 40...47 %. The sand suitability for use according to the grain distribution for concrete is obtained by reference to the sieving curve shown in Fig. 10.1.

It is possible to use sands, obtained by pre-mixing separate fractions to provide the required grain distribution. Additives of large fractions of natural or manufactured sand are expedient when fine sands are used.

Approximately 50 % of all of concrete mass is made up of coarse aggregate (gravel or crushed stone), which forms the stiff framework of concrete.

Gravel is fragile sedimental particles, formed as a result of weathering of dense rocks. Coarseness of the gravel grains ranges from 5 to 70 mm. Round shaped grains and in most cases high content of dust-like particles and grains of weak rocks are often found in the gravel.

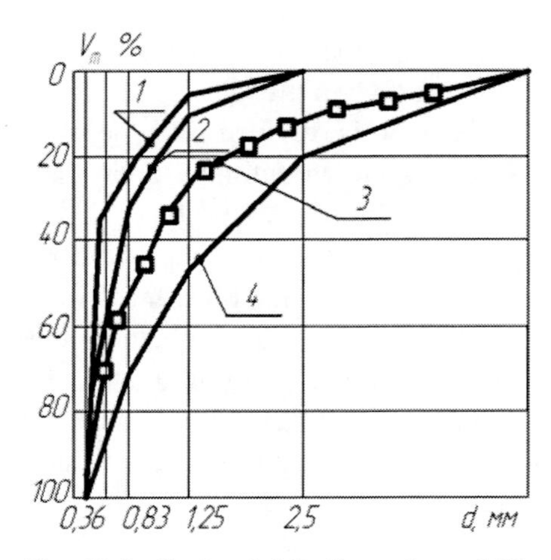

Fig 10.1. Grain distribution of sand for concrete:
1,2- bottom limit of coarseness, which is accepted (M_{fin}=1.5) and recommended (M_{fin}=2); 3,4- top limit of coarseness, which is recommended (M_{fin}=2.5) and accepted (M_{fin}=2.25); V_m- complete residue on sieves by the size d

Crushed stone is a product of rock crushing. The crushed stone is obtained also from gravel, boulders and blast-furnace, steel-smelting and other slags.

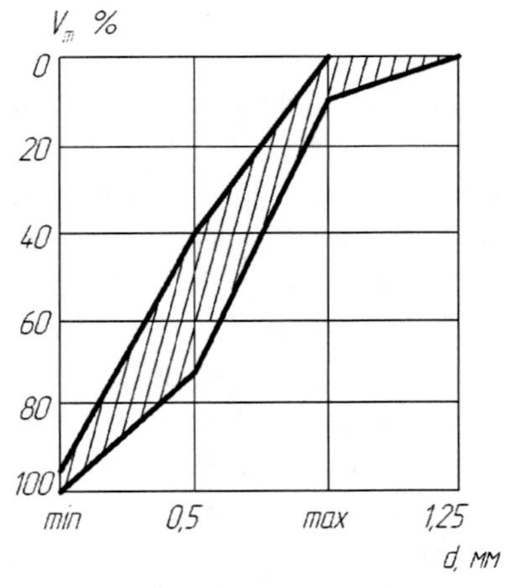

Fig.10.2. Recommended grain distribution of coarse aggregates

Quality of the coarse aggregate, as well as sand, is determined by coarseness and grain distribution (Fig. 10.2), shape, grains surface and admixture contents. The petrographic peculiarities, strength of the initial rock, water- and frost-resistance, have a substantial effect on the quality.

The maximal size of the coarse aggregate should not exceed 1/4 of the minimum cross-sectional area of the structure. The coarseness of the aggregate is normally less than 2/3 of distances between the reinforcing bars in the reinforced concrete structures.

The crushed stone or gravel is divided into separate fractions; mixed up in the recommended correlations to provide an optimum grain distribution (Table 10.3). Fractions 5...10, 10...20, 20...40, 40...70 mm are used as a rule.

Crushed stone and gravel with coarseness up to 150 mm and more are used in the manufacture of the concrete for hydraulic and other massive structures.

One of the important requirements is the strength of the coarse aggregate. The crushed stone can be made of igneous rocks with ultimate strength of more than 80 MPa, metamorphic rocks – more than 60 MPa and sedimentary rocks with more than 30 MPa for heavy-weight concrete. Crushed stone made of natural stone should have strength 1.5...2 times higher than the concrete strength.

Table 10.3 Recommended content of fractions in coarse aggregate, %

Maximum coarseness of grains,	Aggregate fraction, mm				
	5...10	10...20	20...40	40...70	70... 120
20	25...40	60...75	—	—	—
40	15...25	20...55	40...65	—	—
70	10...20	15...25	20...35	35...55	—
120	5...10	10...20	15...25	20...30	30...40

The crushability (Cr) can serve as a parameter of gravel and crushed stone strength:

$$Cr = \frac{m - m_i}{m} \cdot 100\%, \quad (10.3)$$

where m - mass of a specimen, m_i- mass of material which passed through a sieve with mesh size 4 times less than the smallest size of fraction after testing . The higher is the index of crushability of crushed stone or gravel, the lower the expected concrete strength.

It is possible to use aggregates with crushability index of no more than Cr 16, for the manufacture of concrete with compressive strength of 25 MPa and lower; Cr 12, for compressive strength of 30...35 and no more than Cr 6 for 40 MPa concrete.

The special requirements to strength of coarse aggregate are set by its application in hydraulic concrete, concrete for bridge structures, culvert aqueducts, reinforced concrete ties, coverages and bases for highways and aerodromes.

It is required to use crushed stone for hydraulic structures, in which strength is not less than 2.5...3 times higher than concrete strength for the zone of variable level and 2...2.5 times for underwater, internal and upper zones.

Crushed stone with grains density not lower than 2.5 g/cm^3, water absorption not more than 0.5 % is applied for concretes at the variable water level zone, and for other zones - not lower than 2.3 g/cm^3 and 0.8% accordingly. Water absorption of crushed stone made of sedimentary rocks can be a little higher, but not more than 1...2 %.

For abrasive- and cavitation-resistant concrete there are additionally established grades of coarse aggregate for abrasive resistance by testing in a special drum.

Crushed stone, according to the grains shape, is divided into three groups: cube-shaped with content of plate-like (flaky) and needle grains not more than 15 %, improved - 25 %, and ordinary - 35 %. The content of plate-like and needle-shaped grains can exceed 35 %, but on conditions that assigned workability of concrete mixture and density of concrete without the excess of cement content are provided.

The content of dust-like and clay particles allowed in coarse aggregate from igneous and metamorphic rock should not exceed 1 %, for sedimentary rocks with strength 20…40MPa - 3%, and from sedimentary rocks of higher strength - 2 %. The amount of dust-like and clay particles in crushed stone and gravel depending on the type of rock for the concrete of variable water level and upper zones should not exceed 1 %, and for underwater and internal zones - 2 % for hydraulic structures.

The frost resistance of concrete is highly influenced by the frost resistance of the aggregate. For concretes used for hydraulic structures, this will depend on the average monthly temperature of the coldest month in the performance region. If this temperature ranges from 0 to minus 10°C, frost resistance of crushed stone and gravel should be not less than 100 cycles, but from minus 10 to minus 20°C is not less than 300 cycles.

Concrete aggregates can be made on the basis of metallurgical and fuel slags, and other industrial wastes. The reason for using aggregates from metallurgical slags is their resistance to different types of deterioration processes. The sulphur, contained in metallurgical slags can cause corrosion of reinforcing steel. Its content should not exceed 2,5 % of the mass of the crushed stone made of blast-furnace slag. The presence of sulphur should be taken into account during the production of prestressed reinforced concrete structures, where possibility of the use of crushed slag stone should be supported with the special researches.

Porous aggregates. Aggregates with grains density up to 2 g/cm^3 are normally porous. They are utilized for obtaining light-weight concrete and mortars, and also thermal- and sound-proof composites. Porous aggregates are divided into natural aggregate, obtained in the form of sand, crushed stone from porous rocks, and artificial ones. The artificial inorganic aggregates such as: claydite and its varieties (haydite gravel, clay-ash claydite), obtained by burning with expansions of granules from clay rocks, haydite slates, tripoli powder, ash-slag mixture or fly ash from the thermal power-stations. Others include: aggloporite which is a product of calcination of sand-clay rocks, ashes, wastes of coal-cleaning industry; slag pumice-a product of porization during melting of metallurgical and chemical industries slags; granulated slag; perlite and vermiculite-obtained by expansion at the burning-out of effusive rock and hydrated mica grains.

Organic porous aggregates are obtained by the processing of wood and other vegetable raw materials and also polymers.

The type of porous aggregates is often indicated in the name given to the light-weight concrete (claydite concrete, aggloporite concrete, perlite concrete and other). Cement wood is made on the basis of organic aggregates. Claydite concrete is the most widespread among the light-weight concrete.

The porous gravel and crushed stone have a bulk density of 250...1200 kg/m^3 whilst porous sand has a density ranging from 100...1400 kg/m^3. Amongst the porous gravel types and crushed stone, perlite has a maximum density of 600 kg/m^3, aggloporite and slag pumice - 900 kg/m^3, and crushed stone from porous mining rocks 1200 kg/m^3.

The strength grades are also set for the coarse porous aggregates, which are determined by squashing in a cylinder. A frost resistance of the coarse aggregate should be not less than 15 cycles of freezing and thawing, and the maximum possible loss of mass after testing allowed does not exceed 8 %.

For porous, as well as for dense aggregates, grain distribution and content of harmful admixtures is specified. In addition, requirements for the coefficient of grain form, water absorption, humidity, disintegration resistance and other are set.

The selection of porous aggregates is regulated depending on their function and requirements for the strength and density of concrete.

Slag pumice is one of the most effective types of artificial porous aggregates.

The ash-and-slag mixtures can be used as aggregates for heavy-weight and light-weight concrete.

Porous aggregates based on slags and ashes from thermal power stations are also the least power-consuming. The ashes with high content of noncombustible particles which are impermissible for the production of a series of other materials (porous concrete, lime-sand brick, cements and others like that) are suitable for production of porous aggregates.

The porous aggregates, made of the ashes can be obtained by the burning-out of specially prepared mass and by unfired technology. Materials of the first group are obtained by the calcination of granular batch which consists of the ash and little amount of binding additives (clays, technical lignosulphonate and other), in a rotatary kilns (ashy gravel and lumnite claydite) or on the grates of the sintering machines (ashy aggloporite, aggloporite gravel). The basic material of the second group is unfired ashy gravel, used as mineral binder additives.

Admixtures for concrete. The inorganic and organic substances or their mixtures (complexes) introduced to concrete mixture (Table 10.4) to regulate the properties of the concrete, are descried as admixtures. Depending on the basic action and effect, admixtures are divided into the followings groups:

- those, which regulate rheological properties, setting and porosity of concrete mixtures, concrete hardening;
- those, which give special properties to the concrete;
- those, which regulate simultaneously different properties of concrete
- mixtures and concrete (multifunctional action);
- mineral powders - cement substitutes.

The names of some commonly used admixtures of the first three groups and the amounts recommended for use are shown in Table 10.4.

The use of chemical admixtures result in the reduction of labor contents on precast concrete plants and in monolithic construction, where considerable reduction of part of the difficult manual operations, improving concrete quality, its strength, frost resistance, water impermeability, corrosive resistance, and saving cement. By adding admixtures, it is possible to change and regulate the hydration rate and time of concrete hardening, giving it new properties. For example, bactericidalness, hydrophobicity, ability to harden on the frost and others like that.

The plasticizers and superplasticizers, are the most common admixtures used the concrete technology. The use of superplasticizers allows reduction of the labor contents of reinforced-concrete elements casting by 2...4 times and concreting of monolithic structures by 5...7 times. In a number of cases, it is possible to fully eliminate vibration or to replace it with only brief shaking. This cuts down fuel and electric power expenses, and saves up to 20…25 % of the cement.

Table 10.4 Some types of concrete admixtures

Type of the admixtures	Name of the admixtures	Content, % by cement weight	Effect
Plasticizers and superplasticizers	Naphtalene-formaldehyde superplasticizer	0.5...1	Increasing slump of the concrete mixture or reduction of water content.
	Technical lignosulphonates	0.1...0.3	
	Technical lignosulphonates, modified by alkalines	0.1...0.3	
	Plasticizer formate-spirit	0.2...0.8	
Air-entraining and plasticizing	Naphthenate soap	0.01...0.2	
	Synthetic foam-forming admixture	0.05...0.1	
Air-entraining	Neutralized air-entraining oleoresin	0.005...0.035	Increasing of concrete frost-resistance and corrosive resistance by 2 times and more
	Ligneous saponated resin	0.005...0.035	
	Sulfanole	0.05...0.035	
Hardening accelerators	Calcium nitrate	0.3...1	Acceleration of hydration and the hardening process.
	Sodium sulfate	0.3...0.8	
Antifreeze	Potash	3...12	
	Calcium nitrate	3...12	
	Carbamide	3...12	
Setting time retarding	Sugar treacle	0.1...0.2	Increasing the setting time of concrete mixture
Steel corrosion inhibitors	Sodium nitrite Calcium nitrite-nitrate	0.5...3 0.5...3	Inhibits the corrosion process. Improves durability of reinforced-concrete structures.

The large reduction in the water content of superplasticized concrete mixture whilst maintaining constant cement content, results in the production of a high-strength concrete. Along with individual admixtures, complex types also exist with more universal effects and more widespread use. For example, application of complex admixtures containing plasticizers and set accelerators permits both an improvement of the flow characteristics of concrete mixtures and acceleration of the concrete hardening process without thermal treatment. The complex admixtures give an opportunity to eliminate or reduce the negative properties whilst enhancing the positive features of each of the components, or to improve the overall positive effect.

Application of some chemical admixtures requires taking into consideration their possible negative action (setting acceleration of concrete mixture, reinforcement corrosion, equipment and accessories; wall salt formation; the decline of structural stability under the action of ground current; development of deterioration due to interaction of reactive silica of aggregates with alkalines etc). For example, admixtures containing chloride salts can be used as hardening accelerators for only reinforced-concrete structures with nonprestressed reinforcement with the diameter over 5 mm. It is not allowed to use admixture of salt-electrolytes for manufacturing reinforced-concrete structures, if intended for the electrified transport and industrial enterprises which use direct electric current.

Dispersible mineral admixtures (filling agents) are divided into active and inert. Inert admixtures are obtained by fine grinding of sands, limestones, dolomites, loess and other rocks not possessing inherent hydraulic activity. For active mineral admixtures (diatomites, tripoli powders, silica clays, ashes, granulated blast-furnace slags and other) which are normally introduced to the concrete mixture, the same requirements are set, as for those pertaining to Portland cement, Portland-pozzolan cement and blast-furnace cement. The pozzolanic cements should possess similar fineness as the Portland cement.

Increasing the concrete mixture water content is undesirable if mineral additives are to be used.

Fly ash from heat engine generating stations is one of effective admixtures which are introduced to concrete to replace part of the Portland cement. Its can be used to replace up to 30 % of the cement (binder) mass.

Table 10.5 Requirements for fly ash from power generating stations used as concrete admixtures

Indexes	The index value of ash kind		
	I	II	III
Content of $SiO_2 + Al_2O_3 + Fe_2O_3$, % to the mass, not less than, for ash:			
Anthracite and mineral carbon	Not specified		70
Brown coal	70	The same	50
Content of sulfur and sulfuric-acid catenation in enumeration on SO_3 % to the mass, not more than	3	3.5	3
Content of free calcium oxide CaO, % by the mass, not more than	3	5	2
Content of magnesium oxide MgO, % by the mass, not more than: for ash			
Anthracite	15	20	5
Mineral carbon	7	10	5
Brown coal	5	5	3
Humidity % by mass, not more than	3	3	3

The ash is divided into categories, depending on the area of usage: I – for reinforced-concrete structures and elements; II - for the concrete structures and elements; III- for the structures of hydraulic constructions. The requirements, which are requested according to the ash chemistry, are shown in the Table 10.5.

The ash acts, not only as an active mineral admixture in the concrete mix, which increases binder content but also as a microfilling agent which improves the sand grain distribution and actively influences the process of micro-structure-formation in the concrete.

Diminishing of minimum cement content for the non-reinforced elements to 150 kg/m^3, and for reinforced concrete elements –to 180 kg/m^3 is possible if the fly ash is used. Total cement and ash content can be accordingly not less than 200 and 220 kg/m^3. Combined adding of fly ash and plasticizer additives into concrete mixtures is especially effective.

Combined use of the high dispersed silica substances such as microsilica and the superplasticizers is one of the most effective ways of cement content diminishing and obtaining high-strength and durable reinforced-concrete structures

10.2. Fundamentals of Concrete Mixtures Technology

Carefully mixed mixture of binder, water, aggregates and additives, introduced in required proportion is known as concrete mixture. Concrete mixtures are made in concrete mixing units, plants or workshops for reinforced concrete elements.

Concrete mixture can be made in truck mixers to aid the process of transporting to small distances from a factory to the building sites or directly in-situ from dry cement mix with aggregates.

The process of concrete mixture production consists of the followings technological operations: materials preparations, dosage, and mixing (Fig. 10.3).

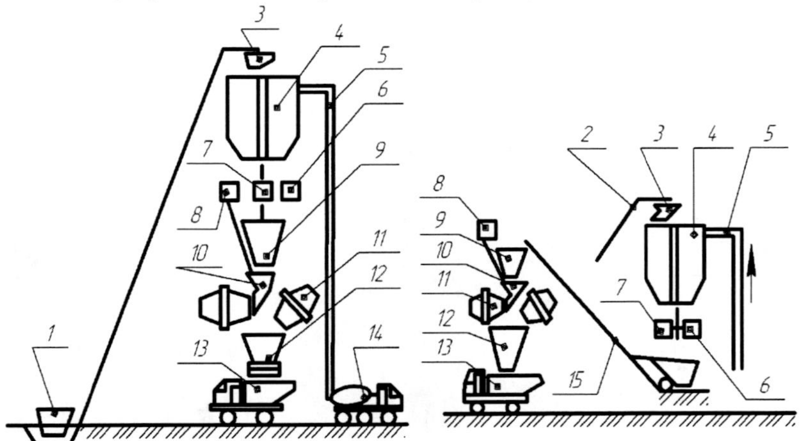

Fig. 10.3. Technological scemes of concrete mixing plants:

a- single-stage; **b-** two staged; 1- conveyer of aggregates; 2-aggregate delivery conveyor to the measuring hoppers; 3,9,10- slewing, bowl and distributive hoppers; 4- measuring hoppers; 5- cement pneumatic pipeline; 6 - cement batcher; 7- aggregate batcher ; 8- water batcher; 11- mixer; 12- hopper; 13- truck mixer; 14- cement truck; 15- auger elevator

The materials preparation for concrete mixture includes aggregates cleaning from harmful admixtures and segregation on factions which is executed on the grinding-sorting factories. The aggregates and water may be heated in the winter concreting. Aggregate and water may be cooled, and also replacement of water part with the ice can take place during concreting to reduce the heat generation in massive hydraulic constructions.

The quality of the concrete mixture would depend on the accuracy of dosage of the components. The dosage of bulk materials is executed according to the mass. Porous aggregates are dosed according to volume with a mass correction. The error of cement, water and admixtures dosage should not exceed ± 2%, aggregates ± 2.,5%. The portioning and continuous batchers with a manual, semi-automatic or automatic control are used on the concrete mixing units. Batchers with a manual control are used only at units with the small capacity.

The main operation during concrete mixture production is the mixing, on which the strength and homogeneity of concrete depend. Depending on the method of mixing, there are two basic types of concrete mixers: free falling and forced mixing.

The free falling (gravity) types of concrete mixers have drums as cylinder or two cones, combined with the bases. Mixing takes place due to fascination of mixture by the blades as the drum rotates with its free falling content.

The forced types of mixing are the most effective for mixing rigid concrete mixtures, and also for porous aggregate-based mixtures. The concrete homogeneity increases due to the rotation of the blade in the stationary drum, or to drum rotating in the opposite direction.

As a rule, the mixing duration of heavy-weight concrete mixtures in gravity batch mixers lasts for 1 to 2 minutes, and in the mixers of the forced mixing - 2...3 minutes. The mixing duration of fine-grained concrete mixtures and concrete mixtures with porous aggregates is increased by up to 3.5 minutes. The mixing duration is also increased in the winter.

Depending on the workability, concrete mixtures are divided into classes accordingly to EN 206-1:2000 (Table 10.6).

Fig. 10.4. Types of concrete mixtures depending on slump

Table 10.6 The workability classes of concrete mixtures

Class	Vebe, secs	Class	Slump, mm
V0	> 31	S1	10...40
V1	30...21	S2	50…90
V2	20...11	S3	100 ...150
V3	10...6	S4	160 ...210
V4	5...3	S5	>220

Also, concrete mixtures can be divided by degree of compaction and spread (flow test) of Abrams cone (Table 10.7).

Table 10.7Compaction and spread of Abrams cone classes

Class	Degree of compaction
C0	>1,46
C1	1,45 ...1,26
C2	1,25 ...1,11
C3	1,10... 1,04

Class	Spread of cone, mm
F1	>340
F2	350 ...410
F3	420 ...480
F4	490...550
F5	560. ..620
F6	>620

Concrete mixtures can also be classified based on slump value (Fig.10.4).

Class S3 and S4 concrete mixtures should be made obligatorily using plasticizing agents.

The homogeneity of the concrete mixture, intensity of hardening, strength and other properties depend on the duration and intensity of mixing.

The concrete mixing plants and factories are batch (batch-type) and continuous action. The plants of batch action are the most widespread; they include operations which repeat periodically: loading of charging feed, mixing and unloading of the prepared mixture. The processes of dosage, mixing and delivery of mixture flow without breaks in automated concrete plants.

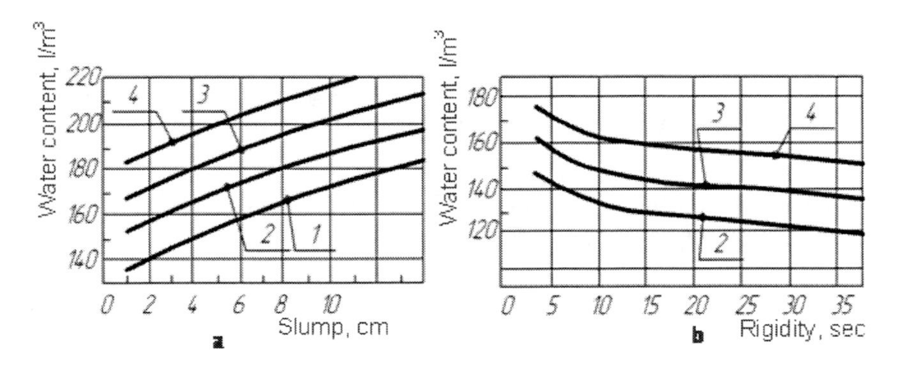

Fig. 10.5. Diagram of water requirement of workable (**a**) and rigid (**b**) concrete mixtures, produced with application of Portland cement, middle-grained sand and gravel of maximum coarseness:

1– 80 mm; 2– 40 mm; 3– 20 mm; 4– 10 mm

For batch gravity mixer, ready-mix batch volume varies from 65 to 3000 litres, depending on the drum capacity. Mixers with a capacity of the ready-mix batch of not more than 165 litres are movable, but those with greater capacity are stationary.

Workability, rigidity and segregation are the basic technological properties of concrete mixtures.

The mixtures which slump values close to zero are classified as rigid. The rigidity of concrete (V) is characterized using the Vebe test (special device), which measures the vibration time required for its full compaction.

The water content (W) is one of the major factors that determine the indexes of workability of concrete mix (Fig. 10.5). The water content, required for achievement of required workability and rigidity of the concrete mixture, depends on the type and amount of cement, specific surface and the aggregates porosity, shape and character of their surface. It also depends on the ratio of fine and coarse aggregate, curing condition and duration before placing, kind and content of plasticisizing admixtures and etc.

Increasing the cement content by 1 % causes an increase in concrete mixture water content of approximately 1.5 to 3 %. Thus, concrete with increased cement constent at constant water content has lower slump and higher rigidity.

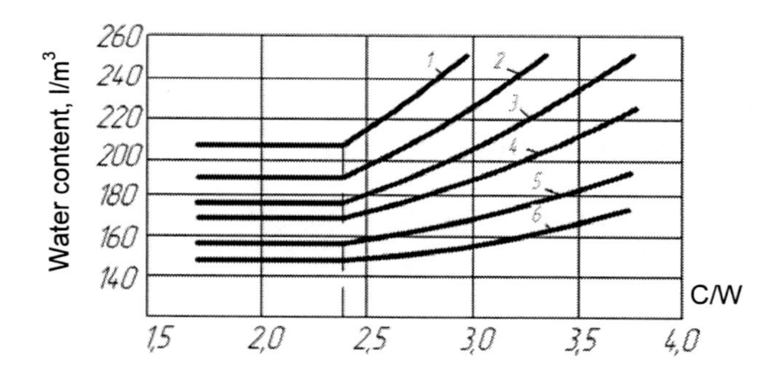

Fig. 10.6. Influence of C/W on the water content of concrete mixture:
1…3- mixtures with slump 10; 5 and 2 cm;
4…6- mixtures with rigidity (by Vebe) accordingly 30; 60 and 100 sec (crushed stone 5…20 mm, sand of average particle size, cement with normal water content 26…27%)

Change in the cement content at constant water content to C/W =1/(1.68$K_{n.c}$) ($K_{n.c}$ is normal consistency of cement paste determined with the Vicat apparatus) does not influence significantly the workability. This statement is called the rule of water content constancy, as it shows that increasing the cement content to some limit with unchanging aggregate/cement ratios, the water content of the mixture does not change (Fig. 10.6). A greater amount of cement paste is required with the increase of the sand content, surface of aggregates, their porosity, and also the content of dust and clays in them. If the cement content remains constant, the workability of the concrete mixture will become reduced. Using crushed stone instead of gravel also leads to an increase of the rigidity of the concrete mixture.

As a result of partial cement hydration, some water evaporation and its absorption by the aggregates, workability of concrete mixtures become reduced during mixing, transporting and before casting. The workability loss of concrete mixtures at transporting should not exceed 50 % of the normal slump or 8 cm. The loss should not exceed 30% in situations where the normal slump is over 8 cm. The rational technological method of workability improvement of concrete mixtures is by the use of plasticizing surface-active substances (SAS). These admixtures are divided into four groups depending on the effect: superplasticizers, high, medium and low plasticizing. Superplasticizers transforms low workability (Slump = 2...3 cm) to concrete mixtures with slump values of 20 cm and more at constant water content without strength reduction. Low plasticizing admixtures increase concrete slump from 2...3 to about 8 cm and less.

Plasticizing substances can be used for the purpose of simultaneous reduction of water and cement content at constant mixture workability and concrete strength, or for reducing the water-cement ratio and increasing concrete strength. Traditional plasticizing agents, such as lignosulphonates, by reducing the water content allow a decrease in cement content of between 8...12 %. The cement economy achieved by the use of superplasticizers can be as much as 20 %.

The workability and rigidity of concrete mixtures is assigned depending on the peculiarities of structures, degree of reinforcement, method of transporting and compaction. A greater workability of mixtures is required for thin-walled elements saturated with reinforcement, than for massive lightly-reinforced structures. By using modern methods of compaction, especially in industrial conditions, the possibility to increase rigidity of concrete mixtures emerges. This leads to cement economy and improves the range of constructive properties of concrete.

Along with the required workability of concrete mixtures it is important to provide adequate homogeneity which is measured by the degree of segregation of mixtures during storage and transporting. Segregation of concrete mixtures develops as a result of the sedimentary processes caused by the different density of the concrete components. Application of segregating mixture leads to the uneven distribution of the aggregates in the fresh concrete and cement stone, and deterioration of the hardened concrete properties. Reduction of segregation is observed as the water content declines, or introduction of fine ground up mineral admixtures, or upon addition of surface-active substances.

Concrete mixtures are transported to construction sites by high-discharge mixers or truck mixers. Transportation of concrete mixtures with dump truck is permitted. The concrete mixture can be delivered with cargo cranes, on runners, band conveyers, or by pipeline transportation to the place of work implementation.

Placement of concrete is carried out in forms or mold following the preliminary placement of reinforcement and embedded items.

The quality of concrete placement is determined by the efficiency of compaction of all the volume of material, in order to prevent segregation. A 1% decrease in the density of concrete causes a reduction in its strength of approximately at 5 %. The compaction ratio(factor) K_c should be not less than 0.97...0.98, and for high density concrete,- not less than 0.99:

$$K_c = \frac{\rho_{0.a}}{\rho_{0.c}} \tag{10.4}$$

$\rho_{0.a}$ - actual average density of compacted concrete mixture;

$\rho_{0.c}$ - calculated average density.

The compaction of concrete mixtures can be carried out by vibration, degassing, pneumatic concreting (shortcreting), and also by rolling, pressing, centrifugation etc.

Vibration is the most widespread and effective method of compaction. The vibration frequency 10 - 50 Hz is used for concrete compaction. The two types of vibration used for cast-in-place structures are - internal and surface vibration. An internal vibration is based on the immersion of the executive device of the vibrator into the mixture and is utilized for the compaction of concrete mixtures with slump values of over 1 cm. Surface vibration is recommended for the compaction of concrete coverings, roads, slabs etc. Concrete layers within a depth of 20...30 cm are well consolidated by surface vibration.

At precast plants, reinforced concrete elements are formed, as a rule, at the special vibro-areas. The surface, internal and external vibrators are used for elements cast in stationary forms. The last(external) vibrators are normally attached to the form.

Vibrostamping is applied for production of elements with complicated configuration. The compaction is achieved by vibrators of certain shape. The progressive method of forming is the powered vibrorolling method- which is a combination of vibration of mobile layer of concrete on a stripe with pressure applied to the concrete by the vibromills of a rolling mill.

The efficiency of the vibration of rigid concrete mixtures is increased by the simultaneous action of small pressure (vibrations with cantledge) – of up to $(1...3) \times 10^3$ Pa at the open surface of the element.

The operating conditions of vibration, intensity and duration of fluctuations, as well as the method and time of their application, influence the compaction quality. Increasing vibration up to the optimal magnitude of compaction increases the concrete strength, but segregation of the concrete mixture is possible if the optimal magnitude is exceeded.

The repeated vibration after 1.5...3 hours after placing (when the setting of concrete mixture did not take place) increases the strength of concrete by 10...15 % or to reduce cement content by up to 20 %, to increase water impermeability by 1.5...2 times, and also to improve frost resistance and adhesion to the reinforcement.

Another effective method of compaction of thin-walled structures made from mixtures with high water content is achieved by v a c u u m i z a t i o n along with repeated vibration. During vacuum compaction by the shields and bushes which houses the vacuum chamber, it sucks off 10...20 % of the total amount of water from the concrete, which improves its quality.

The centrifugation is utilized for pipe forming, bearing supports, piles and other structures of tubular profile, at which distributing and compaction of concrete mixture takes place under the action of centrifugal forces, developing in the revolving forms.

A further effective method of compaction is pressing with the help of which it is possible to get concrete of high strength and durability by using of fine-grained rigid concrete mixtures The different methods of pressing (vibropressing, radial and axial pressing) are used, for example, for making concrete pipes.

The use of superplasticizers allows the production, at moderate water content, of concrete mixtures which eliminates vibrocompaction and ensures the gravity filling of molds. The cast mixes are required for the production of composites and heavily reinforced concrete structures, when high-quality compaction of concrete mixture by vibration is difficult to realize.

The pneumatic methods of concreting - shortcreting and jetcreting are applied in hydraulic engineering and other industries of construction for concreting of channel slopes, dikes, screens and tunnel facing. They are based on repulsion of concrete mixture under the pressure and application of the concrete with dynamic forces at the surface.

The shortcrete does not contain coarse aggregate; its rational compositions range from 1:2 to 1:4. The jetcrete contains a coarse aggregate and allows the creation of a thicker layer(20...30 cm). Shortcrete is also used for repairs, strengthening and reconstruction of various structures, grouting of precast reinforced concrete, and for concreting of thin-walled bearing structures of different shapes. Air cement guns, pneumatic delivers and pneumatic rackers are used for application on the surface of shortcrete and jetcrete.

10.3. Structure and Properties of Concrete

The concrete structure. Dense heavy-weight and light-weight concrete, as a rule, include aggregate framework which actively influence their properties. Structure of light-weight concrete differs from heavy-weight one by the presence of additional pores in the grains of aggregates.

The concrete microstructure represents cement stone structure. Cement stone by itself is a conglomerate of cement hydration products, inclusions of the unhydrated grains of clinker, additives and air bubbles. Pores in the cement stone are presented as connected with each other and canals of capillaries, disconnected by the products of cement hydration (by cement gel). The pores, according to the origin, are divided into the gel pores and capillary pores.

Capillary pores, formed by the excessive mechanically bounded water, impair the basic concrete properties, especially frost resistance. Capillary porosity diminishes as the water content declines and increases with hydration time.

Along with capillary pores, pores and voids, resulting from poor compaction, have negative influence on concrete properties. The total pore volume of the cement stone is 25...40 % of the total volume of the cement stone, and the capillary pores forms a major part of this at early ages. With the increasing of hardening duration, the general porosity and volume of macrocapillaries diminishes. This leads to the improvement of the concrete properties, such as strength, permeability and long-term durability.

Properties of concrete. The basic property of concrete as a structural material is strength. The ultimate strength is defined by control specimens testing, made of the mixture of the set composition or drilled from the concrete structure.

The basic index of concrete strength which is specified is its strength class, (in MPa) which is accepted with the assured probability. As a rule the accepted probability of strength set, is 0.95. This means that the ultimate strength of concrete, which meets the requirements of the numerical strength class, is achieved at not less than in 95 cases from 100. Compressive strength classes for heavyweight and lightweight concrete are listed in Table 10.8.

In practice of making concrete, its strength indexes initially determine the mean strength of the separate specimens tested. After the numeral value of average strength and accordingly class of concrete taking into account the coefficient of variation (variability) determine its strength class.

There is a dependence between the class of concrete (C) and the average strength of the consignment ($\overline{R}$) which is controlled:

$$C = \overline{R}(1 - 1{,}64\,C_v), \quad (10.5)$$

where;

Cv- coefficient of variation of concrete strength. It is defined according to the following formula:

$$C_v = \frac{S}{\overline{R}}, \quad (10.6)$$

where S - average square deviation for concrete consignment;

Table 10.8 Compressive strength classes for heavyweight and lightweight concrete

Heavyweight concrete		Lightweight concrete	
Required grade, i.e. required minimum characteristic cube strength, MPa	Specified compressive strength class	Required grade, i.e. required minimum characteristic cube strength, MPa	Specified compressive strength class
10	C8/10	9	LC8/9
15	C12/15	13	LC12/13
20	C16/20	18	LC16/18
25	C20/25	22	LC20/22
30	C25/30	28	LC25/28
35	C28/35	33	LC30/33
37	C30/37	38	LC35/38
40	C32/40	44	LC40/44
45	C35/45	50	LC45/50
50	C40/50	55	LC50/55
55	C45/55	60	LC55/60
60	C50/60	66	LC60/66
67	C55/67	77	LC70/77
75	C60/75	88	LC80/88
85	C70/85		
95	C80/95		
105	C90/105		
115	C100/115		

In conditions of high quality control, the coefficient of variation of concrete strength varies between 6...8 %, but for poor quality control(-characteristic of insufficient level of technology), Cv varies between 20...25 %. The standardization of concrete strength on a class basis allows the provision of the required design reliability regardless of the coefficient of variation. At the same time, a diminished coefficient of variation decreases the required average strength which in some cases results in a reduction of the required cement content (see formula 10.6),

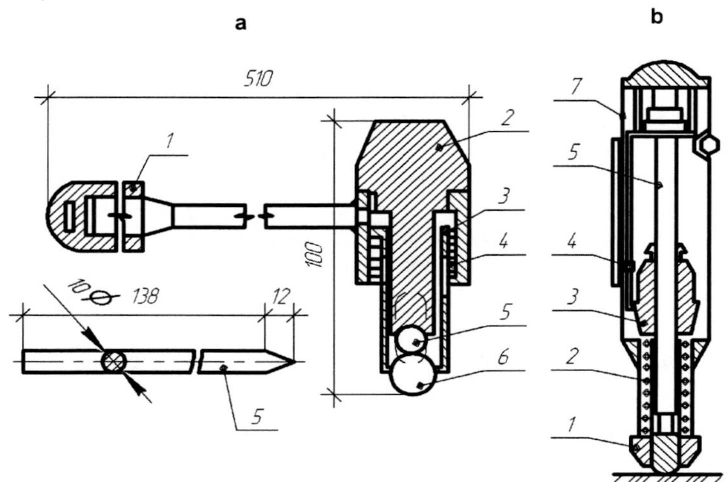

Fig. 10.7. Devices for non-destructive methods of concrete strength control:

a- standard hammer: 1- rubber stick on a steel handle; 2- enclosure; 3- spring cup; 4- spring; 5- standard bar; 6- steel ball;

b- device for the testing of concrete by the method of plastic deformation; 1- shock-worker; 2- shock spring; 3- firing-pin; 4- pointer; 5- bar; 6- sending spring; 7- enclosure

For a transition from the concrete class to the average strength at the normative coefficient of variation 13.5 % (this is accepted as rule at the structural design on the heavy-weight and light-weight concrete) it is possible to use the formula of $\bar{R}$ =C/0.778, where C is the numerical value of concrete class.

An enhanced tensile strength is the characteristic feature of light-weight concrete. The developed surface of aggregates promotes good adhesion with the cement stone. The correlation between tensile and compressive strength for heavy-weight concrete is 0.05...0.1 and for light is 0.06...0.17.

One important performance criterion of light-weight concrete is the coefficient of structural quality - the correlation between the ultimate concrete strength and its average density. The density in the dry condition kg/m^3 for light-weight concrete grades are set from LC 1.0 to LC 2.0 (with gradation through each 200 kg/m^3).

Time interval (age of concrete) after which class is determined, depends on the type of concrete and range of plant conditions. The time accepted, as a rule, equal 28 days of normal hardening, at a temperature of (20 ± 2)°C and relative humidity of air not below 90 %. If it is required age of concrete can be increased to 90 or 180 days that leads to the economy of cement. At the calculation of reinforced-concrete structures next to cube strength it is required to know prism strength which is determined by the prism compressing with the sizes $20 \times 20 \times 80$ cm. The ratio between cube strength of concrete and prism strength is in the range of 0.7...0.8.

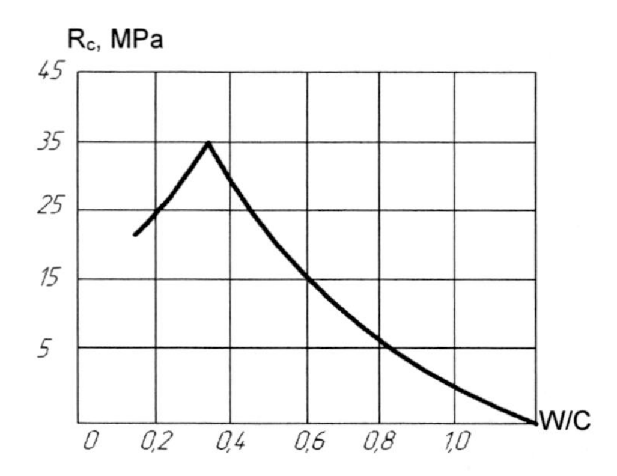

Fig. 10.8. Relationship between concrete strength R_c and W/C

There are various non-destructive methods of control of concrete strength directly in element and structures except of determination of concrete strength as a result of control standards testing (Fig. 10.7). Using of non-destructive methods is caused by the necessity of taking into account production factors: deviations from the set concrete composition, conditions of transporting, casting, hardening and etc. An important advantage of non-destructive methods is the immediacy and simplicity of the strength control.

The mechanical methods of non-destructive control are based on the principles of resilient rebound or impression. Thus concrete strength is determined by resiliency which is characteristics of rebound striking bodies (devices of Shmidt and other) or hardness, measured by the diameter of print (devices of Fizdel, Kashkarov, Gubber and other). The mechanical methods allow defining the strength of surface layer; they give a high error for concrete inhomogeneity in a sectional view.

The impulse method, based on measuring of distribution speed of ultrasonic vibrations in material is spread among the physical methods of non-destructive assessment of concrete strength. The change of speed spreading of ultrasonic vibrations indicates the certain strength changing at the permanent compositions of concrete and the hardening conditions.

The cement strength and water-cement ratio are the main ones among the known factors which influence the concrete strength.

Dependence between compressive strength concrete (R_c), strength of cement (R_{cem}) and water- cement ratio (W/C) can be represented by formula:

$$R_c = \frac{A' R_{cem}}{(W/C)^m}. \qquad (10.7)$$

This dependence expresses the basic law of the concrete strength - rule of water- cement ratio, essence of which consists in the fact that for permanent materials, production technique and hardening conditions, the concrete strength depends only on a water-cement ratio (Fig. 10.8).

The next formula is widely used for calculations in relation to a heavyweight concrete:

$$R_c = AR_{cem}(C/W \pm 0{,}5). \qquad (10.8)$$

The formula (10.8) shows, that in certain situations the concrete strength is inversely related to the W/C, in other words a cement-water ratio (C/W rectilinear dependence). In the two formulae (10.7, 10.8) coefficients A' and A depend on the quality of the initial materials.

It is important to note that the calculated and factual strength can substantially differ, because of the influence of other factors not explained here.

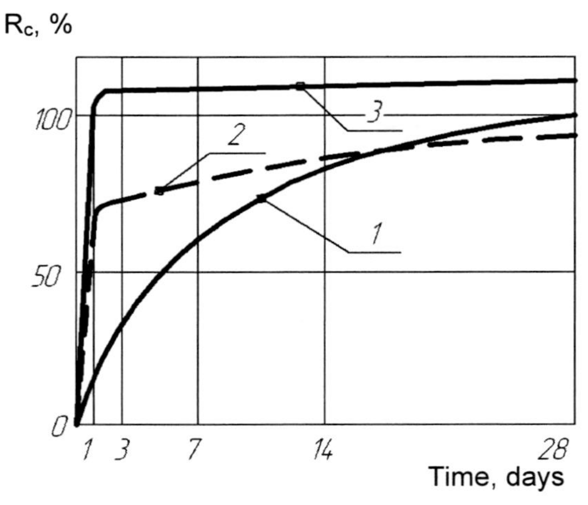

Fig. 10.9. The concrete strength increasing:

1- in the normal conditions of hardening; 2- steaming at atmospheric pressure and temperature of 85°C; 3- steamed in an autoclave at pressure of pair of 0.8 MPa and temperature

In light-weight concrete cement stone and grains of aggregate differ in strength and deformability by a lesser measure, than in heavy-weight concrete. That is why the strength of porous aggregate influences substantially the concrete strength.

For approximate prognostication of the concrete strength increase with time, it is possible to use the logarithmic dependence:

$$R_c^n = R_c^{28} \frac{\lg n}{\lg 28}, \tag{10.9}$$

where R_c^n - the ultimate concrete compressive strength at age n days ($n > 3$); R_c^{28} - 28-days concrete strength.

The increase in the concrete strength based on cement of certain chemical-mineralogical composition as a function of time is determined with the temperature-humidity conditions of hardening (Fig. 10.9). As the temperature increases, the processes of the cement stone hardening (basic structural component of concrete) are accelerated.

The concrete hardening is accompanied with heat release as a result of the exothermic processes of the cement hydration. It is required to ensure high humidity during the process of curing to prevent the concrete from drying out. The factors which influence the hardening intensity, affect also the heat generation (exothermal reaction) of concrete. The mineralogical composition of cement is the major factor which influence the heat generation. For the main minerals contained in normal cement, the order of the intensity of heat generation is as follows: $C_3A > C_3S > C_4AF > C_2S$ (Fig. 10.10).

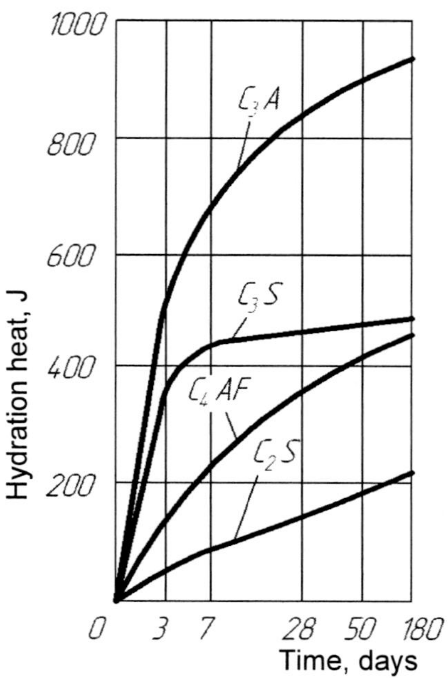

Fig. 10.10. The heat generation during the hydration of clinker minerals

The thermal effect of hydration of C_3A is almost twice higher, than C_3S and in 5 times higher, than C_2S. The heat generation of concrete is increased with increase in the cement content, grinding fineness, increasing temperature, and also by the presence of hardening accelerators. Heat generation reduces – by the introduction of mineral admixtures (slag, ashes of thermal power stations and other) and surface-active substances (SAS) to the cement or concrete mixture.

The positive role of the heat generation of concrete shows up during winter concreting, and also heat treatment with the purpose of accelerating the hardening process.

The heat generation of concrete plays a negative role in massive structures - dams, foundations under generating units and etc. It causes a considerable increase in temperature (up to 50°C and higher) in the core of the mass concrete. This results in the development of thermal stresses and cracking. A reduction in the self-heating of concrete in massive constructions is achieved by the use of low-heat cements with the heat generation in 3-day's age not more than 230 J/g. The reduction of concrete mixture temperature is also effective. Concrete cooling by water which circulates in pipes is also used.

For zero or negative temperatures, hardening of concrete takes place slowly (Fig. 10.11), or does not flow generally, because of free water freezing and stopping the process of cement hydration. Ice formation in concrete is accompanied by increasing of pressure which destroys its structure. After thawing the hydration process resumes, however the strength and other properties of concrete reduce mainly as a result of the delayed structural development.

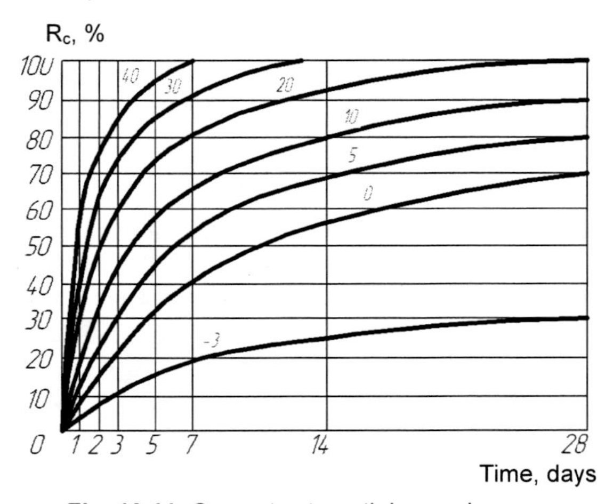

Fig. 10.11. Concrete strength increasing

at temperatures from -3 to to + 40 °C

Winter concreting, as a rule, is executed by the thermostat method and method of electrical warming. The method of steaming treatment is also used and so-called "cold concrete".

The enclosures and large tents are used in some cases, for example, for laying concrete in the blocks of hydraulic buildings at a temperature below minus 20°C.

The thermos method is used for concreting during construction of massive structures and hydrates and hardens due to the heat, which is released during cement hydration. The concrete surface is protected by heat insulator.

The steam- heating is affected in steam jackets or chambers, and also by the steam running through the pipes, embedded in the concrete. The concrete hardening is affected also due to the heat which is introduced in the body of concrete by passing current through metallic electrodes, or as a result of heat-transfer from the heated air by electrical warming. An alternating current of normal frequency is also used for preheating the concrete. Direct current causes electrolysis and that's why it is not used.

"Cold concrete" has an ability to harden at subzero temperatures due to introduction of large amount of chemical admixtures which reduce the freezing temperature of the mixing water. As antifreeze admixtures are used hydrochloric and ammonium salts, sodium nitrite, ammonium water, potash etc.

Concrete is a reliable and durable material if designed to take account of the intended exposure and environmental requirements. The rejection and destruction of concrete can be caused by internal stresses and by external actions. It is possible to resist chemical action of water and deleterious substances, which are contained in it and other such factors. Concrete must be designed specially to resist multiple wetting and drying; freezing and thawing.

The temperature differences, reaction of cement alkalines with some aggregates, crystallization of salts in the pores of concrete and others like that, destroy the integrity of the concrete and cause cracks.

The freeze-thaw resistance of concrete is the ability to withstand several cycles of alternate freezing and thawing in the saturated water state without the substantial strength diminishing.

There are some grades, set according to frost resistance, depending on the amount of cycles of freezing and thawing, which is established by standards. The grades for heavy-weigh concrete are:– F50, F100, F150, F200, F300, F400, F500 and higher and for light-weight – F25 and F35 and higher.

The durability to freeze-thaw action is always enhanced by diminishing the relative volume of capillary pores and increasing the volume of gel pores, within the concrete matrix.

Thus improvement in durability to frost resistance can be assured by reducing the water-cement ratio and the total water content which are the main factors that determine capillary porosity, and high concrete density.(Fig. 10.12). For good frost resistance of the normal concrete, it is recommended, that W/C should not exceed 0.5, and content of water not more than 160 l/m^3.

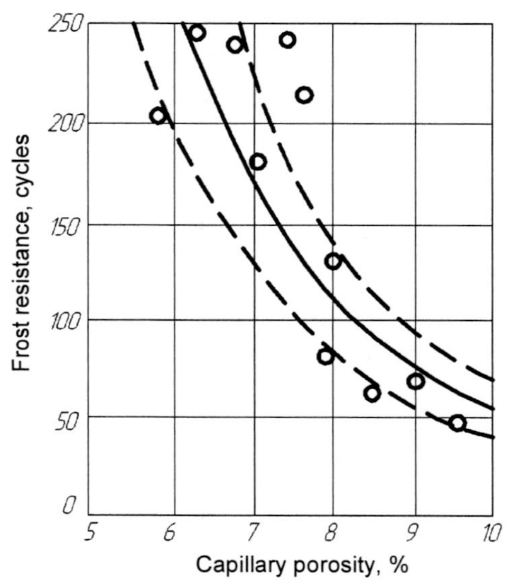

Fig. 10.12. Dependence of frost resistance of concrete on its capillary porosity

The correct selection of cement and aggregates are equally important. The use of cement with minimum content of tricalcium aluminate (C_3A <8 %) have a positive affect on the freeze-thaw durability of concrete. The introduction of increased amount of active mineral admixtures into cement, leads to noticeable decline of freeze-thaw durability

Surface-active substances are widely used to increase the freeze-thaw resistance of concrete. They allow a reduction in the water-cement ratio and provide a fine air pore formation, to which the frozen water is pushed back from capillaries(i.e, they act as pressure release pockets) and they also improve the crystalline structure of the cement stone.

Water impermeability, which is characterized by the maximum pressure at which there is no water filtration through the specimens, is an important index of concrete quality for building structures which work under the action of certain pressure. The extent (grades) of water impermeability are assessed based on the pressure gradient (ratio between maximal water pressure and the thickness of construction) and the type of construction and function. At a pressure gradient of 5, a grade of W4 is recommended; from 5 to 10 - W6; from 10 to 12 - W8; 12 and more than - W12.

Water infiltration into the concrete matrix occurs mostly through the capillary pores and especially sediment capillaries, formed as a result of stratification of concrete mix.

High water impermeability is achieved by reducing the W/C, and by application of expanding, plasticized and hydrophobic cements. Application of pozzolanic and blast-furnace cement will also improve impermeability characteristics of concrete. The ambient relative humidity also affect the permeability of concrete. Water impermeability is substantially increased due to reduction in the volume of opened pores during cement hydration. The water impermeability and water absorption of concrete are the indexes of its density to which the resistance of concrete to different aggressive actions of circumambience is related (Table 10.9).

Table 10.9 Indexes of concrete density

The concrete types according to the compactness	Water impermeability grade	Water absorption, % by weight	W/C not more
Ordinary dense	W4	5.7...4.8	0.6
High dense	W6	4.7...4.3	0.55
Especially dense	W8	4.2	0.45

The crack-resistance of concrete, to a great extent, is related to its deformation properties. Deformation of concrete occurs both as a result of hardening (shrinkage and swelling) and under the action of external forces.

Concrete shrinkage takes place as a result of water evaporating from the pores during drying. It ranges, as a rule, from 0.2 to 0.4 mm/m in annual age and increases with increasing W/C, cement content, at introduction of active mineral additives. Concrete shrinkage changes are approximately proportional to the logarithm of time and exhibits fast increases especially in the first 28 days (Fig. 10.13).

Concrete deformations essentially decrease on introduction of steel reinforcement. The shrinking deformations in concrete lead to crack formation, which may lead to a violation of the serviceability performance requirements. This must be avoided, especially it affects on the thin-walled structures from the prestressed reinforced concrete.

One of the criteria of crack-resistance of concrete is its conditional extensibility:

$$\varepsilon_{c.e} = \frac{R_t}{R_{dyn}}, \tag{10.10}$$

де R_t - ultimate tensile strength; R_{dyn} - dynamic modulus of elasticity.

Creep of Concrete:- This is the process of deformation of concrete without a corresponding increase in applied load. The creep of concrete is affected by loading of structures in early age, increasing of water-cement ratio, cement content, humidity of concrete and duration of imposed loading. It depends on the type of cement and it mineralogical composition, and also on the mechanical properties and content of the aggregates used. Concrete made with high-strength low- aluminate Portland cement is characterized by a lower creep.

The creep of concrete causes significant bends and losses of previous tension in the reinforced-concrete structures. At the same time, creep deformation can compensate the tension, which increases as a result of exothermic processes and shrinkage of the cement stone in massive constructions.

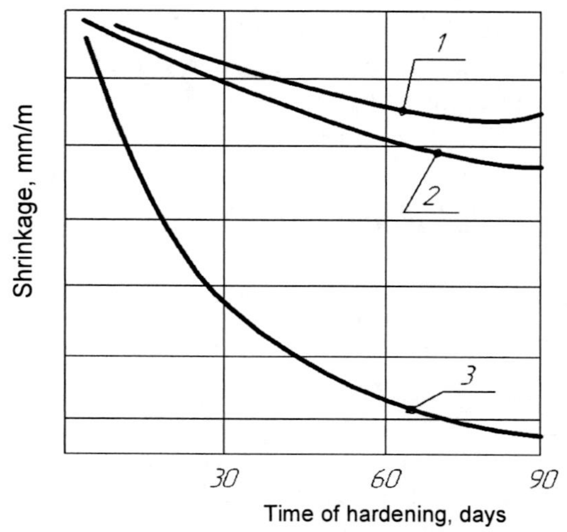

Fig. 10.13. The shrinkage curve:

1- concrete; 2- mortar; 3- cement stone

Light-weight concrete is characterized by greater dimensional instability than heavy-weight concrete. The modulus of elasticity, for example, of claydite concrete is about 20...50 % smaller than the modulus of resilience of heavy-weight concrete of the same class. As the modulus of elasticity diminishes the more deformative is the aggregate and the higher its content is. The maximum extensibility of light-weight concrete is 4...5 times higher than for heavy weight concrete, and is about 0,03...0,4 mm/m. The increased extensibility and low modulus of elasticity of light-weight concrete promotes its crack-resisting and allows its effective use in different buildings.

General shrinkage of light-weight concrete is in15...30 % higher than the shrinkage of heavy-weight concrete. The shrinking deformations reduce with diminishing water and cement contents.

10.4. PROPORTIONING CONCRETE MIXTURES

One of main task of concrete technology is the determination of the optimum composition of concrete. With respect to the optimum composition of concrete, as a rule, the content of materials per 1 m^3 of concrete mixture which provides the required properties of concrete at the least expense of cement is understood.

The composition of concrete for dry materials is called nominal(laboratory) and the materials with certain humidity is described as w o r k i n g one. The nominal composition is found at first, as a rule, and then the working composition is computed, taking into account the aggregate humidity.

For concrete production, the calculation of the concrete composition is added up for determination of the material expenses for the batch of concrete mixer. A volume of batch (V_b) is a less than the total volume of separate friable materials as a result of filling the voidage between grains of coarse aggregate.

$$V_b = \beta(V_C + V_S + V_{C.A}), \quad (10.11)$$

where:

β - coefficient of concrete yield, for a heavy-weight concrete equals 0.55...0.75;

V_C, V_S, V_A - accordingly volumes of cement, sand and coarse aggregate.

Proportioning of the concrete mixtures can be realized by experimental or calculation-experimental methods. For proportioning heavy-weight concrete the required slump or rigidity of concrete and also class of concrete strength are specified. If it is required the grades of concrete are also specified according to the frost resistance and water impermeability and so on.

The experimental methods of concrete mixtures proportioning are expedient at the large volumes of concrete works and heterogeneity of the materials.

Calculation-experimental methods accelerate the mixtures proportioning due to the empirical formulas, accumulated graphic and tabular information used. But the expected composition should be experimentally corrected on required indexes before the recommendation to apply it to production.

The method of absolute volumes is the most common among the calculation-experimental methods of determination of concrete composition.

The calculation of composition of heavy-weight concrete begins with determination of the required C/W or W/C.

The approximate C/W required can be found from the following formula:

$$R_c = AR_{cem}(C/W - 0.5), \quad (10.12)$$

where:

R_c - concrete strength at 28 days; C/W - cement-water ratio;

A - coefficient which depends on the quality of materials.

The formula (10.12) indicates the rectilinear dependence between the concrete strength, cement strength and cement-water ratio. It is just for C/W ≤ 2,5. At C/W > 2,5 ($R_c < 2AR_{cem}$) the next formula can be used:

$$R_c = A_1 R_{cem}(C/W + 0.5) \quad (10.13)$$

Depending on the material quality the next values of coefficients can be used:

	A	A_1
High quality	0.65	0.45
Ordinary quality	0.6	0.4
Satisfactory quality	0.55	0.37

Water-cement ratio should be limited by the action of salt and fresh water on structures and also in other cases when the concrete density is specified (Table 10.10).

Table 10.10 The W/C limitation at the action of salt and fresh water

Condition of exposure	W/C for the non-massive concrete structures		W/C for the external surface of massive concrete structures	
	Salt water	Fresh water	Salt water	Fresh water
Area of variable water level at climatic conditions:				
Especially severe	0.42	0.47	0.45	0.48
Severe	0.45	0.50	0.47	0.52
Mild	0.50	0.55	0.55	0.56
Parts of structures which are constantly under water pressure	0.55	0.58	0.56	0.58
non-pressure	0.6	0.62	0.62	0.62

For concrete of ordinary density W/C can be no more than 0.6; high dense - no more than 0.55 and for specially dense concrete - no more than 0.45.

After selecting the water-cement ratio, the required amount of water can be determined to achieve the required slump or rigidity of concrete mixture. Usually for water content determination, graphic (Fig. 10.5) or tabular data is used, taking into account the influence of the cement type and coarse aggregate and also the water demand of sand.

If the content of water and water-cement ratio are known, it is possible to define the cement content (C) as follows:

$$C=W/(W/C) \text{ or } C = W\cdot(C/W) \qquad (10.14)$$

The cement content should be not less than some minimum values to prepare the homogeneous concrete mixture. The minimum amount of cement content is accepted to be not less than 200 kg/m^3 for non-reinforced concrete. The reduction of cement content is assumed to 150 kg/m^3 for the use of fly ash additive. It is effective to add, also other dispersible mineral additives to provide the minimal binder content.

The contents of sand and coarse aggregate can be found by solving the following system of two equations:

$$\begin{cases} \dfrac{C}{\rho_{cem}} + \dfrac{S}{\rho_s} + \dfrac{C.S}{\rho_{c.s}} + W = 1000 \\ \dfrac{C}{\rho_{cem}} + \dfrac{S}{\rho_s} + W = \alpha V^v_{c.s} \dfrac{C.S}{\rho^b_{c.s}} \end{cases}, \qquad (10.15)$$

where:

C, S, C.S, W - accordingly contents of cement, sand, crushed stone (gravel) and water;

ρ_{cem}, ρ_s, $\rho_{c.s}$ - absolute densities of cement, sand and crushed stone (gravel); $\rho^b_{c.s}$ - bulk density of the crushed stone (gravel);

$V^v_{c.s}$ - voidage of the crushed stone (gravel) $V^v_{c.s} = \left(\rho_{c.s} - \rho^b_{c.s}\right)/\rho_{c.s}$;

α - coefficient of the coarse aggregate grains separation by the cement-sand mortar.

Coefficient α characterizes the mortar surplus for filling the voids between the coarse aggregate grains and for plastic concrete mixes depends on the cement content, water-cement ratio and water demand of the sand (Table 10.11).

Table 10.11 Coefficient α for plastic concrete mixes

Cement content, kg/m^3	Value of α depending on W/C				
	0.4	0.5	0.6	0.7	0.8
250	—	—	1.26	1.32	1.38
300	—	1.3	1.36	1.42	—
350	1.32	1.38	1.44	—	—
400	1.4	1.45	—	—	—

Notes: 1. The values given in Table are true for sands with water demand of 7 %. By increasing the sand, water demand for every percent α diminishes by 0.03, and by diminishing the sand water demand increases accordingly by 0,03.

2. For non-plastic (rigid) concrete (C < 400 kg/m^3) α = 1.05...1.15.

The first equation in system (10.15) state that 1 m^3 of concrete mix (1000 litres) equals the sum of the absolute volumes of four components (cement, sand, crushed stone or gravel and water). From the second equation, it follows that the volume of mortar in a concrete should fill the voidage between grains of coarse aggregate.

The calculation formulae for determination of crushed stone (gravel) and sand contents can be obtained by solving the following system of two equations (10.15):

$$C.S = \frac{1000}{V_{c.s}^{v}\frac{\alpha}{\rho_{c.s}^{b}} + \frac{1}{\rho_{c.s}}}, \qquad (10.16)$$

$$S = \left[1000 - \left(\frac{C}{\rho_{cem}} + \frac{C.S}{\rho_{c.s}} + W\right)\right]\rho_{s}. \qquad (10.17)$$

The determination of optimum water amount as a result of testing of few series of specimens with different water content is the most responsible for light-weight concrete. At the optimum water amount, it is possible to get the maximum strength of concrete for a defined grain composition of aggregates, cement content and conditions of compaction.

There are some other calculation-experimental ways of concrete mixtures proportioning, used in construction practice.

Self-Assessment Questions

1. Give the classification of concrete and describe their importance in construction.
2. What factors are taken into account for the selection of cements for concrete?
3. What are the requirements for the mixing water?
4. What are the requirements for the fine aggregates of concrete?
5. What are the requirements for the coarse aggregate of concrete?
6. What porous aggregates are applied at manufacturing of light-weight concrete?

7. What are the basic admixtures, applied in concrete?
8. Describe the technologies for obtaining concrete mixtures.
9. Describe basic properties of concrete mixtures.
10 Tell about workability of concrete mixtures.
11. Describe the peculiarities of concrete structure?
13. What methods of determination of concrete strength are used?
14. What is the essence of nondestructive methods?
15. Enumerate the various properties of concrete.
18. Explain the methods of proportioning of concrete mixtures.

Chapter 11

VARIETIES OF CONCRETE

11.1. FINE-GRAINED AND POROUS CONCRETES

Along with ordinary coarse-grained concretes there are wide verities of fine-*grained concretes* with maximum aggregate fineness not exceeding 10 mm applied in construction. This is especially so in conditions of coarse aggregate scarcity. The basic and most common variety of fine-grained concretes is the *sand concrete*.

High values of specific surface of aggregates and in some cases intergrain air space cause necessity in the increased cement paste content in the fine-grained concretes.

Fine-grained concrete has more homogeneous structure and greater tensile strength, than coarse-grained. The change of W/C, grain composition, form and other features of sand influence substantially the quality indexes of fine-grained concrete. The reduction of the cement content for the obtaining of fine-grained concretes is arrived at by the improvement of the aggregate grading, for example, due to introduction of additive- enlargers, like fly ash and other effective filling agents and plasticizers. Fine-grained concretes are used for the various thin-walled structures, arranging of road carpet etc/.

By reinforcing fine-grained concrete having a maximal fineness of 2,5...3 mm with woven metallic nets, *ferrocement* is obtained. It is a material with high bearing strength, which is used in thin-walled spatial constructions with complicated configuration.

The non-sand concrete is characterized by a deficiency of sand in the composition of concrete mixture and limited cement content, sufficient only for enveloping of grains of coarse aggregate. In such non-sand concrete, the average density goes down to 1500...2000 kg/m^3 and the thermal conductivity diminishes to 0.64...0.93 W/(mK). The porous structure of non-sand concrete results in strength decline. The compressive strength of such concrete at 28 days ranges between 5...10 MPa. The properties of non-sand concrete allow applying it as a walling material.

Non-sand concrete possess sufficient filtrational properties to warrant its use for drainages and as filters in ameliorative construction as a more rational and economic option compared with drainages and filters from the other materials.

Another variety of light-weight concrete is *cellular concrete* with a large amount (upto 85 % of the total volume) of artificially formed pores as meshes, filled with air or other gas. The strength of these types of concretes relates to their porosity (Figure 11.1).

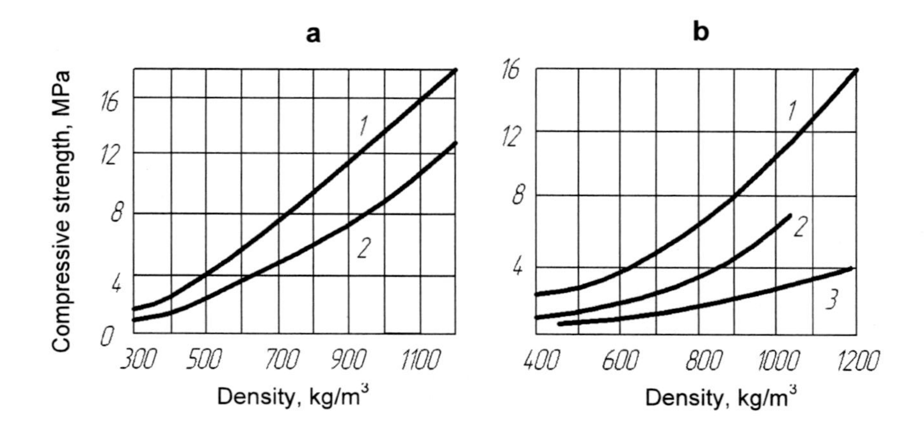

Figure 11.1. Relationship between strength of cellular concretes and density:
a- autoclave aerated concrete; 1- cement strength 40 MPa; 2 - cement strength 30 MPa; b- aerocrete; 1- autoclave; 2- steam-cured; 3- normal hardened.

Also distinguishable are aerated concrete (foam-concrete) - obtained by mixing a binder, water and silica component with foam; gas concrete - obtained by mixing similar mixture with a gaseous substance (aluminum powder, hydrogen peroxide).

Cements, lime, or mixtures of lime with cement or ash or ground slag can be used as binders for the production of cellular concretes. It is also effective to apply low-alumina cements and lime with high activity for cellular concretes (content of CaO > 80 %).

Lime-silica binders used to produce cellular concretes are called gas-or foam silicates.

Ground sand or fly ash from thermal power-stations are used as silica component, as a rule. The ground sand should contain not less than 90 % SiO_2. Comparatively with the ground sand, ash has higher hydration ability, needs considerably less expense for grinding and results in the production of concrete with less average density. In general, fly-ash contains not less than 50 % of the glassy and alloyed particles and not more than 5 % non-fired particles of coal.

For the acceleration of cellular concretes hardening, autoclave or non-autoclave methods of thermal treatment - in steam-curing chambers, special moulds with electric curing are applied. Much higher mechanical properties of concrete (Figure 11.1) are achieved by the autoclave method of hardening involving the application of a vapor pressure of 0.8...1.2 MPa,

Autoclave hardening allows a wider use of low grade binders such as those based on slag-ash material. The duration of the thermal treatment are reduced to 10...12 hours, and properties such as frost- and crack-resistance of cellular concretes become highly improved.

Cellular concretes, is categorized as light-weight concrete and is made with porous aggregates. It is divided into insulating ($\rho_0 \leq 500$ kg/m^3), constructively insulating (900 kg/m^3 $\geq \rho_0 >$ 500 kg/m^3) and constructive (1200 kg/m^3 $\geq \rho_0 >$ 900 kg/m^3). Cellular concrete has an ultimate compressive strength in the range of 2-15 MPa and frost resistance in the range of 15...100 of freezing and thawing cycles.

To obtain a dense matrix, speed mixers, dispersible reinforcement and fibrous additives are added to increase the density and strength of cellular concretes.

The properties of cellular concretes are substantially improved by the vibration action during preparation of mixtures and forming. Vibration is instrumental for intensification of air expulsion from the matrix, and the formation of a fine-pored and homogeneous structure.

The cellular concrete has high values of sorption humidity, vapour permeability and air permeability (5...10 times higher than that of heavy-weight concrete). The walls and roofs built using cellular concretes should have special vapor sealing on the internal surfaces (layer of heavy-weight concrete, lacquer coating) of buildings, exposed to high humidity and/or aggressive gases. The high permeability of cellular concretes stipulates a necessity to protect the reinforcing steel with the help of special coatings (e.g., cement-casein, cement- latex, bituminous coats).

Cellular concretes, especially those produced by non-autoclaved hardening, are characterized by considerable deformations (up to 1 mm/m and higher) including shrinkage and swelling. In order to significantly reduce such deformations, it is possible to introduce coarse porous aggregates into the composition of the cellular concretes.

The use of cellular concrete as panels and fine wall blocks in structures could be instrumental in the reduction of cost, labor expenditures and materials content in the construction of domestic and industrial buildings. Thus use of cellulose concrete normally reduce the general mass of houses by 25...30 %.

11.2. Hydraulic and Road Concretes

Hydraulic concrete is the type on concrete used for construction of structures constantly or periodically in contact with water. Depending on the location of the hydraulic concrete in structures and on the waterline, the concrete can be categorized as:- underwater (stays in the water constantly), variable water level, above-water. The concrete in underground hydraulic structures is classified as underwater. The hydraulic concretes are also divided into massive and non-massive. Massive concretes are mainly used for construction of dams. *Such* structures need special measures for adjustment of temperature stresses that arise during heat evolution in early-age concrete.

Requirements for hydraulic concretes are differentially specified taking into account the zonal location of concrete in structures (Table 11.1). This complex requirement for the concrete of variable water level and external surface of massive structures is strictly applied.

Compressive and tensile strength are the basic indexes of durability for concrete used on modern hydraulic structures. The ultimate strength of concrete is determined at the age of 28, 60 or 180 days depending on the terms of construction. The indexes of maximum tensile strength, shrinkage and swelling of concrete are specified to provide the required crack-resistance. The maximum tensility of hydraulic concrete obtained for specimens after 180-day's must be not less than $5 \cdot 10^{-5}$ for concrete in internal areas and not less than $7 \cdot 10^{-5}$ for the concrete on external areas of buildings. It is improved with the increasing of concrete strength, application of cements without mineral additives, by introduction of surface-active additives and polymers to the concrete mixture.

Table 11.1. The requirements for hydraulic concrete according to location

Requirements for the concrete	Massive structures						Non-massive structures		
	External area			Internal area					
	Areas of relative water level								
	Underwater	Variable level	Above-water	Underwater	Variable level	Above-water	Underwater	Variable level	Above-water
Waterproofness	+	+	+	+	+	—	+	+	+
Water resistance	+	+	+	+	+	—	+	+	+
Frost resistance	—	+	+	—	—	—	—	+	+
Low heat generation	+	+	+	+	+	+	—	—	—

Note: "+" means the presence of the proper requirement.

Linear shrinkage of hydraulic concrete at a relative humidity of 60% and 18°C, at 28 days should not exceed 0.3 mm/m. At 180 days shrinkage should not exceed 0.7 mm/m comparatively with their initial length. Swelling must be accordingly not more than 0.1 and 0.3 mm/m comparatively with the sizes of the dry standard specimens.

Concrete shrinkage exerts sharp and negative influences on the construction of storage facilities and prestressed concrete reservoirs. The dampproofing.-resistance of hydraulic concrete of massive constructions is directly related with temperature stresses, arising from the heat of hydration generated during hardening.

Frost resistance, water impermeability and resistance to chemical corrosion in a water environment are the major properties which influence the durability and life-span of hydraulic concretes. The abrasive action of suspended sediments in water and cavitation also aggressively influence the performance of hydraulic concrete.

One of methods by which concrete can be protected from the influence of cavitation is by imparting to the structures such form which ensures non-separated flowing of the water stream. For this purpose a smooth cross section of parabolic character is given to the structures, for example, to the spillways. The cavitation and abrasive resistance of concrete, is well characterized by the compressive strength. For concrete revetments of waterworks, it is required to apply concrete with compressive strength of 45...50 MPa to ensure cavitation firmness.

For protection against cavitational-abrasive wear can be used stone, metal, rubber, polymeric concretes revetments.

Road concrete differs from ordinary one by the high tensile and compressive strength, enhanced deformability and frost resistance.

Road concrete is divided into concrete for one- and two-course pavement, and also for bases of improved pavements. The ultimate compressive strength of road concrete is in the range of 15-50 MPa and tensile strength is 1.5-5.5 MPa.

The requirements for road concrete are predefined by the severity of its service conditions: by the actions of static and dynamic transport loadings, and by the variability of humidity, temperature etc.

The strength and frost resistance of concrete are specified depending on the category of road, layer of pavement and climatic conditions in service.

Special Portland cement, as well as plasticized, and sulphate-resistant cements are applied in the manufacture of road concrete. The content of the tricalcium aluminate in the cement used is limited to not more than 10 %. The initial setting time of cement is not less than 2 hours. Surface-active substances can be added to improve the properties of the concrete mixtures and to improve resistance against the simultaneous aggressive action of chloride salt solutions and frost action.

In addition to strength determination, the abrasiveness of the crushed stone and gravel used are usually tested and checked.

In road construction, the high quality of concrete pavements is insured by proper curing of the concrete by application of different film-forming materials.

11.3. Cast, Self-compacting and Rigid Concretes

For monolithic concreting of reinforced structures, it is important for concrete mixtures to have high flowability and to be compacted without application of significant vibratory actions. Concretes which correspond to these requirements are called *cast and self-compacting concrete*.

Cast concrete is the concrete, made from mixtures with slump 15…22 cm. Such mixtures are able to flow freely and easily flow around the reinforcement and embedded items and fill fully the molds for structures of irregular shape.

Due to high water content, obtaining the cast concrete which meets the necessary properties is a more difficult task at industrial conditions, than normal consistency. The basic problems which occurs during cast concrete production and application are water bleeding and segregation, preventing qualitative casting and compacting of structures; raised cement content because of the high water content; high flowability lost with time; high shrinkage, creep; low quality at casting of densely reinforced structures.

To ensure the homogenity of cast concrete mixtures special measures are normally recommended, to increase water-retaining ability; to increase sand content in concrete mixture up to 40…50% by aggregate mass; to limit water-cement ratio to prevent water bleeding and segregation of cement paste; to apply superplasticizers, air-entraining or special water-retaining admixtures (bentonitic clays, silica gel); to use cements with sufficient water-retaining ability.

Introduction of thermal power-station fly ash in the production of cast concrete mixtures is a comparatively effective means. Application of fly ash allows obtaining cast concrete at the same cement content and as a moderately more flowable mixtures. Effectiveness of cast concrete mixtures increases significantly when superplasticizers or high-effective plasticizers are applied. Application of such admixtures permits the production of cast concrete with reduced water content and provides the required water-retaining ability, rapid hardening and flowability for a period of 40…50 minutes and more.

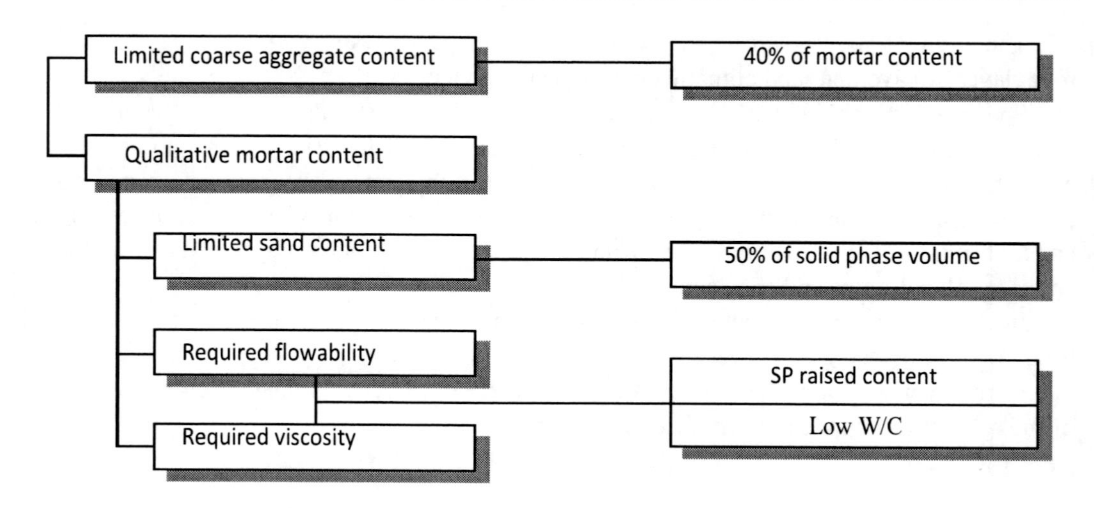

Figure 11.2. Methods of self-compacting concrete mixtures obtaining (by H.Okamura, K.Ozawa, M. Ouchi)
SP – superplasticizer.

Self-compacting concrete (SCC), is a high-technology type of cast concrete, that is concrete based on mixtures, able to fill the forms and frameworks without need for any vibration resulting in enhanced uniformity and absence of microcracks forming ability. Such concrete mixtures are characterized by the slump more than 22 cm. Production of self-compacting concrete is useful for complex applications and is made possible by modern and very active superplasticizers addition (Figure 11.2.).

The main advantages of self-compacting concrete are: to reduce energy and labor contents, and to improve work conditions at concrete works by decreasing of noise and vibration influences. It is also characterized by improved pumpability, reduced pressure requirement and reduced losses.

Rigid mixtures are concrete mixtures, for which standard-measured slump is not observed (Slump (SL) equals 0 cm). As opposed to cast and self-compacting mixtures rigid concrete mixtures are characterized by low water content and high viscosity. That is why for production of elements based on rigid mixtures, intensive methods of compacting should be applied. Concrete from rigid mixtures can be manufactured at plants as it is difficult to provide the required level compacting in other cases. For concrete mixtures requiring Vebe consistency grades V4 and V3, the elements can be manufactured with application of ordinary vibration (vibration amplitude – 0.35…0.5 mm, frequency – 50 Hz). Mixtures with higher rigidity are compacted by vibration at raised amplitude and frequency, cantledge vibration, vibrostamping, vibrorolling, vibrocompression and other methods. The ability of rigid mixtures to flow and fill the required shape is apparent only on application of vibration action and some pressure. Normally the compaction of rigid mixtures is implemented in several stages: at the initial stage ordinary vibration is conducted, resulting in regrouping of aggregate particles and filling the voids with mortar (matrix). After that, maximum density is reached by intensive influence of pressure vibration.

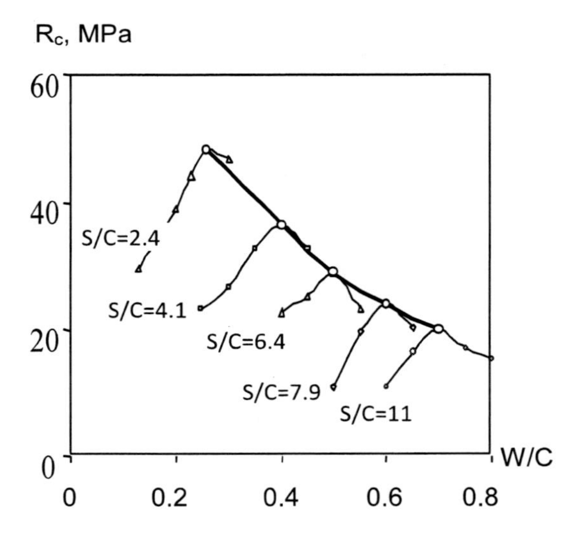

Figure 11.3. Relationship between strength of vibrocompressed concrete and W/C at different sand-cement ratios (S/C).

The increase in concrete mix rigidity is connected with increase in aggregate content and decrease of cement matrix content. The concrete strength is determined mostly by structural role of aggregate because it is the aggregate framework takes up the loads, therefore cement stone strength is less influential.

Rigid mixtures can produce concrete with strength increased by 30…80% depending on the rigidity and reduce cement content by 20…40%. Concrete based on this type of mixtures exhibit increased modulus of elasticity, tensile strength, reduced shrinkage and creep.

The increased resistance to abrasion, water-tightness and corresponding durability enable the wide application of such concrete in road construction. Rigid mixtures are also used in the manufacture of cellular and ribbed slabs and in tubular structures.

Application of rigid mixtures gives the possibility for instant formwork removal from the cast elements, causing increase in forms turnover, reduction of heat treatment duration of concrete and set of strength. However application of rigid mixtures based concrete requires significant capital investments for necessary equipment installation.

For rigid mixtures the most widespread method of compacting among the known ones is vibrocompression at present. This method achieves compaction of mixture by short-term influence (5…15 sec) of intensive vibration and dynamic cantledge. This results in the attainment of the required initial strength; products are instantly demolded and transported for further curing/maturing.

V*ibrocompressed concrete*, are mostly manufactured using, fine-grained aggregates(less than 10mm). The water-cement ratio of concrete depends on the cement content (Figure 11.3) and peculiarities of the aggregate. An optimal water content exists for every cement content which permits the production of concrete with the maximum density. At a cement content of 400…500 kg/m^3 W/C is normally within the range 0.35…0.25. That allows frost resistant concrete with strength of 50…60 MPa to be obtained, which is widely used in road, roofing and other elements.

At low cement content (150…200 kg/m^3) for vibrocompressive concrete usually necessity for application of mineral fillers appears since increasing binder volume, is insufficient for obtaining dense concrete. In that case W/C is 0.6…0.7, and concrete strength varies as rule from 15 to 20 MPa. Such concrete is applied for hollow masonry blocks.

11.4. High-strength and High-performance Concretes

Up till the present time, there is no uniform definition for the type of concrete considered to be h*igh-strength concrete.* Conditional boundary between ordinary concrete and high strength concrete changes with the development of concrete technology. In the early part of the 20th century, concretes with ultimate compressive strength 30-50 MPa, were considered to be high-strength. In the 60s, concretes ,more than 60-70 MPa were considered to be high-strength. At present High-Strength Concrete is defined as concrete with compressive strength at 28 days age 70...150 MPa. European standard EN206 provides the possibility of manufacturing and application of concrete including class C100/115. This has been made possible mostly due to the application of the effective modifiers such as- superplasticizers and silica fume admixtures. Also the industrial technology of concrete production with strength in the range indicated above has been developed, and corresponding norms have been introduced. Such concrete are widely used for load-bearing structures of monolithic frameworks of high-rise building (Table 11.2), bridges, offshore structures, and vibrohydro-compressed pipes. In laboratory conditions it has become possible to obtain concrete with strength up to 200 MPa and more.

Strength and other properties of concrete get better as it incorporates to saturation some organic and inorganic substances. For the obtaining P*olymer-concrete*, elements are dried, vacuumized and saturated with liquid monomers (methyl methacrylate, styrene etc) which then polymerize directly into a concrete by the radiation or catalytic thermal methods. Thus strength of concrete, water impermeability and frost resistance, resistance to the aggressive environments, abrasive resistance increases several times; creep also diminishes many times. The concrete-polymer products have properties similar to cast-iron, steel, reinforced concrete. The use of polymer-concrete pipes instead of reinforced concrete provides an essential economy of steel. Polymer-concrete has substantially improved durability performance, there is therefore a positive effect derived from using the concrete in severe conditions especially in chemically aggressive environments.

High Performance Concrete is also considered to be the type of high-strength concrete, which have compressive strength of 30...50MPa, at the age of 2 days; 60...150 MPa, at the age of 28days; frost resistance – F600 and more, watertightness – W12 and more, water absorption – less 1...2%, abrasiveness to 0.3...0.4 g/cm^2, and regulated parameters of deformability.

Obtaining of high-strength heavy-weight concrete at high-strength aggregates is possible by increasing density and strength of cement stone (cohesive factor) and transition zone strength (adhesive factor). The main way of obtaining of high-strength concrete is providing ultimate low water-cement ratios at comparatively high hydration degree and required compacting of concrete mixture. At low W/C values obtaining optimal ratio between crushed stone and mortar content influences positively on concrete strength.

Table 11.2. Examples of high-strength concrete application at high-rise buildings construction

City	Year of building construction	Number of floors	Concrete strength, MPa
Montreal	1984	26	119,6
Toronto	1986	68	93,6
New York	1987	72	57
Toronto	1987	69	70
Paris	1988	36	70
Chicago	1989	82	78
Guangzhou	1989	63	70
Chicago	1990	65	84
Frankfort	1990	58	45
Seattle	1990	58	133
Frankfort	1991	51	112

With W/C increasing at other constant conditions not only the porosity of cement stone increases but differential porosity of concrete, that is pore distribution per volume unit of concrete by their radii substantially changes. It was found out that even at moderate W/C increasing capillary radius increases.

Traditional methods of W/C reduction at constant cement content are water content reduction and conversion to rigid mixtures, application of superplastizing admixtures, increasing of aggregate purity and conversion to the aggregates with lower specific surface.

Drastic method of W/C reduction without significant flowability loss of concrete mixture is introduction of superplasticizers (SP). In comparing with ordinary plasticizers reducing water content up to 10...15%, SP permit to decrease water content at 20...30% and more and increase correspondingly concrete strength (Figure 11.4, 11.5). By regulation of SP content and W/C high early strength concrete can be obtained. Considering the possibility of air-entraining increasing due to SP introduction especially at raised dosages, admixtures with defoaming additives were identified, which permit the achievement of greater superplasticizing effect.

Introduction of SP allows achieving the required strength of steam-cured concrete at reduction of temperature of isothermal heating up to 30^0C or decreasing at 40...50% of steam curing duration due mostly to the duration of isothermal heating duration.

Reduction of W/C within a certain range is possible also due to the increasing of cement content, however this method has restrictions, taking into account increased shrinkage and crack resistance reduction of concrete. High-strength concrete is preferred to produce at cement content not more than 550...600 kg/m^3.

At equal W/C increasing in cement content above the optimal value causes usually strength reduction. This conclusion has been repeated by researchers, what can be explained by the necessity in certain optimal thickness of cement paste as adhesive interlayer at aggregate grains, at which maximum adhesive effect take place.

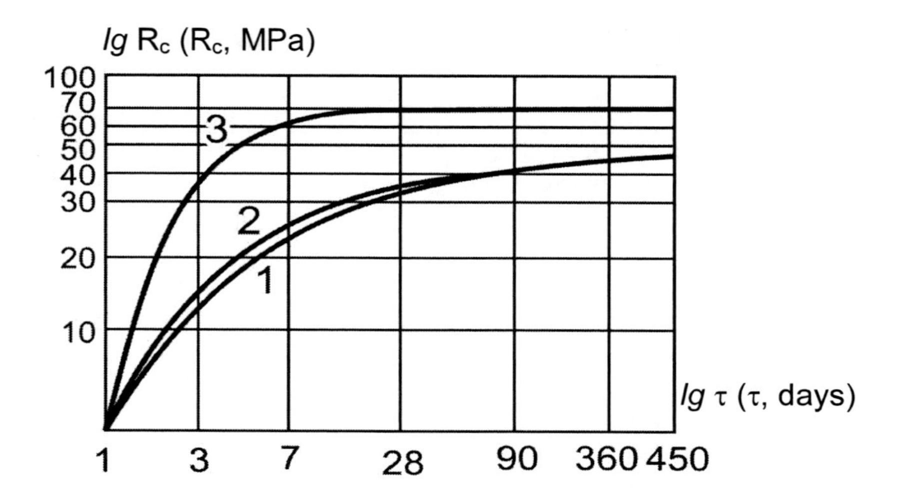

Figure 11.4. Kinetics of hardening of concrete with naphthalene- formaldehyde superplasticizer (SP): 1 – without SP; 2 – SP at constant W/C; 3 – SP at reduced W/C.

Concrete strength changes almost linearly with increasing of cement content. High-strength concrete of classes C50/60-C80/95 can be manufactured at reduced W/C, applying cements with compressive strength 70…80 MPa. There is an experience of such cements production in the industry mostly due to the increasing of alit content in clinker up to 60...65% and increasing of specific surface of cement up to 4000...4500 cm^2/g. Concrete at mentioned cements gains in 1 day of hardening in normal conditions 20...40%, and in three days – 50...70% of grade strength. However, manufacture of such cements is characterized by high energy content and is not very common.

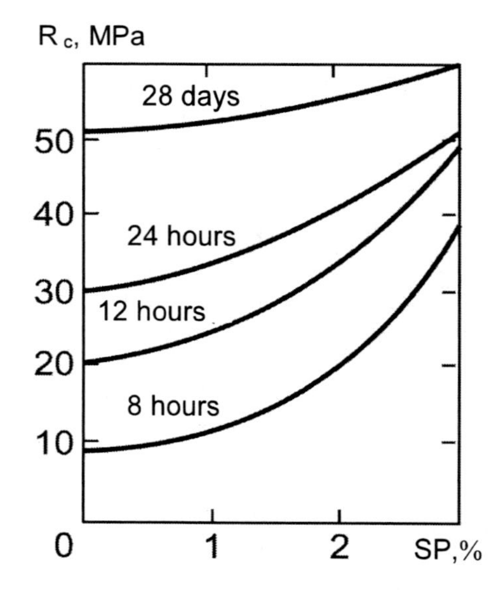

Figure 11.5. Relationship between compressive strength of concrete and the SP dosage at different age fo hardening.

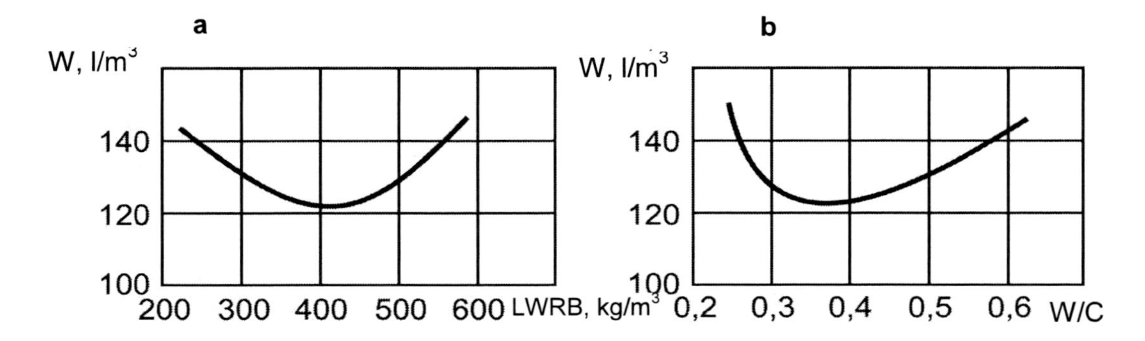

Figure 11.6. Dependence of water content of concrete mixtures (Slump equals 1...4 cm) on the LWRB content (a) and W/C (b).

Low water required binders (LWRB), which are obtained by fine grinding of Portland cement clinker and mineral admixture with introduction of powder-like superplasticizer belong to the effective binders for high-strength concrete. The binder is characterized by high specific surface (S=4000...5000 cm^2/g), low water requirement (16...20%) and strength up to 100 MPa. Water requirement of concrete mixtures based on LWRB is on 35...50% lower, than on ordinary Portland cement (Figure 11.6). Their peculiarities are: raised sensitivity of the parameters of flowability to water content change; absence of segregation; high viscosity and improved thixotropic properties. Concrete based on LWRB is characterized by intensive strength acceleration, even after a few hours, and in 1 day its strength can reach up to 60 MPa.

The processing of the experimental data permits suggestion of the formula for strength determination of concrete based on LWRB:

$$R_c = 0{,}4R_{cem}(C/W - 0.4), \tag{11.1}$$

where C/W– cement–water ratio; R_{cem} - cement strength at 28 days, MPa.

In the 50s of the 20th century in Norway it was suggested to improve concrete properties, by introducing ultra-fine wastes from metal manufacturing industry – *silica fume* (SF), and since the middle of the 70s wide manufacturing of concrete with SF admixture has been going on. The most effective admixtures among the silica fume are the manufacturing wastes of crystalline silica and ferrosilicone. They consist mostly of amorphous silica (85...95% SiO_2) with an average particle diameter 0.1 μm and dispersivity of 1500...2000 m^2/kg. For SF transportation it is subjected to granulation or briquetting. SF is supplied also as water paste with application of stabilizing admixtures to prevent further gel forming.

The use of SF containing 7…15% of dry superplasticizer is effective.

Increasing of concrete strength due to SF introduction is caused by the complex of special properties of this material and first of all by its raised pozzolanic activity, ability to densify the microstructure of cement stone especially in the transition zone with aggregates. Adding SF to concrete increases drastically the water demand of mixture. Superplasticizer addition is very effective in improving the consistency of the mix and making possible easy placing and compaction of the concrete. The presence of SF increases the concrete strength both by increasing the degree of hydration of the binder and by reduction of concrete porosity, especially capillary porosity. The cement gel, consists mostly of low-alkaline hydrosilicates C-S-H(I) type. The strength of this hydarate is almost twice higher than the strength of the hydrosilicates C-S-H(II), formed in concretes without modifiers. The strength increase of concrete with silica fume containing modifiers and the reduction of water segregation constitute a positive influence which increases the adhesive interaction of the cement stone and aggregates.

Kinetics of concrete hardening with SF at ordinary conditions is characterized by intensive increasing of strength within the period from 1 to 28 days. In the conditions heat-steam curing at effective regimes of treatment the strength is obtained up to 90 % from 28 days one. SF admixture content usually recommended is 5…15% by cement mass. At application of modifiers including SF and superplasticizer, strength increasing is 30...60%. The researchers from different countries estimate contribution of 1 kg SF in concrete strength to be equivalent to the contribution of 2…5 kg of Portland cement. Other ultrafine silicious and aluminosilicious materials can also be effective when composed with superplasticizer.

High values of elastic properties and upper limit of microcracking are a characteristic of high-strength concretes with modifiers. Shrinkage deformations can be almost negligible or have extremely low values. Creep of such concrete is normally significantly lower than the required ones. Concrete is characterized by comparatively low water permeability (water pressure is about 1...2 MPa and more), high sulphate resistance, acid resistance, alkaline corrosion resistance caused by interaction between active silica and aggregate.

High-strength concrete obtained at low W/C, which correspondingly provide low porosity of concrete, has high frost resistance. Application of complex modifiers, including superplasticizer and silica fume admixture, resulting in good frost resistance of concrete F600...F700 and more. Additional effect can be reached by introduction air-entraining surface-active substances (SAS) to modifier composition.

Ultra High Performance Concrete (*Reactive Powder Concrete (RPC))* is a fiber-reinforced, superplasticized, silica fume-cement mixture with very low water-cement ratio (W/C) characterized by the presence of very fine quartz sand (0.15-0.40 mm) instead of ordinary aggregate. The absence of coarse aggregate was considered by the inventors to be a key-aspect for the microstructure and the performance of the RPC in order to reduce the heterogeneity between the cement matrix and the aggregate.

11.5. Heat-Resistant and Decorative Concretes

For the casing of fireboxes, breechings, chimney flues at construction of cogeneration plants, in the elements of protective walls, structures of the atomic power plants and other structures which are heated, *heat-resistant concrete* is applied. Ordinary heavy-weight cement concrete is suitable for making building structures which are tested with the continuous influence of temperature only to 200°C.

The heat-resistant concretes according to the maximum possible temperature of application are divided into classes from 300 to 1600°C.

The hydraulic (Portland cement, blastfurnace cement, alumina cement) and non-hydraulic (caustic magnesite, liquid glass) mineral binders can be used in the heat-resistant concrete. Cement stone acquires the heat-resistant properties due to introduction to it of the different fine milled mineral additives, resistant to the action of high temperatures and binding the free calcium oxide. The fly ash, blastfurnace, fuel slags and other can be used as such additives (Figure 11.7).

Ordinary aggregates, which consist of the quartz, have polymorphic transformations, which can cause the destruction of concrete. Unsteady to the action of high temperatures are also carbonate rocks.

Aggregates from basalt, andesite, tuffs and other outpoured effusive rocks which do not contain free quartz can be used in concrete which work at temperatures to 700°C. Chamotte as a product of fire-clays burning is the most spread as the aggregate for heat-resistant concretes to 1300°C. Magnesite, chromite, corundum and other aggregates are used for the obtaining of concretes with higher heat-resistance.

Decorative concretes and mortars are applied as finishing for walls and other elements of buildings, constructions, and flooring arrangement. Binders that can be applied include ordinary cement as well as, lime, gypsum, caustic magnesite, dolomite etc. The binders could be colored or white. Obtaining cements of light color is possible by mixing of ordinary cement with whitening admixtures (chalk, ground limestone, marble etc.).

Aggregates for concrete of such type are crushed stone and sand, obtained by grinding of decorative rocks. Along with dense materials natural and artificial porous materials are used with softening coefficient more than 0.8. Fine-ground quartz sand, marshalite, stone powder and others are used as fillers.

For obtaining light facing coatings limestone and dolomite aggregates are added to the decorative concrete.

Grain distribution of decorative aggregates mixture depends on the desirable texture of a material. Maximum coarseness of the aggregate is selected depending on the thickness of structure or facing layer, it can vary from 5 to 40 mm. The strength of coarse aggregate for decorative concrete should not be less than 40 MPa, water absorption less than 4% by mass, and its frost resistance should provide frost resistance of concrete.

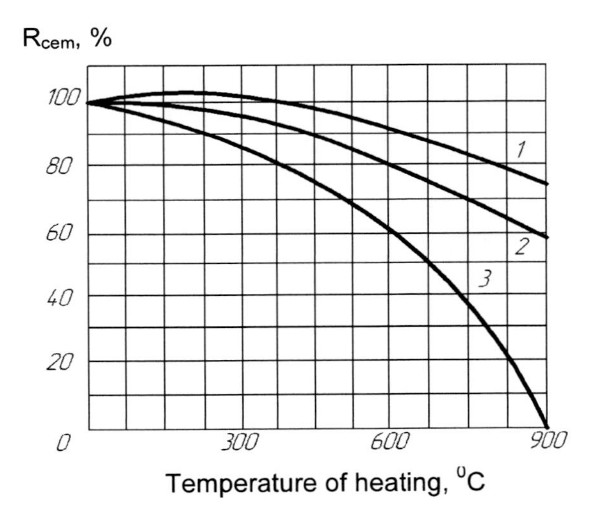

Figure 11.7. Relationship between temperature and cement strength:
1- Portland cement 70% + tripoli powder 30%; 2- Portland cement 70% + pumice 30%; 3- Portland cement.

Pigments for colored cements and decorative concrete should have high alkaline, solar and atmospheric resistance. Pigments content in cement and concrete mixture depends on their tinting power (color value), spreading capacity, and influence on cement strength. Comparatively intensive mineral pigments (chrome oxide, red oxide and ultramarine) are introduced in quantities less than 10% by cement mass. Pigments of medium color value like ochres, mummies, and sienna can be introduced into concrete mixtures in higher quantity and whitening admixtures content can reach up to 25%. It is desirable to add red iron oxide admixture of not more than 5%, because concrete strength decreases significantly if high content of the iron oxide is used. Organic pigments, for example, blue and green phthalocyanic are introduced into the colored concrete and mortars in quantity of 0.15...0.35% by cement-sand mixture mass. Organic pigments can be used both as basic and as admixtures to basic pigments for obtaining more intensive coloring in quantity of 0.05...0.1% by cement mass.

During manufacturing of colored concrete, it is desirable to mix pigments with cement preliminarily. The most uniform homogenization of the mixture is achieved by application of vibration or ball mills. Colored cement paste is obtained by processing of binders and pigments in acoustic or aerohydrodynamic activators. Instead of mixing water at concrete mixtures manufacturing there are also applied colored stabilized suspensions obtained by mixing water with pigments and SAS admixtures.

Colored concrete is known as *mosaic concrete*. It is applied for construction of floors, production of construction elements such as staircases, windowsills and others. Mosaic mixtures are produced by using broken stone able to be polished (marble, granite, basalt etc.). Sizes of broken stone chips can be in the range of 2.5...15 mm.

Approximate compositions of mosaic mixtures are given below:

Components in parts by weight	Strength of mosaic concrete, MPa		
	20	25	35
Portland cement	1	1	1
Water	0.65	0.5	0.4
Sand	2	1.4	1
Broken stone	3.4	2.4	1.7

Along with ordinary concrete, composite mosaic facing, which imitates decorative rocks, is also produced. Facing elements composed of separate pieces of rocks of given colors are placed by pattern with application of cement-sand mortar. Hardened coating is polished.

Polymer-cement concrete and mortars have improved decorative properties and incresed adhesion to different underlayers. They are used for flooring in premises with intensive movement and frequent requirements to be cleaned. Polymer-cement mortars are applied for gluing of facing materials, plasterworks and for facades treatment.

The decorative effect of concrete, especially at prefabrication of structures can be achieved by special treatment for obtaining required texture (Figure 11.8). Required pattern of concrete surface is obtained by rolling, relief matrix etc.

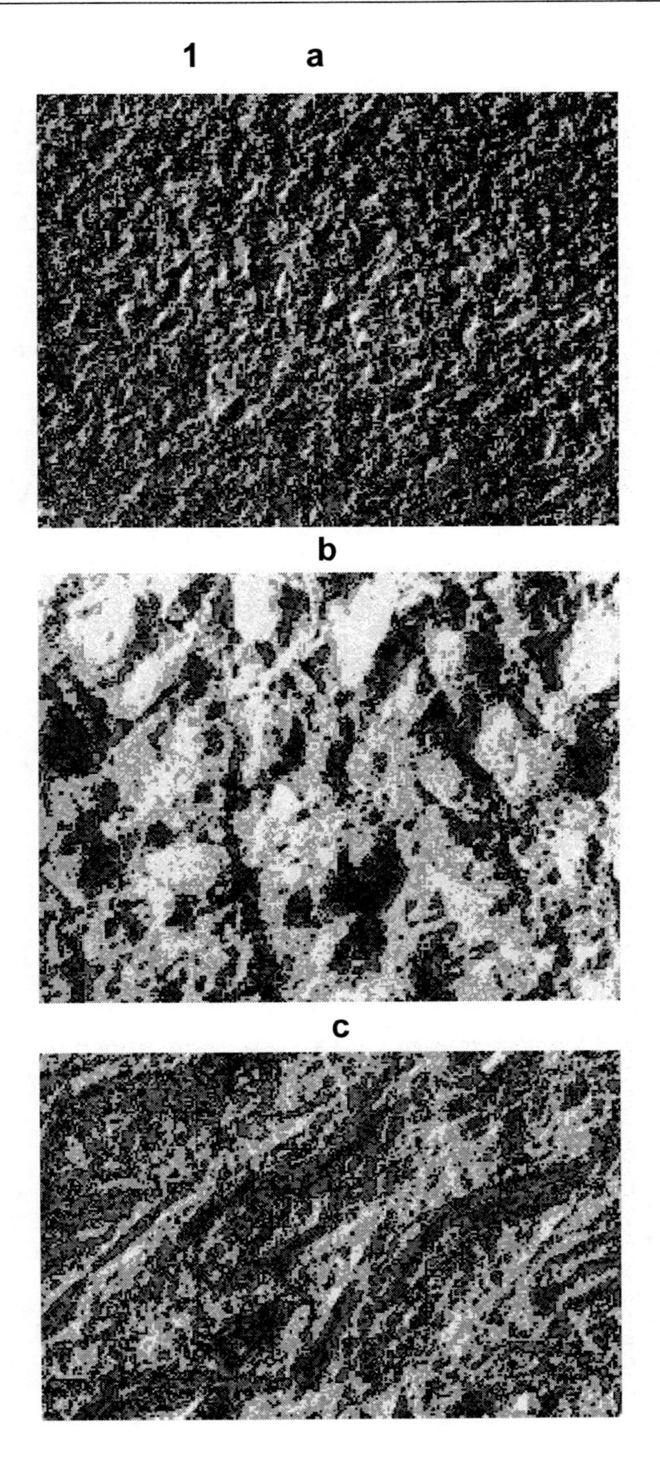

Figure 11.8. 1. Samples of concrete surface treatment by pneumatic or electrical hammers with nozzle: a – bush hammer, b– groove, c- boaster.

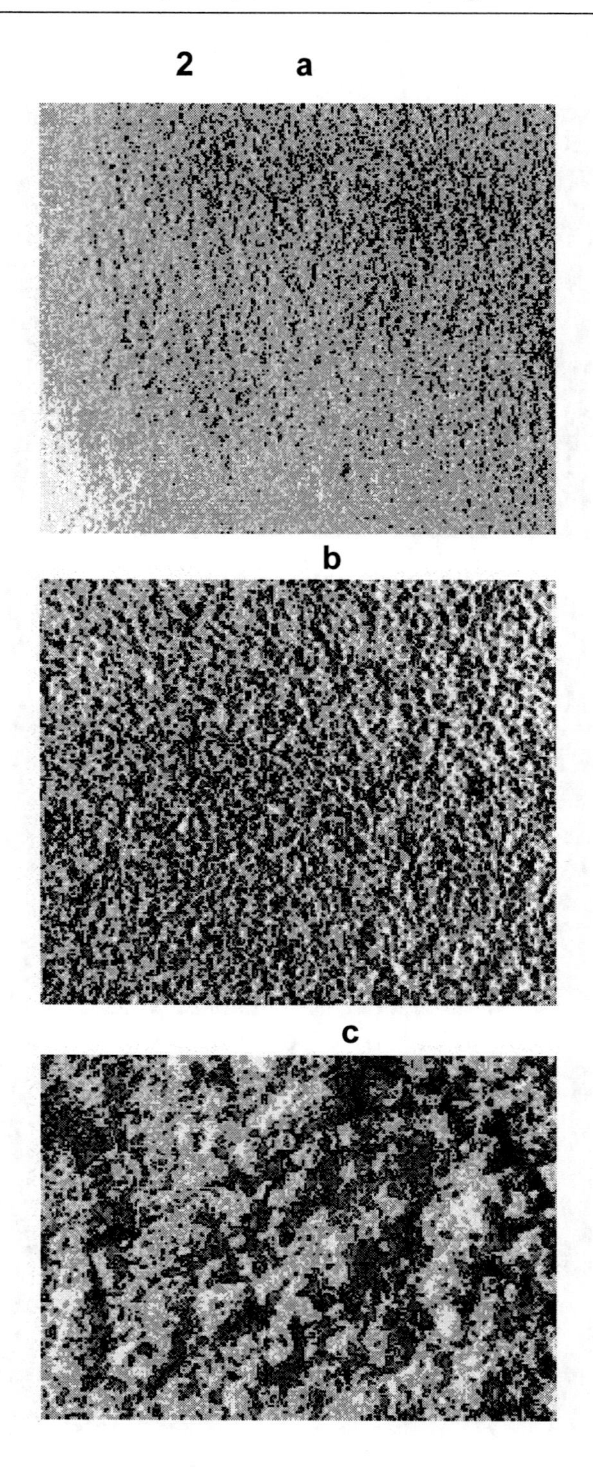

Figure 11.8. 2. Samples of plaster surface treatments: a – scrapping, b – spraying with fine relief; c – spraying with coarse relief.

11.6. Electrotechnical and Radiation Protective Concretes

Electrotechnical concrete is a special concrete with specified electrical properties. Such concrete can be used for production of elements of different dimensions and complicated shape, able to carry mechanical and electrical loads. Electrotechnical concrete is divided into two groups – electrical insulating and electricoconductive.

Electrical insulating concrete has high specific electrical resistance, with low value of dielectric losses, high dielectric permeability, and electrical strength. It is applied for the manufacturing of current-limiting reactors, traverses of power lines and other high-tension insulating structures.

Ordinary cement concrete in certain temperature-moisture conditions conducts electric current. However electrical resistance of concrete is comparatively unstable, at season changes of temperature and humidity it varies by 60…80 times. For obtaining electroconductive concrete with the required electrical properties it is possible to apply conducting admixtures – special carbon black, graphites, fine-ground coke, metal slimes etc. Electroconductive concrete can be applied for supports of high-voltage lines, manufacturing electrical heating elements of buildings. On the basis of metal-saturated electroconductive concrete there are created radioprotective materials, which permit a reduced level of electromagnetic fields inside the buildings. It is well-known that the electromagnetic influence of the environment grows with development of modern electronic technologies and systems.

Electroconductance of clinker minerals can be related to ionic type. Its value largely depends on the orderliness of the ions in the crystalline lattices. Value of *electrical resistivity* (measured in ohm-metres, Ω·m) of Portland cement clinker minerals are given below:

$\beta\text{-}C_2S$	C_3S	C_3A	C_4AF
$1.5 \cdot 10^7$	$1.2 \cdot 10^6$	$6.0 \cdot 10^5$	$6.4 \cdot 10^5$

Electroconductance of cement stone is closely connected to the moisture content (Figure 11.9). Moisture in concrete pores is the electrolyte, containing the ions of soluble products of hydration and hardening of cement and the ions of environment. Electrical resistivity of the moisture in concrete pores (ρ_m) is defined by:

$$\rho_m = \frac{k}{W^n}, \qquad (11.2)$$

where W– concrete humidity; n–parameter, which depends on the temperature; k– coefficient, which depends on the parameters of porous space.

The simplest method of obtaining concrete with improved dielectric properties is reduction of its humidification by density increasing. When cement content and correspondingly volumetric concentration of cement stone in concrete increase up to certain degree then its electrical resistivity reduces. For obtaining concrete with improved dielectric properties application of aggregates with minimal porosity is a compulsory condition.

Reduction in concrete electroconductance under conditions of natural humidity is reached by introduction of waterproofing and polymer admixtures. The most reliable stabilization of electroinsulating properties of concrete is provided by volumetric impregnation it in dryed state by dampproofing materials. There is a widely used petrolatum for such purposes. The technology of concrete impregnation by petrolatum is comparatively simple and can be done both in special chambers under pressure and in ordinary infiltrating bathes. Volume electrical resistivity of concrete (ρ_V), impregnated with petrolatum stabilizes at the level $10^{12}...10^{14}$ $\Omega{\cdot}m$ (Figure 11.10).

For stabilization of electroinsulating properties of cement concrete it is also effective to impregnate the dryed cocnrete with monomers with subsequent polymerization by thermocatalytic or radiative method.

Mechanical loads, causing the microcracks formation and therefore facilitate moistering of the material influence on electrical properties of cocnrete. Generally change of electrical resistance in time of concrete, under the influence of mechanical loads, is described by equation:

$$\rho_c = \rho_0 \tau^{-m}, \tag{11.3}$$

where ρ_0– specific resistivity of concrete at initial state; τ– time; m– parameter, depending on the character of loads and environmental conditions.

The important characteristic of concrete is its *electric strength,* which is characterized by the values of breakdown voltage. All the factors, which facilitate improving mechanical strength, make positive influence on its electric strength. When aggregates are introduced into cement stone, electric strength reduces. The weakest element of concrete structure through which destruction occurs is contact zone between cement stone and aggregates. Minimal values of electric strength are inherent in concrete with application of coarse aggregate.

Electroconductive concrete differs from the ordinary one by the presence of disperse conducting component that should have required electroconductance, temperature stability, ability to resist the oxide processes during manufacturing and heating. Refractory elements made of electroconductive concrete should have stable thermomechanical properties at the temperature up to $100...200^0$C. There are different technologies used for elements and structures manufacturing based on electroconductive concrete, which are based both on plastic molding and static or dynamic pressing of rigid mixtures.

The availability of electroconductive component in concrete determines the possibility of its application for production of electric heating units, domestic, civil, agricultural buildings, antistatic floor, protective screens for electromagnetic radiation protection, X-radiation and gamma radiation, cathode grounding elements etc.

Radiation protective concrete can be used for protecting against radioactive substances and ionizing radiation. The most penetrating among all radiations are gamma-radiation and neutrons. Ability of material to absorb gamma-radiation is proportional to its density. Elements with low atomic mass, e.g., hydrogen, must be present conversely in materials for neutrons flow absorption. Concrete is an effective material for biological protection of nuclear reactors, because they are successfully combined with concrete high density and certain content of hydrogen in chemically bound water at comparatively low cost. For the reduction of the thickness of protective screens during the erection of atomic power stations and at the isotopes-producing enterprises along with ordinary *extra-heavy-weight concrete* with average density from 2500 to 7000 kg/m^3 and hydrated *concrete* with high content of chemically bound water are applied. For this purpose, heavy-weight natural or artificial aggregates: magnetite, hematite or limonite iron ores, barite, metallic scrap, lead fraction etc are applied. Below are shown approximate compositions of some types of extra-heavy-weight concrete in Table 11.3.

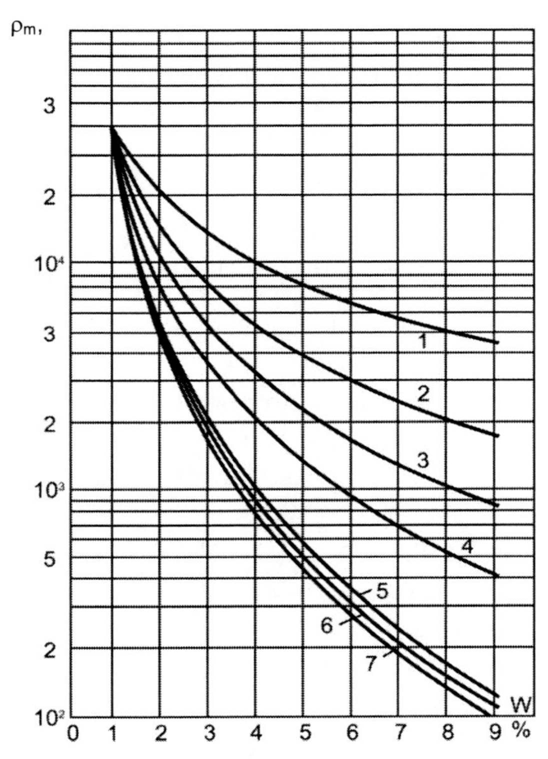

Figure 11.9. Relationship between electrical resistivity of concrete (ultimate compressive strength at 28 days equals 15 MPa) and humidity at different temperature:
1. -10^{0}C; 2. -7^{0}C; 3. -5^{0}C; 4. -3^{0}C; 5. 0^{0}C; 6.+5^{0}C; 7.+10^{0}C.

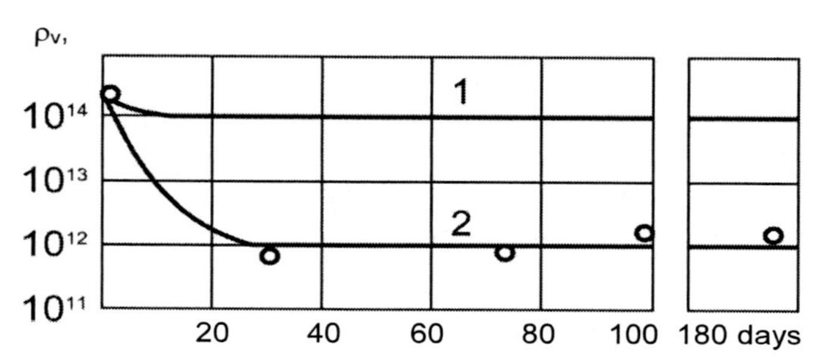

Fig.11.10. Change of volume electrical resistivity of sand concrete impregnated by petrolatum:

1 – dry air conditions;

Figure 11.10. Change of volume electrical resistivity of sand concrete impregnated by petrolatum: 1 – dry air conditions; 2 –raised humidity and temperature.

Table 11.3. Compositions of extra-heavy-weight concrete

Type of concrete	Materials content kg per 1 m^3 of concrete				Average density of concrete, kg/m^3
	Cement	Fine aggregate	Coarse aggregate	Water	
Magnetite	389	1365	1762	184	3700
Hematite	300	1100	2140	195	3735
Barite	395	1352	1800	193	3740
Metallic aggregates	395	2637	2637	170	5839

Limonite, serpentinite, and other materials, with high density and significant content of chemically bound water, are effective for obtaining of hydrated concrete. As binders there are applied Portland cements and blast-furnace cements in extra-heavy-weight and hydrated concrete. It is possible to use cements of special purpose, which provide on hardening, an increased content of hydrosulfuraluminate binding large amount of water. Particularly in hydrated concrete it is permitted to apply alumina and gypsum-alumina cements, which bind more amount of water than ordinary Portland cement. For improving protective properties, admixtures are introduced which increase hydrogen content such as: – carbide, boron, lithium chloride and other admixtures, containing light elements into hydrated concrete

Among the improved protective properties concrete applied for arrangement of protective screens of nuclear reactors should also have other peculiarities, such as raised temperature resistance, high heat conductance and low values of shrinkage, thermal dilatation and creep.

Extra heavy-weight concrete mixtures are disposed to segregation because of significant difference between cement paste and aggregates density. To prevent segregation it is recommended to transport such mixtures in truck mixers, and to apply methods of separate concreting etc.

For neutrons flows of high intensity,(characteristic of some reactors based on fast neutrons), it could be necessary to apply *radiation protective concrete*.

Qualitative changes can result in ionizing radiation to occur in a concrete structure, the character and depth of which depend on the concrete properties, type of initial materials and the radiation dosage. For determination of radiation resistance of materials, consideration must be taken of the density of particles flux, intensity and absorbed radiation dosage. The density of particles or quanta flux is characterized by the ratio of amount of particles, penetrating into the sphere of the elemental volume per unit time to the cross-sectional area of the sphere (quantum per sec on sq. m - $sec^{-1}m^{-2}$). As opposed to the density, radiation intensity is a specific value of energy (W/m). Absorbed radiation dosage is equal to the ratio of absorbed energy to the mass of the environment irradiated (J/kg). For example, the density of neutron flux, emitted by a nuclear reactor reaches $5 \cdot 10^{17} sec^{-1}m^{-2}$, by isotope source – $10...10^{8} sec^{-1}m^{-2}$. Radiation intensity is correspondingly 10^4 and 10^{-6} W/m^2. The radiation dosage, absorbed by the concrete of structures placed behind the enclosure of nuclear reactor after 30 years of its service life is $10^{11}...10^{12}$ J/kg.

Irradiation causes thermal shrinkage of cement stone, which increases when radiation dosage increases. As the temperature rises up to 350^{0}C and its partial dehydration occurs. The deformations due to radiation of cement stone significantly exceed the deformations caused by water evaporation resulting from its heating. Radiation-chemical reactions promote shrinkage as if their possible result is formation of chemically active particles, reacting with each other. During radiation, there occurs radiolysis of chemically bound, adsorption and free water of cement stone, resulting in the release of oxygen and hydrogen in gaseous-like state.

Radiolysis of water is accompanied by a reduction in strength of the cement stone and development of creep deformations.

The permissible intensity of irradiation for concrete depends on the type of coarse aggregate, 10^{19} neutr-cm^{-2}:

Granite, diorite, syenite, gabbro, labradorite	2...5
Dunite, basalt, diabase, andesite, olivinite	5...15
Serpentinite	10...50
Limestone	10
Magnetite, hematite	10...100
Chromite	200

Concrete irradiation is characterized by decreasing density and increased linear dimensions of the aggregates. Minerals transition from crystalline to amorphous state, which is supplemented by expansion deformations, is also possible. As irradiation occurs, the formation and pilling up of different defects of minerals in the crystalline lattices which compose the aggregates, also occurs. The largest changes due to radiation influence are representative for hypogene acid magmatic rocks. When the amorphous phases content in the rock structure increases and crystals sizes reduce, the radiation resistance of rocks becomes higher.

Modulus of elasticity of concrete reduces when irradiation dosage increases because of the structural defects accumulation in the aggregates and cement stone. It was found out that at high dosages of irradiation, compressive strength of concrete reduces by 4 times and the tensile strength reduces by more than 2 times.

For radiation-resistant concrete, it is usual to apply high-silica Portland cements with low content of aluminates and alumoferrites. Under the irradiation conditions application of fine-grained concrete is effective.

Self-Assessment Questions

1. Discuss the peculiarities of the fine-grained and non-sand concretes.
2. What are the peculiarities and methods of manufacturing of cellular concrete?
3. Outline the peculiarities of the hydraulic concrete?
4. What requirements are set for the road concretes?
5. What are the peculiarities of cast concrete?
6. Outline the properties and methods of self-compacting concrete making.
7. Outline the peculiarities of rigid concrete mixtures.
8. What are the peculiarities of vibro-compressive concrete?
9. Explain the technology of high-strength concrete.
10. Outline the properties and technology of high-performance concrete.
11. What are the peculiarities of the heat-resistant concrete?
12. Discuss the technology of the decorative concrete.
13. Outline the properties of electro-technical concretes.
14. What are the peculiarities of the radiation protective concrete?.

Chapter 12

BUILDING MORTARS AND DRY MIXTURES

Building mortars unlike the concrete do not contain the coarse aggregates and can be placed, normally, on porous base without intensive mechanical action. The building mortars, by virtue of their range of structural features and properties, are similar to the fine-grained concrete. Comparatively low strength permits the successful use of local binder materials and various industrial wastes (ash and slag, lime and gypsum containing products etc) in the production of mortars..

Lime, gypsum, cement and composite mortars (cement-lime, cement-clay etc) are mainly applied in construction. Cements are used as a binder for mortars which are exploited at high humidity. Heavy-weight mortars have an average density over 1500 and light-weight less than 1500 kg/m^3. For light-weight mortars natural or artificial porous sands can be used as aggregates. Depending on the purpose, building mortars are divided into: masonry(intended for arrangement of brick, large panel and other walls structures), finishing and special mortars. Among the special mortars grouting, injection, waterproofed, heat-resistant, chemically proofed and others are used in building.

Modern construction is characterized by wide application of dry building mixtures - carefully batched and industrially mixed mortar and concrete mixtures to be prepared by adding water on a building site.

The efficiency of dry mixtures consists in increasing the mechanization level, essential reduction of the period of construction, reduction in labor and industrial expenses, high quality supply.

Dry building mixtures are classified by:

- basic destination (kind of works);
- type of binders in a mixtures;
- modification degree by mix admixtures;
- the most typical property after hardening;
- application conditions.

By destination dry building mixtures can be categorized as masonry, facing, seam, spackling, plastering, gluing, sealing mixtures, etc.; by type of the main binder as - gypsum, anhydrite, lime, magnesia, cement, cement-lime, polymer, etc.; by modification degree - economic, standard, high-quality; by characteristic property after hardening - adhesive, weather-proof, rapid-hardening, water-proof, frost-resistant, high-strength, self leveling, elastic, etc.; by application conditions - manual and machine placing, for porous materials, etc. The same mixture can be often used at production of different types of works.

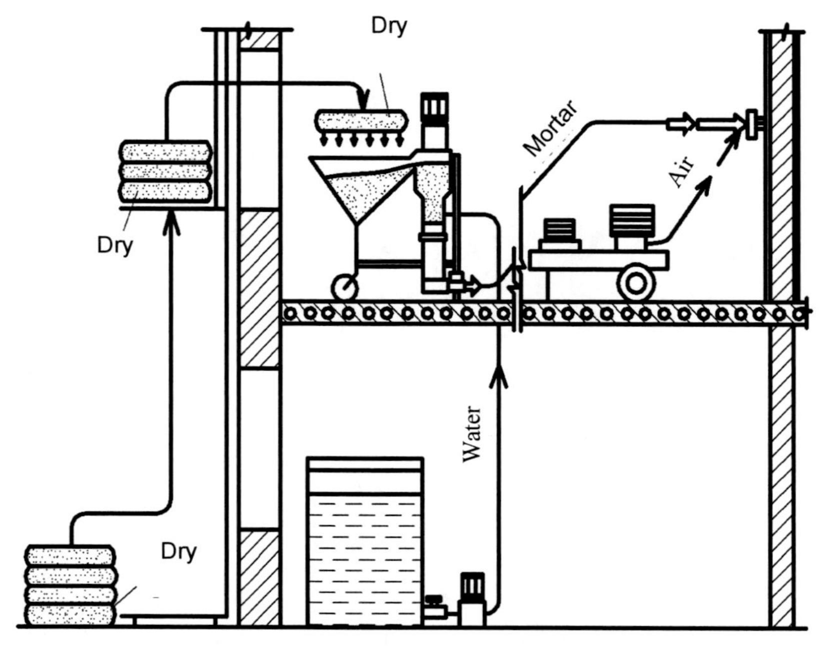

Fig. 12.1. Scheme of plaster works with the use of dry mixtures

The dry mixtures contained in single type of binder, are called simple mixtures; mixed binders - composite mixtures.

12.1. TECHNOLOGY OF MORTARS AND THEIR PROPERTIES

Preparation of mortars. Building mortars are prepared in movable and stationary mortar-mixing plants in the form of ready-mixed mortar mixtures with required plasticity or as dry mixtures which are mixed with water before use (Fig. 12.1). It is effectively to use the hydrophobic cements which provide long-term of their storage at the production of dry mixtures.

The technological process of manufacture of the ready-made mortar mixture includes preparation of the initial materials (screening-out of the gravel impurities, warming up in winter etc), dosage and careful mixing.

The automated mortar-mixing machines with productivity of 18-20 m^3/hour are widely applied for the centralized manufacture of mortars.

The compositions and modes of mortar mixtures making, graphics of mortars delivery on the building objects can be determined with the help of computers on modern plants.

The forced mixing of mortar components can be realized in mortar-mixers with cyclic or continuous mode. The productivity of widespread mortar-mixers is 30-1200 l/hour of the prepared batch. The duration of the mortar mixing should be appropriate for obtaining sufficient homogeneity of the mix. The duration of mixing should be, usually, not less than 1 minute for heavy-weight mortars, for light-weight mortars – 2 minutes.

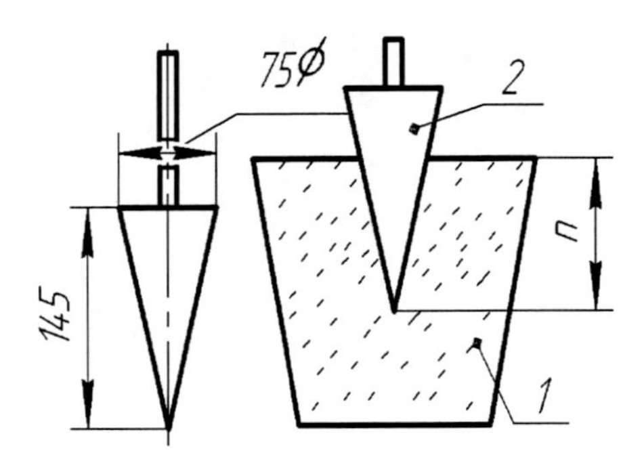

Fig.12.2. Determination of workability of mortars by a cone:

The mortars are manufactured with antifreezing admixtures in winter.

The truck mortar-mixers are used to transport the mortar mixtures. Transporting of mortars to the building site is carried out with the help of the mortar pumps.

Properties of mortars. Workability and water-retaining ability are the basic properties of mortar mixtures.

Workability describes the ease with which the mortar is placed and compacted into the moulds as a homogeneous layer. It can be measured by Abrams cone test or by penetration test with a special cone (Fig. 12.2). It is required by specification to take into account the type of mortar, method of its application, humidity and porosity of base, and temperature of air. For example, the depth of mortar cone penetration specified for placing the ordinary brick is – 9-13 cm.

It is possible to attain the required workability of mortar mixtures without segregation by application of plasticizers and fillers. Two types of plasticizers are used for mortar mixtures: inorganic (lime and clay paste) and organic - surface active substances. The mechanism of action of inorganic plasticizers consists in the creation of adsorption layers on the surface of the particles making them mutually repulsive and the resulting dispersive action on the grains gives a feel of enhanced workability. –Unlike the inorganic plasticizers, the organic plasticizers are added to mortar mixtures in considerably less amount (0.03-0.3% by weight of binder).

Water-retaining capacity prevents segregation of the mortar mixtures. The use of reduced water – binder ratio (due to correct proportioning of composition); addition of fine mineral fillers and plasticizing agents can achieve this purpose.

Mortars are placed on a porous base in most cases. The process of normal cement hardening is impeded by the superfluously intensive sucking of water by base. At the same time fly ash can be used in mortars as an active admixture, microfilling agent and/or plasticizer. Its presence improves plasticity, water-retaining ability and other properties of mortars, substantially allows reducing the cement and lime content. Fine-grain fly ash that is taken away from the last fields of electrostatic collectors is the most effective.

The compressive strength of mortars is in the range of 0.4-30 MPa. Mortars with ultimate strength 0.4-1 MPa are made mainly of lime. The strength of mortars can be determined on cube specimens after 28 days curing. At the time of making the specimens, it is important to take into consideration the real conditions of mortar hardening.

Strength of mortars, as well as concretes, depends, mainly, on binder content and water-binder ratio. When placed on a dense base, the strength of mortars (R_m) can be calculated by the following formula:

$$R_m = 0.25\, R_{cem}(C / W - 0.4), \qquad (12.1)$$

where

R_{cem} - cement strength;

C/W - cement-water ratio.

Regarding the water suction by the porous basis; strength of mortars can be calculated depending on binder content (B):

$$R_m = KR_{cem}(B - 0{,}05) + 4, \qquad (12.2)$$

where K – sand quality coefficient.

The frost resistance of building mortars is determined by the number of cycles of freezing and thawing, which the specimens can withstand without destruction. The mortar grade of the frost resistance depends on operating conditions. For wall laying and external plaster grade is determined, usually, in the range of F10-F50, and at the humidity conditions in the range of F100-F300. The frost resistance of mortars, as well as frost resistance of concrete, is determined by the type of binder and admixtures, water-binder ratio, quality of aggregates and conditions of hardening. It reduces in the presence of leakage, cavities and large voids, conditioned by insufficient workability.

The structure of mortars is modified by admixtures containing organic surface-active substances. Thus water absorption and capillary suction go down, a freeze-resistance and watertightness is enhanced.

Compositions of building mortars are proportioned according to tables or calculation and in both cases are corrected by experimental data on the given initial materials. Typical compositions of the composite masonry mortars (measured in volume parts - Cement: Lime paste: Sand) are presented in Table 12.1.

Table 12.1 Compositions of composite masonry mortars

Ultimate binder strength, MPa	Compressive strength, MPa			
	2.5	5.0	7.5	10.0
40	–	1:0.9:8	1:0.5:5.5	1:0.4:4.5
30	1:1.4:10.5	1:0, 6:6	1:0.3:4	1:0.2:3.5
20	1:0.8:7	1:0.3:4	1:0, 1:2.5	–

12.2. Masonry, finishing and special mortars

Masonry mortars. Masonry mortars are applied during brick and stone laying, arranging of leveler layer at installation of reinforced-concrete structures, grouting columns etc. The composite cement-lime and cement-clay mortars are used mostly for wall construction. Cement mortars without inorganic plasticizing agents are expedient only for structures which are erected below groundwater level. Lime mortars and mortars on different lime-containing binders (lime-slag, lime-ash and other) are applied in mass low-rise building. Thus an air-hardening lime is used in mortars for the laying at a relative humidity in apartments up to 60%, and hydraulic lime-containing binders – in mortars which perform in the moist conditions. Portland cement is widely applied for the different building mortars. The special cements for building mortars, which contain active mineral admixtures or microfillers, are also used.

Along with the coarse, middle and fine sands it is allowed to apply very fine sand with a fineness module of 1-1.5 for production of building mortars. Application of fly ash is rational for the production of effective mortars for ordinary stonework and also for the walls for large-size elements. But it is undesirable to use the mortars with ash additive in winter time due to slow speed of their hardening at a decreased temperature.

If the fly ash is used in the cement mortars than, usually, the cement content goes down on 30-50 kg/m^3 at the simultaneous improvement of workability of mortar mixture. Excess of cement content at a complete replacement of sand by the ash can be removed with adding of lime in little quantity. The admixtures of cement kiln dust and other active microfillers are effective in building mortars in addition to the fly ash.

It is possible to provide masonry mortars with increased early strength in winter by application of chemical admixtures (Table 12.2). Sodium nitrite, potash, urea, calcium nitrite with urea, sodium and calcium chloride, and also some other salts are applied as antifreezing admixtures for making mortars in winter. Thus, account should be taken of the range of limitations: mortars with the chlorides of calcium or sodium can be applied only for the non-reinforced underground structures and also for wall laying of non-residential buildings. It is forbidden to apply any of the types of admixtures on building constructions which are exploited at a temperature higher than 40°C or at humidity over 60%, in aggressive conditions, in the areas of variable water level and under water without special insulation. For constructions on the surface of which the formation of salt efflorescences are not permissible, mortar with admixtures should be preliminary tested for salt efflorescences.

Finishing mortars. Finishing mortars are intended for plastering. Depending on the purpose, they are divided into the mortars for ordinary and decorative, external and internal plasters.

Plaster coverage consists of two or more layers. Bond with the base surface is provided by preparatory layer with thick not more than 5 mm on brick and concrete surfaces and 9 mm on wood. Basic layer 5-7 mm thick is used for the plane surface making. The types of the improved plasters include also covering (decorative) layer not more than 2 mm thick.

Plaster machines can be used for production of plastering mortars, their delivery to the workplace and placing on a processing surface.

Table 12.2 Approximate strength of mortar with admixtures depending on the temperature of hardening

Admixture	Average daily temperature, °C	Amount of admixture into mortar, % by weight of cement	Approximate strength of mortar at frost hardening, % of 28-day strength of normally cured concrete		
			7 days	28 days	90 days
Sodium nitrite	0...-5	5	10	40	55
	-6...-9	8	5	30	40
	-10...-15	10	5	30	40
Potash	0...-5	5	25	60	80
	-6...-15	10	20	50	65
	-16...-30	15	10	35	50
Sodium nitrite + potash	0...-5	2.5+2.5	20	55	75
	-6...-15	5+5	15	40	60
	-16...-30	6+6	5	35	45
Sodium chloride + calcium chloride	0...-5	3+0	35	80	100
	-6...-10	3.5+1.5	25	45	70
	-11...-15	3+4.5	15	25	50

All the types of plaster mortars should have specified flowability, ability to not disintegrate during production and also to provide the required strength and good adhesion to the base.

The required flowability of the mortar mixture depends on the location of the layer in plaster coverage. The most plastic mortar which fills all of surface voids is applied for a preparatory layer. The basic layers are produced from comparative viscid mortar which forms the required thickness of plaster. Mortar for decorative or covering layer has the consistency, which allows smoothing its surface. The maximum sand fineness also depends on the type of plaster layer. For preparatory layer, this is equal to 2.5 mm, for the covering layer - 1.2 mm.

The mortars for ordinary plasters are categorized into: cement, lime, cement-lime, lime-gypsum and gypsum.

Cement mortars are used for external plasters which are subjected to systematic moistening and internal plasters - in apartments with relative humidity of air over 60%.

Waterproofing admixtures (for example organic silicon liquids) can be used to increase water resistance. The following ratio between cement and sand is recommended:- for preparatory layer 1:2.5-4, basic layer 1:2-3, covering (decorative) layer 1:1 -1.5.

Cement-lime mortars are used for plastering of both facades of buildings and internal apartments. Application of lime sharply increases the plasticity of mortars. The content of lime paste depends on the function of the layer. The volume content of lime can be for preparatory layer, usually, 0.3-0.5; for basic layer 0.7- 1; for covering (decorative) layer 1-1.5. The cement-lime mortars can be manufactured with two methods: by dry sand-cement mixture which is moistened by lime milk to the required workability; or by adding of the cement to sand-lime mortar.

The mortars, based on the air-hardening lime are used for plastering of surfaces in apartments with relative humidity of air up to 60%. They can be used for external plasters which are not subjected to systematic moistening. It is better to add lime in the form of lime milk into the mortar. Slow hardening is the main lack of lime mortars. Gypsum plaster can be added to accelerate the hardening of lime mortars. The lime-gypsum mortars are the most appropriate for plastering wooden surfaces into apartments. The hydraulic admixtures: tripoli powder, diatomit, slags, fly ash and others are added into mortars to increase their water-resistance.

Gypsum mortars are applied in areas, mainly dry by climate, for plastering of wooden and gypsum walls. Dry gypsum plaster mixtures are intended for the mechanized placing of plaster coverage. They are made in factory conditions with the purpose of the centralized supply to building site.

Decorative mortars are applied to the facing of external and internal walls. The ordinary and coloured Portland cements are used as binder in decorative mortars. The coloring ingredients and fillers from natural or ground up materials also can give the color to mortars. The mica or ground up glass can be added to impart the surface brilliance of decorative layer.

The lime-sand, terrazite and stone-like decorative mortars are used for decorative plasters.

Lime - sand decorative mortars are the most economic. Quartz, marble or limestone sand is added to their composition depending on the required color. The different methods of placing and finishing the covering (decorative) layer can be used.

The facades of buildings are plastered with coloured lime - sand mixtures. Mortars with enlarged cement content are used for plastering of the surfaces of dense concrete and structures, which are subject to intensive moistening.

Terrazite plasters, besides binder and quartz sand, include stone chips of different fineness and certain amount of mica and coloring agents if it is required. They are usually manufactured from pre-proportioned dry mixtures. Terrazite plaster can be treated with the help of sand blasters, brushes and other tools for obtaining a finished surface, which imitates sandstone, tuff or other rocks. The terrazite finish depends on maximal fineness of filler grains. Fine-grained fillers in terrazite mixtures have grain size of 0.15-2 mm, medium-grain size of 2-4 mm, and coarse-grains of 4-6 mm in size. Coarse-grained terrazite for facade finishing can be used for plastering of socles and ground floors, middle-grained for wall surfaces, and fine-grained for the cornices.

Stone-like mortars imitate different rocks, depending on the type of stone chips and method of surface treatment.

Stone-like plasters are applied mainly for finishing of facades and socles. Portland cement is the main binder for these mortars. Lime paste is added in an amount of 10-20% to provide the required plasticity and colour (brightening) of the mixture. The grain composition of the stone chips has a great importance and can take into account the method of surface treatment of the hardened mortar and the required finish.

Polymer-cement mortars have improved decorative properties, reduced weatherability, high tensile strength, better adhesion to different bases. They are made from mixtures which consists of complex binders (Portland cement and the plasticized polyvinyl acetate dispersion or latex), fillers, aggregates, coloring agents and water.

Polymer-cement mortars can be applied to the construction of floors in apartments with intensive movement of people and special requirements for the cleanness (public places and buildings, shops, stores etc).

Polymer-cement mortars also can be applied for gluing of different finishing materials, finishing of facades etc.

Special mortars. Special mortars include the mortars for waterproofing, injection, tamping, acid-proof, heat-resistant, heat-protection, acoustic, X-ray plasters and others.

Waterproofing mortars are applied for creation of waterproofing coverages of reinforced-concrete pipes and other structures, for dampproofing of tubings seams, etc. The selection of cements for waterproofing mortars is a very important activity. As the cement strength incease, the water impermeability of mortars improves. The sulphate resistant cements can be applied for the production of waterproofing mortars, intended for service in conditions of aggressive waters. The main influence on water permeability of mortars is the water-cement ratio because the system of voids and capillaries in the mortar,which influence permeability, increase as the water-cement ration increases. As a rule, for waterproofing mortars, water-cement ratio in the range of 0.3-0.5 is recommended.

The water impermeability of mortars can be improved by the use of admixtures like: micro-fillers, plasticizers, polymer etc. Usually, inexpensive materials such as stone powder, fly ash and others can be used as microfillers. Microfillers must compact and create a fine-grained structure for the mortars matrix. Chlorides of iron, aluminium and calcium are widely used for mudding. Liquid glass, sodium aluminate, calcium nitrate and other admixtures can also be used for waterproofing mortars.

Plasters, which are placed by spraying method, perform reliably well. They are used for coverage of pool walls, pipelines, tunnels, basements which are subjected to the action of ground water.

Injection mortars are used for filling of channels of prestressed reinforced concrete structures, operations for provision of reinforcement in concrete and protecting it against corrosion. The compressive strength of injection mortars should be not less than 30 MPa.

Grouting mortars are used for closing (tamping) of water-bearing cracks and voids in rocks and structures. The method of cement mortar injection in cracks and voids of rocks and structures is called cementation. This method is widely used, for example, in hydraulic construction for soil compaction.

Also, cementation is successfully applied for repairing the defects and damages in concrete structures. The complex requirements for the grouting mortars are achieved by the application of plasticizers, special cements and cement-clay mixtures processed by the special reagents, using of the high-speed turbulent mixers etc. Oil-well Portland cement, Pozzolanic cement, Blast furnace cement are applied in the conditions of pressure and aggressive waters.

The plasticized and hydrophobic cements are the effective types of binders for grouting mortars. Water-cement ratio in such mortars is dependent on the rock fissuring.

Acid-proof mortars are produced from a mixture of acid-proof aggregates and dispersible fillers with the fluorsilicate sodium that are mixed with liquid glass.

Quartz sand and also sands which are produced by crushing granite, andesite and other rocks are used as the aggregates for acid-proof mortars. Fine-grain quartz sand, marshalite and other siliceous powder materials can be used as fillers. Fluorsilicate sodium content which is needed for the accelerated hardening of liquid glass is 5-8% by it weight.

The active mineral admixtures – tripoli powder, diatomite and others can be added to increase the waterproofness of acid-proof mortars. Polymer admixtures in composition of acid-proof mortars increase their density and the impermeability.

Mortars with light aggregates are applied for thermal-insulation plasters.

Sound absorbing light mortars with an average density of 600-1200 kg/m^3 are used for acoustic plasters. Sand grains having a fineness between 3 to 5 mm from porous materials like: pumice, slags, expanded pearlite, claydite and others are used as aggregates in such mortars.

Heavy-weight mortars with an average density of 2200 kg/m^3 and higher are applied for plastering of x-ray cabinets and apartments in which the work with radioactive isotopes are carried out. Portland cement or Blast furnace cement is applied as the binder materials in such mortars, as aggregates are used barite or other heavy-weight rocks.

12.3 DRY BUILDING MIXTURES

Initial materials and admixtures. The type of the binders defines the conditions of hardening and operational properties of dry mixture, including humidity conditions of work, frost resistance, resistance to cyclic moistening and drying. The hydraulic binders are applied in the mixtures used for dry and damp conditions; air-hardened binders used for dry conditions.

Depending on the fineness of aggregates and fillers, dry mixtures can be categorized as concrete, mortars and mixtures for thin-layer technologies with filler fineness no more than 1.25 mm(also called disperse). Disperse mixtures are also divided into coarse, small and fine-dispersed.

For manufacturing of dry mixtures in most cases ordinary and white Portland cements are used. Application of Portland cement without mineral admixtures that provides the required stability of chemical-mineralogical composition and properties of the binders is the most preferable. To provide accelerated hardening of mixtures is required to apply high dispersion cements. White cement is used for spackling manufacturing, decorative plasters and special kinds of glue.

For rapid-hardening mixtures, which are used for repair works; and also for expanding compositions, alumina cement is used as binder.

The gypsum binders are applied as the admixtures in cement mixtures and as the basic binders in spackling and other finishing compositions.

Hydrated lime can be added in cement mixtures to increase plasticity and for regulation of some other properties. Humidity of hydrated lime should be no more than 2 %, also completeness of slaking of lime must be guaranteed.

Polymer binders in dry mixtures are applied in the powdered state.

Quartz sand, chalk, limestone, microsilica, kaolin, dolomite, etc are applied as fillers for dry mixtures.

The widest nomenclature of dry mixtures is made with application of quartz sand. The content in sand clay (more than 0.5-1.5 %) and powder-like impurities, which sharply increase water requirement and shrinkage, reduce strength and adhesion with the base, decrease frost resistance is undesirable. At the same time some increase in sand of the fine-grained fractions improves the water-retaining ability of mixtures and their density.

The stone crumb (granite, marble, etc.) with fineness no more than 2.5 mm can be applied in the protectively-finishing compositions along with quartz sand.

The type of filler depends on the use of the dry mixture. In the dry mixtures used as high water resistance coverings the bentonitic clay is applied as filler. In dry mixtures used as organic fibrous fillers: polypropylene, acrylonitrile, cellulose are used. They have positive influence on strength of materials, increase their crack resistance and reduce shrinkage.

The humidity of dry mixtures should not exceed 1% by weight.

Mortars are obtained by mixing the dry mixtures and required volume of water specified in the standard documentation on a dry mixture or in the instruction of it application. During manufacturing and application of mortars the harmful substances should not escaped in environment in the amounts exceeding maximum permissible concentration.

Regulation of technological and operational properties of dry mixtures can be realized by application of various chemical admixtures. For the mixtures placed by a thin layer on porous surfaces, it is important to provide required water-retaining ability. Fast absorption of water by the base leads not only to loss of the plasticity of the material, but also to insufficient hydration of the binders, decrease of strength, adhesion and frost resistance.

Modification of dry mixtures for improvement of their water-retaining ability, plasticity suitability, is by application of admixtures based on cellulose esters - hydroxyethyl cellulose and hydroxypropyl cellulose (0.05-0.5 % by weight).

Methylcellulose has wide application. In the form of cellulose methyl ether, with a density of 1.29-1.31 g/cm^3, and bulk weight of 0.3-0.5 g/cm^3, it is very soluble in water. On swelling in water the volume of methylcellulose increases by many times. Macromolecules of methylcellulose, even in diluted water solutions, are inclined to aggregation. Water solutions are stable at pH values between 2 to 12. Methylcellulose has good surface-active properties in water solutions. However, on heating to temperatures between 35- 36°C it is observed to undergo a gelling process. At lower than the gelling temperature, the gel collapses. Methylcellulose is capable to keep the properties after dissolution in water, repeated drying and dispersion.

To increase the water-retaining ability of dry mixtures, bentonitic clays(-natural aluminum silicates) can be added, which are characterized by high dispersion, swelling capacity, ion-exchange ability. These clays consist of montmorillonite minerals with the general formula $Al_2O_3{\cdot}4SiO_2{\cdot}nH_2O$. Bentonitic clay, modified by the organic compounds is recommended to use for dry mixtures manufacture. They make less considerable affect on the water-retaining ability of the building mortars, than methylcellulose. Montmorillonite clay, microsilica, etc are used as stiffeners, along with bentonitic clay.

Ethers from starch (5-20 % by weight of cellulose ethers) can often be added into dry mixtures as stiffener alongside cellulose ethers.

Higher level of modification is achieved by adding redispersible polymer powders in the composition of dry mixtures, which can represent itself as the admixture or the polymer binder. Redispersible powders are obtained by drying of water dispersions of polymers usually by a spray drying method. Redispersible polymer powders (RPP) are capable at mixing to create water dispersions similar to the initial material.Mineral binders do not always work well at bending and tensile loadings, have insufficient adhesion especially to such materials as glazed ceramic, plastics, metals, expanded polystyrene, etc. A dosage of redispersible polymer powders of 0.5-5% considerably improves adaptability to mixture manufacture, adhesion to the base, decrease water absorption (Fig. 12.3, 12.4), raise bending strength, water resistance and frost resistance. At a dosage 5-7 % redispersible powders start to work as a polymer binder. The materials modified by them start to have elastic properties;

higher resistance to load deformation; and better resistance against abrasion. The general characteristics of RPP is presented in Table 12.3.

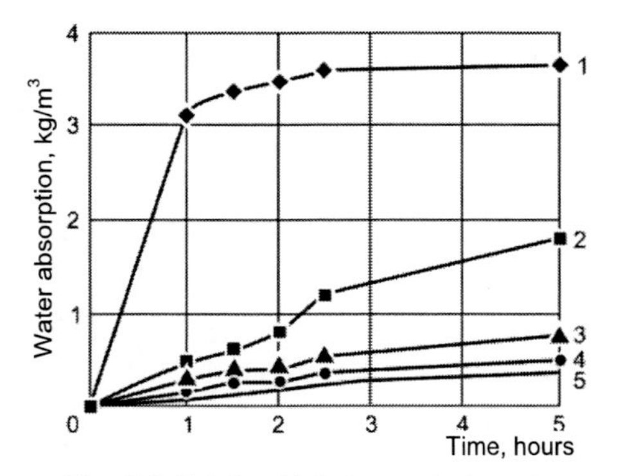

Fig. 12.3. Relationship between plaster water absorption and Winnapas powder content: 1 – powder content 0.5%; 2 – 1%; 3-1.5%; 4-2%; 5-5%

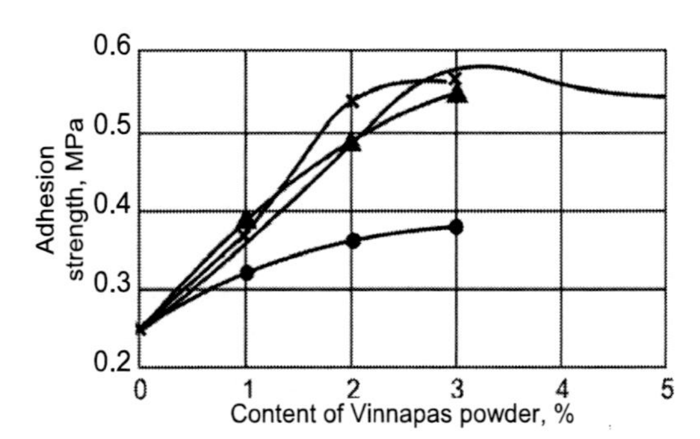

Fig.12.4. Relationship between decorative plasters adhesive strength to a concrete surface and content of various trade marks of Vinnapas powder

A well-known typical redispersible powder is the thermoplastic polymer known as Vinnapas (commercial name). In combination with mineral binders, Vinnapas redispersible powders are successfully used for dry mixture manufacture. The basic fields of application are glues for tiles and thermal protection systems; self levelling compositions; repair compositions for concrete; all kinds of plasters; dispersive and powder cement-lime paints; waterproofing compositions; filling paste for seams, and also spackling compositions.

Table 12.3 RPP general characteristic

Characteristic	Value
Chemical nature(homo-, copolymers)	Vinyl acetate, vinyl acetate – ethylene, vinyl acetate – acrylate, acrylate, butyl acrylate – styrene, styrene -acrylate
Bulk density, kg/m^3	140-500
Anticoagulant content, %	10-15
Content of anticaking agents, %	5-15
Skinning effect minimal temperature, °C	0-25
Average particle size, μm	50-250
Ash content, %	0.5-13
Characteristic of redispersible product:	
Solid substance content, %	2-50
Viscosity (by Brookfield), m·Pa·s	500-5500
Average particle size, μm	0.01-0.5
pH of 30% dispersion	4-12

Unlike liquid and paste-like products, the mixtures modified by polymers are resistant to the influence of low temperatures and bacterial pollution.

A number of chemical admixtures can be added into dry mixtures to increase thixotropy and diluting effect; regulate setting time; accelerate hardening, etc.

The most effective diluents are super- and hyper plasticizers - products of polycondensation of naphthalene and melamine formaldehyde, polycarboxylate, polyethylene glycol. Some effective types of super- and hyper plasticizers are shown in Table 12.4. Super plasticizers can be added to composition of dry mixtures in quantities ranging from 0.05-1.5% by weight of binder to increase of mortars fluidity; decrease of water content and consequently, increase the final strength, density and uniformity of the hardened mortar. Such admixtures are particularly recommended in self levelling mixtures as plasticizers, dispersants and shrinkproof agents.

Admixtures (citric acid, etc.) which delay the setting time of gypsum binders are applied. They enable an increase in the operating time of gypsum proportional to their dosage (0.01-0.08 % by weight). Effective hardening-accelerating admixtures are calcium formates and their modifications. Their content is 1-4 % by weight of the binder.

Dispersants, porophores, defoaming and water-repellent agents, etc. can be added into compositions of special chemical admixtures.

Principal types of dry mixtures. Gluing dry mixtures are widely applied for plaster works, surfacing preparations, floor making and filling of seams.

Gluing mixtures are the polymer-mineral systems containing mineral binders, fillers and polymer admixtures, for regulating physical-mechanical and rheological properties of mortars.

Table 12.4 Types of super- and hyper plasticizers

Type and mark	Chemical base	Field of application
Melment F 10	Melamine formaldehyde	Superplasticizer and diluent for dry Portland cement and gypsum mixtures
Melment F 156	Melamine formaldehyde	Superplasticizer and diluent for dry Portland cement and gypsum mixtures
Merflux 1641 F	Polycarboxylate	High effective superplasticizer, diluent and shrinkproof agent for dry cement mixtures
Merflux P 100 F	Polycarboxylate	High effective superplasticizer, diluent and shrinkproof agent in self levelling floor compositions
Peramin SMF 10, SMF 30	Melamine formaldehyde	Superplasticizer and shrinkproof agent for mortars
Peramin-Conpac 149 S	Polycarboxylate	Superplasticizer and shrinkproof agent for dry mixtures
Bevaloid 36	Naphthalene formaldehyde	Superplasticizer and shrinkproof agent for dry mixtures and other building materials

Gluing mixtures of all groups should:– provide strength of jointing of facing materials with the various bases – concrete, stone, cement-sand, gypsum plaster, foam concrete and wood-particle plates – not less than 0.5 MPa. They should also keep this indicator from the influence of various operational factors, such as – negative temperatures, the water environment, static and dynamic loadings. They should:

– have long term serviceability of mortar mixtures – not less than 60 min;

– have high fixing ability (resistance to displacement of the tile put on the mortar mixture) – no more than 0.5 mm;

– have high adhesion to pasted facing material and the bases;

– provide duration of correcting of tiles position of not less 10-15 min.

Basic compositions of gluing mixtures and the basic indicators of quality of trademark "Ceresit" mixtures are shown in Tables 12.5 and 12.6.

Mixtures for plaster works represent the polymer mineral systems containing mineral binders, aggregates, fillers and various admixtures, raising vapor permeability and reducing shrinkage of mortars.

Depending on the type of main binder, the dry mixtures are divided into cement, cement-lime and gypsum.

Table 12.5 Basic compositions of gluing mixtures mark "Ceresit"

Component	The content of components, %, depending on a composition variant		
	1	2	3
Portland cement	25-40	30-45	25-40
Alumina cement	-	-	0-3
Quartz sand fineness to 0.63 mm	25-75	55-70	40-70
Limestone powder	0-50	-	-
Lime hydrated	-	0-3.0	0-3.0
Cellulose ester	0.25-40	0.25-0.6	0.25-0.6
Redispersible powder	-	0.7-3.0	4.0-8.0
Defoaming agent	-	0-0.2	0.3-0.5
Cellulose filaments	-	0-0.3	0-1.0
Hardening-accelerating admixture	-	0-1.0	-
Starch ether	-	-	0-0.1
Superplasticizer	-	0.7	0-1.0

Depending on filler fraction, they can be categorized as: fine-grained (not more 0.315mm), medium-grained (0.315-1.2 mm) and coarse-grained (1.2-2.5 mm and more).

Cement mixtures should have adhesion with the base of not less than 0.5 MPa, and vapor permeability of not less than 0.1 mg / (m·h·Pa).

Upon termination of the gypsum hardening process, there should be no cracks formed on the surface.

Cement mixtures of all groups should:

- ensure high performance of external works by providing high strength covers against the influence of moisture and various climatic factors, including the negative and high temperatures. They should also have a factor of water absorption of no more than 0.2 kg/(m^2·h);

- be easily placed and leveled.

The basic characteristics of cement mixtures of different groups are presented in Table 12.7.

Gypsum mixtures are intended for finishing of internal surfaces of premises. Into their composition can be added: lime, fillers, polymer modifying components, retarders of set and other admixtures.

Gypsum mixtures should:

– not collapse due to periodic short-term humidifying effects;

Table 12.6 The basic quality characteristics of gluing mixtures (trademark "Ceresit")

Properties and features of mortars	Facing tile, destination					
	Ceramic, inside and outside of structures		Marble, inside and outside of structures	For deformated base	Large-dimensional for floors	Natural stone
Content of water for 25 kg of a dry mix (liters)	6	6.5-7	6.75-7.0	8.75	5-6	7-8
Serviceability	for 2 h	for 30 min	for 30 min	for 2 h	for 2 h	for 2 h
Position correcting time	for 10 min	for 10 min	for 10 min	more than 10 min	nearly 15 min	nearly 15 min
Slipping	less than 0.5 mm					
Filling of seams	in 24 h	in 3 h	in 24 h	in 24 h	in 24 h	in 24 h
Adhesion of a mortar to the base, MPa:						
Air-dry hardening	not less than 0.8	not less 0.8 than	not less than 0.8	not less than 1.6	not less than 1.0	not less than 1.1
Water hardening	the same			1.3	the same	Not less than 1.1
Freezing and thawing conditions, 50 cycles	the same			not less than 1.0		Not less than 0.7
Operation temperature, °C	from –50 to +70					

– provide possibility of regulating the setting process over a wide time range.

The basic characteristics of gypsum mixtures are:

Serviceability, min	Not less than 30
Thickness of a layer, mm	5 - 30
Compressive strength, MPa	Not less than 2.5
Bending strength, MPa	Not less than 0.5
Adhesive strength, MPa	Not less than 0.5

Mixtures of polymer-cement and polymer decorative plasters have a wide application in painting and decorating. They are intended for external and internal decorative finishing of surfaces of buildings built with concrete and brick bases.

Table 12.7 The basic physical and mechanical characteristics of cement mixtures for plaster works

Indicators	Mixture group		
	1	2	3
Color	Should correspond to the standard		Is not regulated
The suitability, min., not less than	60	60	60
Thickness of a layer, mm	1.5-5	1.5-5	3-30
Compressive strength, MPa, not less	5	10	10
Bending tensile strength, MPa, not less	1.5	2.5	2.5
Bending tensile strength, MPa, not less	Is not regulated	50	50

The compositions and specified properties of polymer-cement decorative plasters obtained from dry mixtures are presented in Table 12.8 for example.

Polymer-cement decorative plasters are classified as:

- A – coarse-disperse, for internal works, without water protective properties, possesses low resistibility to abrasion and adhesion to the bases;
- B – coarse-disperse, for external works, it is recommended for finishing of constructions from a brick and light-weight concrete;
- C - intended for finishing of building structures from the high density materials (heavy-weight concrete, asbestos-cement sheets, etc.);
- D - improved, placing on glazed tile is possible;
- E - fine-dispersed, with water protective properties, for internal and external works.

Table 12.8 The basic technical characteristics of polymer-cement decorative plasters Ceresit

Parameter	CT 35	CT 36	CT 137
	Mixture with mineral fillers and modifiers		
Content of water for mortar manufacturing for 25 kg of dry mixture, liters	5.0-5.6	5.0-5.6	Grain 1.5 mm-5.0-5.6 Grain 2.5 mm-4.3-4.7
Serviceability of mortar, min.	till 60		till 90

Table 12.8 (Continued)				
Temperature of the base at plaster application, °C	from+5 till +35		from +5 till +30	
Adhesion to all bases, MPa, not less	0.5			
Mortar amount, kg/m^2	Approximately 3.2	From 2.0 to 5.0 depending on the tool and a way of formation of invoices	2.0 4.2	Grain 1.5 mm- Grain 2.5 mm-

Mixes for surface preparation represent the systems containing mineral binders, fillers and various admixtures, including, reducing sticking cement mixtures to the tool.

They should:

– have adhesion to the base of not less than 0.5 MPa;

– Vapor permeability should be not less than 0.1 mg / (m·h·Pa);

– be frost-resistant (for external works) and withstand not less than 50 cycles of freezing and thawing;

– have firmness to crack formation and minimum shrinkage (no more than 0.2%);

– be ground and painted, including paints on organic solvents;

– not flow down from vertical surfaces;

– be waterproof (on the cement base).

For surface finish preparation for wall-paper, paints, decorative coverings, spackling and priming mixtures are applied. Unlike plaster they are applied as a thin (0.2-3mm) layer. Compositions for surface finish preparation should provide adhesion to the bases –of not less than 0.5 MPa, vapor permeability –of not lower than 0.1 mg/(m·h·Pa) and high crack resistance.

After hardening of mixtures covering should be ground and painted well, if required to have sufficient water- and frost resistance, to attain the required colors and shades.

The basic physical-mechanical characteristics of mixtures for surface finish preparation and some of their compositions are shown in Tables 12.9 and 12.10.

Table 12.9 Composition of some spackling mixtures

Component	Componential composition of mixtures, mass. %			
	Mixtures on the basis of cement		Mixtures on the basis of gypsum	
	General purpose spackling mixtures	Finishing spackling mixtures	General purpose spackling mixtures	Spackling mixtures for filling cracks and joints
Portland cement/ alumina cement (proportion depending on quality of cements and required rate of setting)	38-42	-	-	-
White cement	-	35-40	-	-

Gypsum hemihydrate	-	-	30-45	70-88
Hydrated lime	0-10	8-12	-	2-5
Quartz sand fineness to 0.4mm	44-48	40-45	-	-
Crushed mica, chalk	0-10	5-10	50-65	15-25
Crushed quartz	0-10	5-10	-	-
Polymer fibres	0-1.5	0-1.5	-	-
Polymer redispersible powder	1-4	3-5	1-3	1-3
Cellulose ester	0.15-0.5	0.15-0.5	0.5-0.8	0.5-1.0
Setting retarder	-	-	0.01-0.05	-

Dry mixtures are widely applied to obtaining the mortars intended for the making of floor coverings.

Mixtures for the making of floors are polymer mineral compositions, containing mineral binder, aggregates, fillers and various admixtures. They help to increase crack resistance, abrasive resistance, frost resistance and water resistance of mortars.

Mixtures for the making of floors should:

– harden quickly (maintain small loads through some time);

– have high abrasive resistance;

– be frost-resistant (not less than 75 cycles of freezing and thawing) in non-heated premises and outside premises;

– provide required firmness of facings, layers and coverings with high adhesion between layers.

Floor coverings are applied when it is required to level a surface, to hide communication pipelines, to distribute loadings on heat - and to a sound-proof layer, to create required biases.

Table 12.10 The basic technical characteristics of spackling mixtures Ceresit

Indicator	CT 29	CT 225
	A cement-lime mixture with mineral filler and organic admixtures	Cement with finedisperse fillers and a complex of modifying admixtures
Content of water for manufacturing cement mixtures for 1 kg of a dry mixture, liters	0.24	0.35- 0.37
Serviceability of mortar, min.	To 60	To 60
Temperature of the base for mortar application, ^{0}C	From +5 to +35	
Thickness of layer, mm	From 2 to 20	To 3
Adhesion to all bases, MPa, not less	0.5	
Amount of mortar for a layer thickness of 1 mm, kg/m^2	1.8	

Coverings applied mainly consist of the mixtures on the basis of cement, gypsum and the anhydrite, forming self-leveling floors. Characteristic compositions of such mixtures are presented in Table 12.11.

For filling seams for tile facings of walls and floors, special mixtures are applied which are carried out simultaneously with decorative and protective functions.

Mixes for filling seams are represented by polymer-mineral systems containing mineral binder, filler and reinforcing mineral binder admixtures. The admixtures also increase the adhesive and deformative characteristics of mortars (the elastic modulus) and reduce shrinkage.

Mixes of all groups for filling seams should:

- give the required decorative properties to surfaces together with facing materials;
- protect constructions from moisture penetration;
- possess resistance to atmospheric, shrinkable or mechanical influences;
- harden quickly;
- have frost resistance of not less than 50 cycles of freezing and thawing;
- be characterized by wearability no more 0.7 g/cm^2 and shrinkage no more than 2 mm/m;
- have strength of adhesion to the base of not less than 0.5 MPa.

Table 12.11 Mixtures for floors

Component	Cement covering for primary alignment of floors (from 10 mm)	***Self levelling*** floors (to 5 mm) with casein	***Self levelling*** floors (to 5 mm) without casein	***Self levelling*** floors (to 10 mm) on the basis of gypsum	Spreading anhydrite covering
Mineral components, %					
Portland cement	30-40	35-45	35-45	1-4	1-8
Alumina cement	-	3-7	3-7	-	-
Hydrated lime	0-2	1-4	1-4	1-2	0-3
Anhydrite	-	-	-	-	40-60
Gypsum hemihydrate	-	0-3	0-3	45-55	0-10
Limestone powder (40-100 μm)	0-10	5-10	5-10	20-25	-
Quartz sand (0.1-0.4 mm)	15-25	38-42	38-42	15-30	15-25
Quartz or limestone sand (0.4 - 2.0 mm)	30-40	-	-	-	20-30
Gravel (2 - 8 mm)	8-12	-	-	-	-
Admixtures, %					
Tylose	0.01-0.03	0.05-0.1	0.05-0.1	0.05-0.1	0.02-0.1
Redispersible copolymer powder Mowilith Pulver	0.5-2.5	1.5-2.5	1.5-2.5	1.5-2.5	-
SuperplasticizerMelment F10	0.5-1.0	-	0.5-0.7	0.5-1.0	0.4-1.0
Casein	-	0.5-2.0	-	-	-
Antifoaming agent	0.1-0.2	0.1-0.3	0.1-0.3	0.1-0.2	0.1-0.2

Some basic mortars compositions for filling seams are presented in Table 12.12.

Technological lines on production of dry mixtures provide storage of binders and fillers in a silage warehouse which is rational for placing over the equipment for dispensing, mixing and packing of components. Weighing of components is performed on batch scales. Mixing of dry loose materials is carried out in the mixers providing uniform distribution of admixtures and dispersion of components, inclined to agglutination and to formation of lumps.

Table 12.12 Basic mortar compositions for filling seams between facing tiles

Component	The content in the mixture, %	
	for seams <5 mm	for seams >5 mm
Ordinary or white Portland cement	25-50	33-35
Alumina cement	1-2	0-1.5
Quartz sand (0.2-0.5 mm)	50-65	-
Marble powder (up to 0.1 mm)	5-10	5-6
Hydrated lime	0-1	0-1
Polymer fibers	-	0-0.2
Pigments	0-2	0-1
Water-retaining agent	0.05-0.1	0.05-0.1
Redispersible powder	0.5-2	0.5-2
Waterproofing agent	0-0.4	0-0.2

The optimum mode of mixing is characterized by high speeds of shift and ensures thorough mixing of the suspension state of the components. Mixes are packed in valve bags by packing machines. Technological lines are equipped with computer controlled systems.

Self-Assessment Questions

1. What are the peculiarities of building mortars? Give the classification of mortars?
2. Explain the peculiarities of mortar technology.
3. Describe the basic properties of mortar mixtures.
4. Discuss the technology and properties of masonry mortars.
5. Discuss the technology and properties of plaster mortars.
6. Describe the basic types of decorative plaster mortars.
7. Discuss the technology and properties of special mortars.
8. Give a classification of dry building mixtures.
9. What are initial materials and admixtures for dry mixtures?
10. Discuss the properties of gluing and plaster dry mortars.
11. Explain the technology of mixtures for surface preparation and for making floors.

Chapter 13

PRODUCTS BASED ON MINERAL BINDERS

Products based on mineral binders form the largest group of construction materials. Concrete and reinforced concrete products take a lead place among them. Reinforced concrete products are presented in chapter 14. Un-reinforced concrete is used for a limited range of products which are subjected to relatively low tensile loads.

Cement stone, reinforced with the asbestos fibers is used for asbestos-cement products, which have high compressive and tensile strength, frost-resistance and low water permeability. Asbestos-cement products are heat-resistant, have reduced thermal conductivity. Under the influence of moisture, their strength increases over time and they do not corrode.

Using clinker-free binders like lime-silica binders, which include a variety of raw materials from waste products is effective for a group of building products. Autoclave technology, based on formation of hydrosilicates and other binding compounds in conditions of water vapor impact at a temperature of 175 - 300°C and pressure of 0.8 -3 MPa, opens ample opportunities for using lime-silica binders. The most popular products that are produced using autoclave technology are lime-sand bricks and stones, and other products from dense and porous silicate concretes.

Gypsum products have acquired more importance in modern construction. They are fire-resistant and biostable, chemically neutral, have low thermal conductivity, and low density. They possess an inherent architectural and decorative expressiveness. Ability to change in a wide range of properties of gypsum and concrete mortars allows the use of highly-mechanized and automated methods of product formation such as: casting, pressing, rolling and extrusion.

13.1. ASBESTOS - CEMENT PRODUCTS

Asbestos is a natural mineral which is mined mainly in Russia, Canada and South Africa. Once mined the asbestos rock is crushed producing fibres which can then be woven into fabrics and used to reinforce cement and plastics.

Some asbestos fibers, when inhaled, constitute a health hazard leading to asbestosis, a form of lung cancer. These health risks prompted the establishment of strict environmental regulations with regards to working with asbestos.. Health risks were shown to be greatest during the mining and production processes, but minimal during installation and use of asbestos-cement products. According to the Environmental Protection Agency (EPA), a material containing asbestos is deemed potentially hazardous only in a friable state, which means when it can be crumbled, pulverized, or reduced to a powder by hand pressure. Asbestos-cement is not considered friable, and therefore not hazardous, because the cement binds the asbestos fibers and prevents their release into the air under normal use conditions. However, asbestos-cement products are classified as friable when severe deterioration disturbs the asbestos or when mechanical means are used for chipping, grinding, sawing, or sanding, therefore allowing particles to become airborne.

The most propagated asbestos-cement products are corrugated and flat sheets, pipes and connective sockets. Asbestos-cement products have relatively good processabilty. They are much lighter than the metal and concrete. Average density of asbestos-cement products changes from 1400 to 2100 kg/m^3, ultimate compressive strength 40-60 MPa, tensile strength - 8 - 15 MPa, and absorption of water is in the range of 10…30%. The main imperfections of asbestos-cement products are the low shock strength and the ability to buckling.

The source components for the production of asbestos-cement are asbestos, Portland cement and water.

There are three common types of asbestos: Crocidolite (blue asbestos), Amosite (brown asbestos) and Chrysotile (white asbestos) and each have different characteristics. All three forms of asbestos can be used to make asbestos-cement but the most effective is Chrysolite.

Due to the strong linear and very weak transversal ties crystals asbestos has a high tensile strength (300 MPa) and the ability to split into thin fibers.

For asbestos-cement products a special type of Portland cement is used, the properties of which contribute to the filtering process of the solid phase in asbestos-cement suspensions and accelerating of products hardening. Fineness of grinding of the cement is characterized by specific surface area in the range of 2200 - 3200 cm^2/g. The content of free CaO in the clinker should not exceed 1%, C_3A - 8%, and C_3S should not be less than 52%. The initial set of cement paste should occur no earlier than 1.5 hours after mixing with water. Cement used for the production of asbestos-cement products, does not usually contain mineral additives.

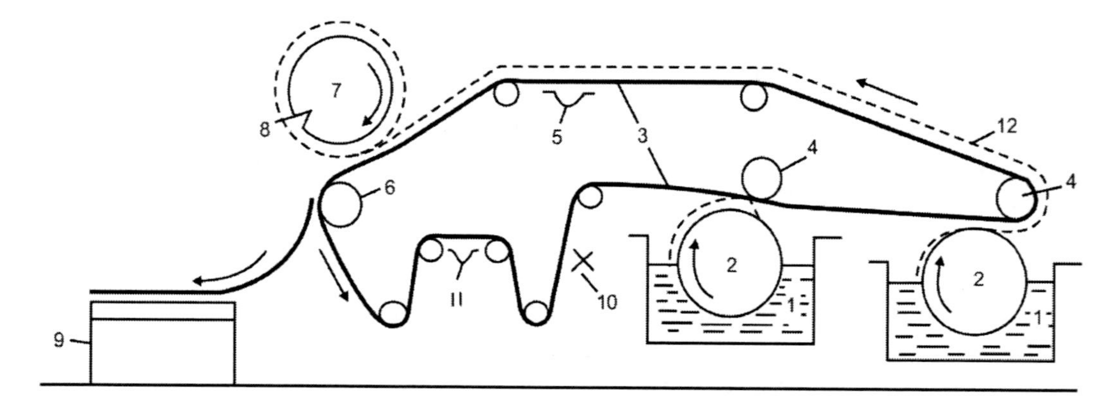

Figure 13.1. Simplified representation of a Hatschek machine for asbestos-cement production
1 - slurry vats, 2 - sieve cylinders, 3 - running felt, 4 - rolls, 5 - vacuum box, 6 - breast roll, 7 - formation cylinder, 8 - cutting wire, 9 - take-off conveyor, 10 - whipper, 11 - suction box (felt drying), 12 – lamina.

The industrial manufacture of asbestos-cement products was made possible by the invention of an Austrian engineer, Ludwid Hatschek. The Hatschek machine was a wet transfer roller. It was used to produce the initial asbestos-cement sheets (Figure 13.1), while two other manufacturing processes - the Mazza process, and the Magnani semi-dry process were used for producing pipes and corrugated sheets respectively. After being formed, most products were steam cured to achieve the optimum microstructure for strength and durability. Asbestos-cement building products have many desirable material characteristics, such as being lightweight, impermeable to water, durable, tough, resistant to rot, termites, soiling, corrosion, warping, and fire, and easy to clean and maintain. Asbestos-cement also possesses low thermal conductivity and is therefore a good electrical insulator.

Processing of asbestos-cement products as the most widespread wet method of production involve: asbestos fission; mixing of fine fibers with the cement and water and formation of suspension; followed by formation of products on the sheet- and pipe-molding machines with subsequent thermal treatment. Molding is the formation from the suspension in meshed cylinder of a thin layer of asbestos-cement, their dehydration and compacting. The resultant products of the fibrous sheet are subsequently subjected to profiling, and pipes - turning. Composition of asbestos-cement mass can be the following: 13-17% asbestos, Portland cement 83-87% (content of asbestos, is higher at the production of pipes than the sheet materials). In some cases, finely powdered quartz sand (30-40%) can be added into the Portland cement. Corrugated sheets are made of the different profiles characterized by height and pitch of corrugations (Table 13.1). Corrugated sheets are manufactured presently usually with length up to 2.5 m, width of 1.15 m and thickness of 7.5 mm. These sheets are assigned for roofs of industrial and agricultural buildings and structures. They are supplied along with the details, which are used for arrangement of walls angles, deformation joints, etc.

Roofs from corrugated asbestos-cement sheets are easily constructed and does not require placement of rigid foundation.

The indicator of the profile fabricability is the ratio between the internal radius of the corrugation and the sheet thickness.

Table 13.1. Characteristic of corrugated asbestos-cement sheets and details

Index name	Value for sheets of profile			
	40/150*	54/200** thickness mm		Details
		6.0	7.5	
Concentrated load, kN, not less	1.5	1.5	2,2	–
Ultimate strength, MPa, not less	16	16.5	19	16
Average density, kg/m^3, not less	1600	1650	1700	1600
Impact resistance, kJ /m^2, not less	1.5	1.5	1,6	1.5
Water impermeability, hour, not less	24	24	24	–
Frost-resistance:				
number of cycles of alternate freezing and thawing without destruction;	25	25	50	25
retained strength,%, not less	90	90	90	90

*40 mm – height, 150 mm – pitch of corrugation; **54 mm – height, 200 mm – pitch of corrugation.

Flat sheets may be pressed or not pressed. They normally measure about 3.5 m in length; have a width of 1.5 m, and thickness of 10 mm. Pressing improves the strength properties of products. The bending strength of pressed sheets is not less than 23 MPa, the average density - not less than 1800 kg/m^3; whilst non-pressed sheets have average strength and densities of 18 MPa and 1600 kg/m^3 respectively.

Flat sheets are intended mainly for construction of prefabricated structures of wall panels, sanitary cabins, as well as for arranging acoustical ceilings, air pits, etc.

Flat sheets are often incorporated into composite products. 'Transitop' is a typical composite board consisting of an integrally impregnated insulating board core, faced on both sides with asbestos-cement board. Waterproof adhesives are used to laminate the insulating core as well as to bond the noncombustible asbestos-cement faces to the core. This combination of materials provides for structural strength, high insulation values, and maintenance-free interior and exterior finishes in a single fire-resistant panel.

The painted sheets and the textured finish by the pattern, can be used for the external facing of walls and panels inside buildings. There is a widespread method of asbestos-cement sheets painting with perchlorovinyl enamels. Decorative coatings, based on the silicate paints are also effective. Asbestos-cement sheets can be finished with wood veneer.

The *asbestos-cement panels* are widely used for the erection of the industrial buildings in particular for construction of their roofs. It is permitted to erect the walls and roofs of industrial buildings with dry and normal operating regimes from asbestos-cement panels. Porous gaskets can be used for construction of joints between panels.

The application of asbestos-cement panels for construction of walls and roofs decreases the weight of structures several folds, and reduces the labour-intensiveness of building works, in comparison with the reinforced concrete,

Asbestos-cement pipes are widely used in the construction. Depending on its purpose, they are divided into:- pressure pipes for arrangement of water-supply systems and non-pressure pipes for arrangement of ventilation and sewerage systems, cable laying, etc.

A maximum working pressure of between 6-15 MPa can be used for the construction of pressure conduits, and asbestos-cement pipes.

Different types of clutches in particular with rubber gaskets (Figure 13.2) can be used for connection of asbestos-cement pipes.

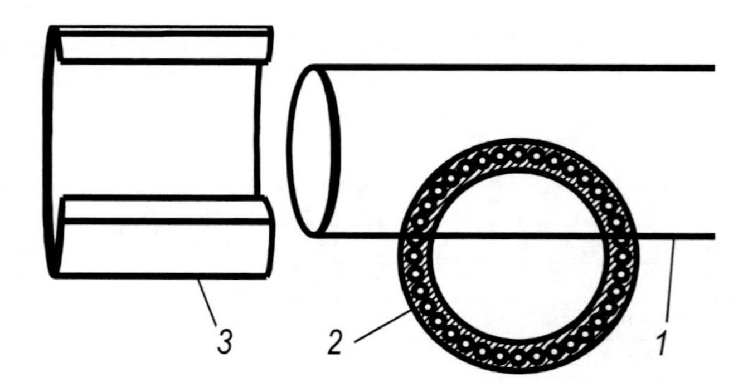

Figure 13.2. Asbestos-cement pipes connecting piece:
1 - pipe; 2 - gasket; 3 – clutch.

Asbestos-cement clutches provide the containment of the connections between pipes. Pipes and clutches should be straight, cylindrical shape, without cracks. Hydraulic test of pipes and clutches should not reveal any signs of water infiltration on their surfaces.

Asbestos-cement pipes should withstand the hydraulic pressure of not less than 0.4-0.6 MPa, and the minimum crushing load in the water-saturated state is in the range of 3.5-5 kN. These products have high corrosion resistance and do not collapse under the influence of electrical current. Low capillary porosity provides water impermeability and freeze resistance of asbestos-cement pipes. They are much cheaper and durable than steel and cast iron pipes.

Asbestos-cement pipes in comparison with the metal pipes are about 3-4 times lighter. Their imperfections include fragility and low acid resistance.

13.2. Products of Autoclave Hardening

Products of autoclave hardening are one of the most widespread wall construction materials aside from the ceramic bricks and precast concrete. The silica bricks and wall products from cellular concrete are the most used autoclave products. The abundance of autoclave products is explained by the prevalence of raw materials, simplicity of technology, high quality and low cost.

The main raw materials for manufacture of autoclave products are the binders used for autoclave hardening and aggregates. Binders of autoclave hardening at normal temperature and moisture conditions and steaming are characterized by relatively low strength. However, their strength increases significantly as the temperature and pressure in the autoclave increase. This results in obtaining an artificial stone of high strength.

Binders of autoclave hardening are divided into clinker-free, lime-based with silica and aluminosilicate components (lime-silica, lime-ash, etc.) and composite, based on Portland cement or Portland cement with lime and aluminosilicate components (sand, pozzolanic and slag cements). The strength of these binders depends on the ratio between components, their chemical and mineralogical composition and fineness of grinding. Components of binders should be characterized by a reduced water demand for production of durable and frost-resistant products.

Lime – silica binders are the most widespread of the production of autoclave products. Their main components are air-hardening lime and powdered quartz sand. Fly ash and others industrial wastes can also be used as the silica component of these binders along with fine-grained sand.

Autoclave hardening occurs as the result of chemical reactions between components of the binder in the presence of water under high pressure and temperature. The main chemical process during autoclaving is the interaction between calcium hydroxide, silica and water. As a result of this chemical process calcium hydrosilicates, are formed. Thus an artificial conglomerate can be obtained. The speed of chemical reactions and strength of the conglomerate are increased if the initial materials have the higher dispersibility.

Lime-sand bricks are non-fired wall material made from a mixture of quartz sand and lime-silica binder, by pressing and subsequent autoclaving under the action of steam of high pressure.

Manufacture of lime-sand bricks involves making lime-silica binder by screening of sand, lime burning and grinding in ball mills along with the sand. This is then followed by mixing of sand with lime-silica binder and slaking of the mixture produced, pressing the bricks and steaming it in the autoclave.

Lime content in the raw material mixture is in the range of 7-10% (by the CaO content). Quartz sand with grains in the range of 2-0.2 mm can be used; content of clay impurities in sand should not be more than 10%. Duration of full cycle of autoclave processing is 8-12 hours under the pressure of 0.8-1.6 MPa.

Solid or hollow lime-sand bricks based on heavy-weight or light-weight aggregates can be manufactured. Average density of light-weight bricks is not more 1400 kg/m^3, and ordinary solid bricks – not more 1650 kg/m^3. Products may be manufactured as white or colored.

Effective lime-sand products have thermal conductivity of not more than 0.486 W/(m·K), they can reduce the thickness of structures on comparison with the ordinary ceramic bricks.

Ultimate compressive strength of lime-sand bricks and blocks is in the range of 7.5-30 MPa and freeze-resistance range between 15-50 cycles of freezing and thawing.

Water absorption of lime-sand products should not be less than 6%.

Along with ordinary bricks, face lime-sand bricks and blocks are produced.

The lime-sand bricks are used along with ceramic brick for masonry work of overground structures. As the result of insufficient waterproofing, lime-sand bricks cannot be used for foundations of buildings. Application of these bricks at high moisture and also in conditions of high temperatures without special protection is not permitted.

The technical economic indexes of lime-sand brick are higher than those of ceramic bricks due to the lower fuel and energy consumptions, and lower labour-intensiveness.

Lime-slag and lime-ash bricks are the varieties of lime-sand bricks.

Also, cellular concrete blocks can be manufactured by autoclaving. Such blocks can have different dimensions. Ultimate strength of blocks is in the range of 2.5-10 MPa. The average density of blocks varies depending on the strength. For example, density of blocks having a strength of 2.5 MPa equals 500 -700 kg/m^3; for a strength of 10 MPa, density ranges between 900 -1100 kg/m^3.

Blocks from cellular concrete are used for masonry work of structures involving dry operating conditions. Vapor sealing is needed if there are moist operating conditions.

Exterior walls of cellular concrete blocks are covered with cement plaster. Application of cellular concrete blocks, reduces the weight of walls by up to 4-5 times, compared with the lime-sand brick.

13.3. Gypsum Products

Gypsum binder is used for the production of gypsum sheathing, slabs and panels, wall stones, facing, architectural and ornamental products, air channels, etc. Gypsum products are light-weight, have a low density, and are non-combustible. However, they have reduced strength when moistened and are characterized by plastic deformation under the influence of loads. Their water-resistance increases when 5-25% of lime, granulated blast furnace slag, at impregnation with the carbamide resins, organic-silicon fluids, etc are added. In many cases, water-resistance of gypsum products is increased when produced in composite forms useing gypsum-cement-pozzolanic binders and gypsum-cement-slag binders.

Gypsum products are used for facing bricks, reinforced concrete, wood surfaces of dwelling houses, public and industrial buildings with dry and normal moisture operating conditions. Increasing the strength of gypsum products with reduced average density is achieved by their reinforcement with fibrous materials. One of the most effective ways to reduce the weight of gypsum products is through creating a porous or cellular structure by introduction of foam into the gypsum mixture.

High aesthetic properties and simplicity of technology give gypsum products a very good visual effect when used as facing for structures of different types.

The most widespread are *gypsum paper boards*, the decorative appearance and sound-absorbing properties of which are provided through perforations, facing with film materials and application of drawings on the board. Gypsum paper boards consist of a hardened gypsum core, firmly connected to the carton envelope (Figure 13.3). The process of gypsum paper boards manufacturing includes making of the molding mass, carton preparation, forming of the continuous sheet, cutting it into separate boards after curing and drying in the tunnel dryers. The humidity of the boards should not be more than 1%.

The density is about 950 kg/m^3, and the failure load is in the range of 250-520 N, depending on the thickness. The boards usually have a rectangular shape and come in different dimensions.

Gypsum fibre boards, unlike the gypsum paper boards, are reinforced not by carton, but with uniformly positioned fibers. Loosened waste-paper is used mainly as a fibrous initial material. The main advantage of gypsum fibre boards, as compared with gypsum, is higher strength. They keep the shape during drying and wetting, provided a good microclimate exists in hot and cold weather. They also have high sound insulating ability, are noncombustible, have good processability, are easily glued by paper and colored plastics. Veneered. Gypsum fibre boards are applied in the same ways as gypsum paper boards. Along with gypsum fibre boards, gypsum fibre slabs can be manufactured, which are used as window stools, elements of built-in furniture, etc.

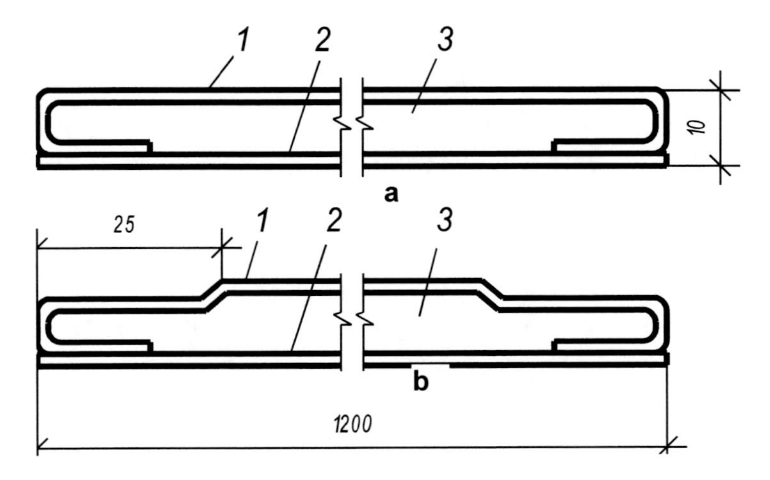

Figure 13.3. Gypsum paper boards:
a - smooth edge; b - cogged edge; 1 - carton envelope; 2 - front-face surface; 3 - gypsum core.

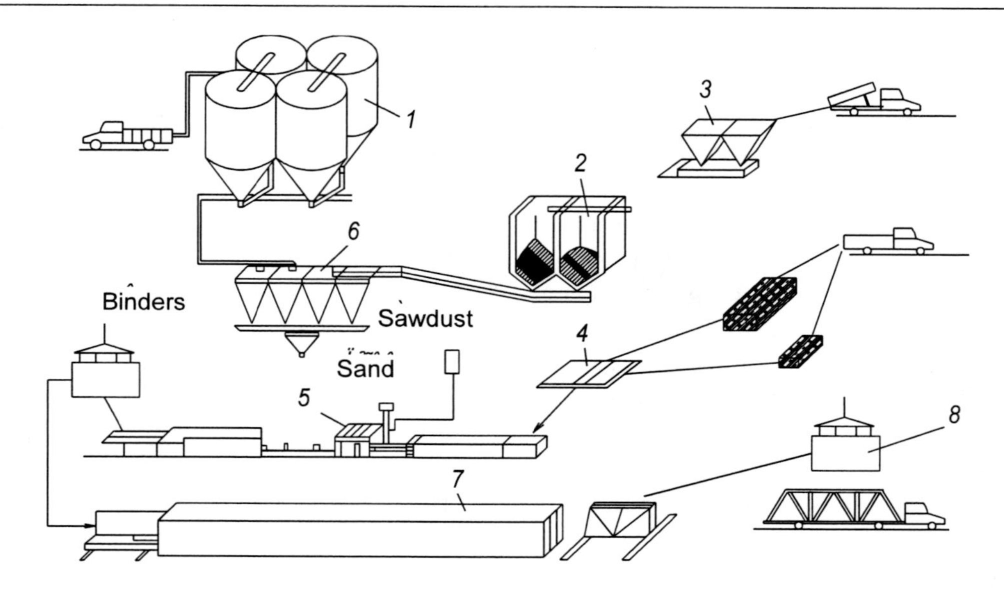

Figure 13.4. Scheme of processing line with rolling mills for production of the gypsum-concrete panels: 1 - gypsum silos; 2 - sawdust bins; 3 - sand bins; 4 – frames; 5 - rolling mill; 6- batchers; 7 - tunnel drier; 8- transportation of panels.

Gypsum decorative boards are produced usually as cast, smooth or corrugated, with different patterns on the surface, reinforced and non- reinforced, with fillers or without them. Gypsum decorative boards are produced, usually square with different dimensions.

Glass cloth, galvanized iron, aluminum and other materials are used for reinforcement of boards.

Front-face area of boards can be covered with water-repellent compounds to reduce the water absorption.

Pressed gypsum boards (*artificial marble*) can also be manufactured. They are obtained from water-gypsum mixture at specific pressure 7-10 MPa with removing of water surplus and curing under the pressure till the beginning of gypsum hardening. Products obtained after 1.5 hours of the formation have compressive strength values ranging from 25-28 MPa and 10% humidity. They are characterized by glossy a surface that imitates the color and texture of natural marble. The products can also be polished to get a smooth surface.

Artificial marble can be used for internal facing of shops, markets, bus terminals, halls, etc.

Considerable amount of gypsum products are used as acoustic materials to build acoustical ceilings and walls ashlaring.

Sound-absorbing boards consist of gypsum perforated screen and absorbent materials, which are heat-insulating plates based on glass fiber, mineral wool, pearlite sand, etc.

Basic types of *gypsum-concrete products* are: blocks and panels for walls, ventilation panels and sanitary cabins. Gypsum concrete are reinforced with different fibrous materials, metallic and wooden elements. Their density range between 800 -1900 kg/m^3, but the ultimate strength depends on the type of mineral aggregates used. For heavy-weight mineral aggregates, the ultimate strength is between 8-12 MPa; light-weight porous aggregates – 3.5 to 7 MPa, organic – 1 to 3 MPa.

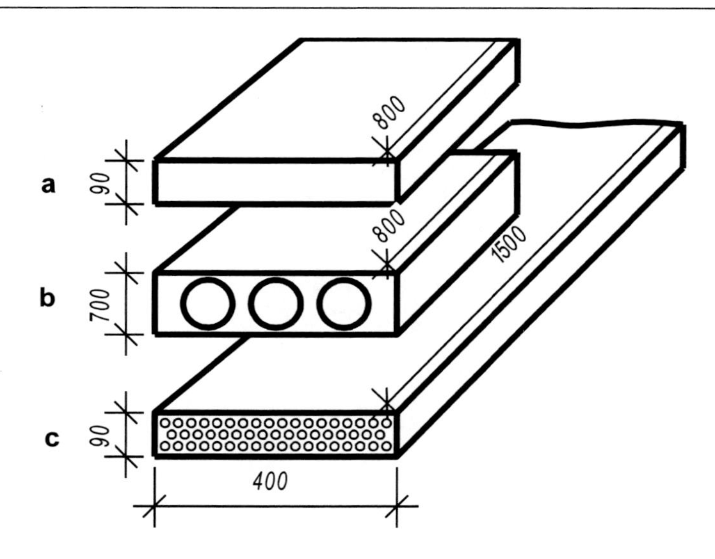

Figure 13.5. Gypsum-concrete panels:
a - solid; b – hollow; c – reinforce.

Processing lines with rolling mills are widely used for mass production of the gypsum-concrete panels (Figure 13.4, 13.5).

Extrusion is an advanced method of gypsum products manufacture. Extrusion allows forming of products (Figure 13.6) from the plastic gypsum mixtures with low water content of any cross-section with the high voidage, low weight, and compressive strength up to 30 MPa.

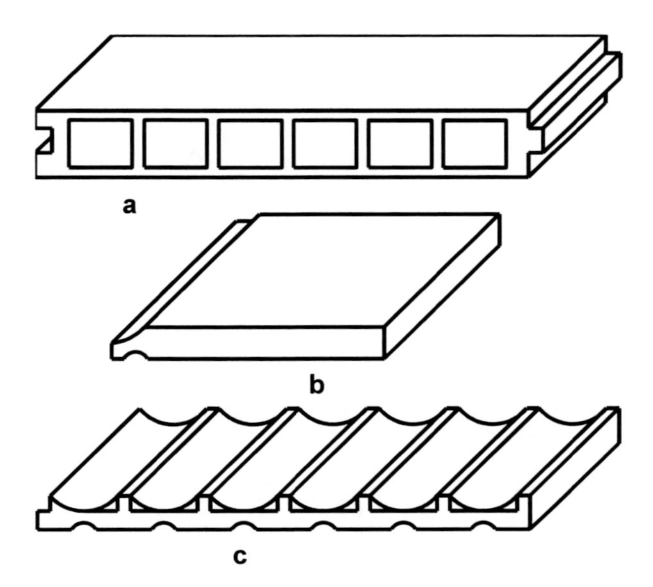

Figure 13.6. Extrusion gypsum products:
a – panel for partitions; b – window stool; c – decorative board.

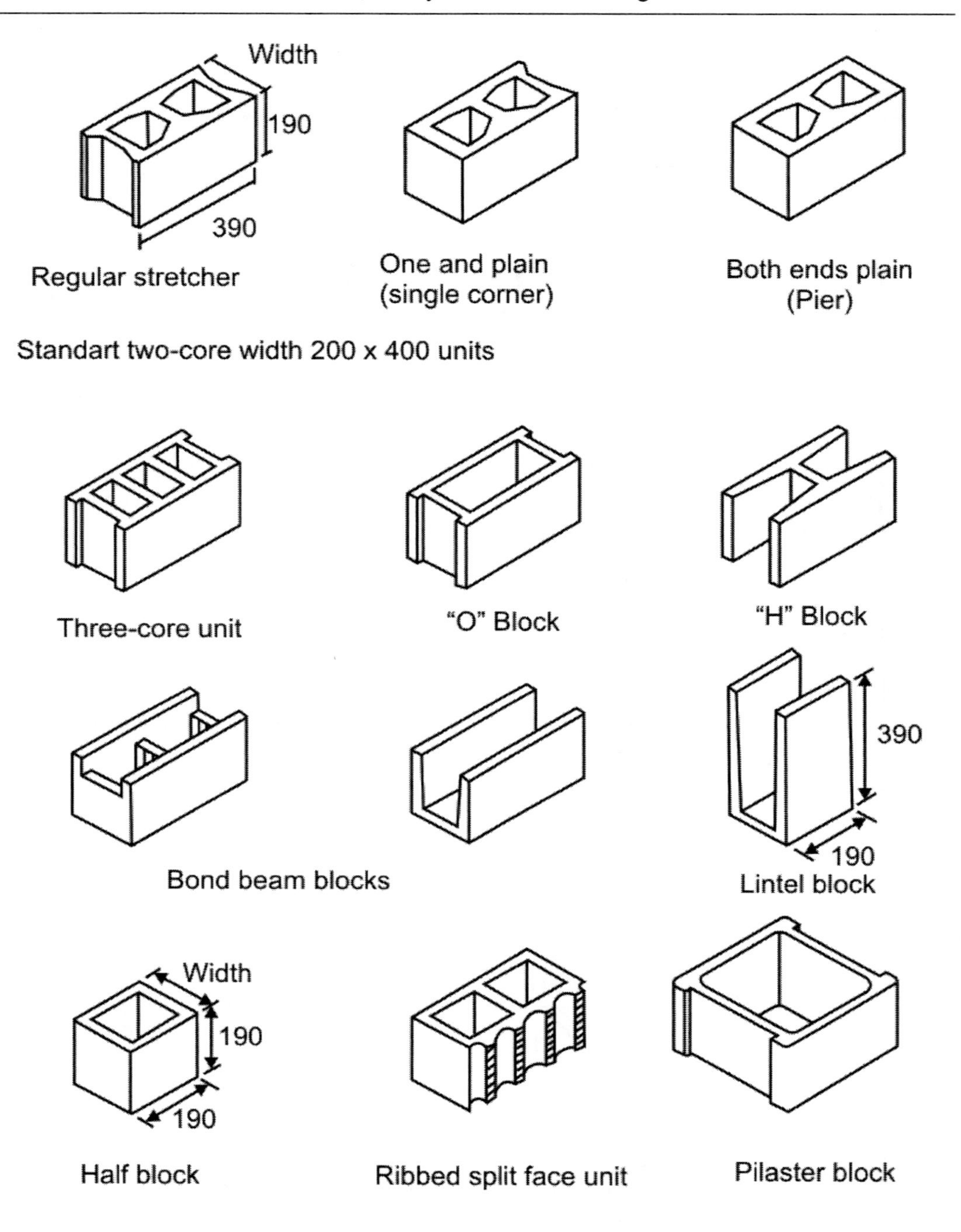

Figure 13.7. Typical concrete blocks.

Wall gypsum blocks are produced by casting or other methods and can be used for frame fillings of structures. Blocks are manufactured on conveyor lines. Gypsum blocks are of 100% fire safe. Once installed, they can be quickly finished. Paintwork, wallpaper, tiles and sprayed finishes can be applied. These products have high mechanical resistance.

13.4. Non-Reinforced Concrete Products

The main types of non-reinforced concrete products are: concrete blocks, sidewalk tiles, concrete roofing tile, decorative concrete products and etc.

Wall concrete blocks (Figure 13.7) based on cement, lime and slag binders can be manufactured of light-weight or heavy-weight concrete by vibration compacting, casting or other methods. Blocks can harden in natural conditions, steaming or autoclaving. They are used for production of bearing and fencing structures of dwelling houses, public and industrial buildings. There are manufactured blocks and in different in sizes. Compressive strength, usually, is in the range of 2.5-20 MPa. Frost-resistance is within the range of 15-50 cycles of freezing and thawing

Depending on the purpose, there are ordinary and facing blocks. Facing blocks can be manufactured undyed or dyed. The average density of hollow blocks is not more than 1650 kg/m^3, and for solid blocks - not more than 2200 kg/m^3.

Sidewalk tiles are made of square, rectangular or figured forms. These tiles are based on heavy-weight, including sand concrete. Tiles can be moulded by vibration, pressing and vibrostamping.

The ultimate compressive strength of concrete tiles should be in the range of 30-50 MPa; the frost-resistance range between 150-250 cycles of freezing and thawing; the abrasiveness no more than 0.6 g/cm^2, and water absorption is no more than 5%.

Colored cement or pigments can be used for the manufacture of colored tiles.

Self-Assessment Questions

1. List the raw materials and technological peculiarities of asbestos-cement products manufacture.
2. What are the technical characteristics of asbestos-cement sheets?
3. What asbestos-cement panels and pipes are used in construction?
4. Explain how the autoclave technology is applied to construction materials.
5. Explain the technological peculiarities of lime-sand bricks manufacture.
6. What are the technical characteristics of lime-sand bricks?
7. What are the peculiarities of gypsum products?
8. Explain the peculiarities of non-reinforced concrete products.

Chapter 14

REINFORCED COMPOSITE MATERIALS

The materials of this group are characterized by the presence of reinforcing component in their composition, which are mainly subjected to stretching and bending stresses, during the operation of the products. The reinforcing elements of materials are distinguished by their chemistry and substantial composition, form, sizes, and degree of orientation in the matrix phase.

Typical reinforced composite materials include: reinforced concrete, glass-cement and glass-fiber plastics. Fiber concrete and asbestos-cement can be related to the fiber-reinforced materials. Particle boards and fiberboards, other laminated plates and roll materials also belong to the fiber composites. They are considered in other chapters.

Composite materials, reinforced by continuous fibers, are characterized by the composite interaction of all the constituent elements. For a given orientation of the uniform continuous fiber rigidly coupled with the matrix in one direction, the relationship between the loads, perceived by both components of materials, will be determined by the ratio between their modules of elasticity and the volume content of fibers. The deformation behavior of composites, reinforced by discontinuous fibers, differs from those where unequal tensile stresses act in the segments of fibers along their length: at the ends of segments they are equal to zero, but achieve maximum values in their effective part.

Most building structures are subject to the action of compressive and/or tension stresses. Concrete has a relatively low tensile strength. For complex resistance of compressive and tensile stresses *reinforced concrete* constitute a composite material solution in which concrete and steel reinforcement are rationally combined and co-work. The combination of concrete and steel in one material is possible due to their high adhesion and similar values of thermal expansion coefficients. Furthermore, in the alkaline environment existing in hardened concrete, steel reinforcement is well protected against corrosion due to the formation of a dense protective concrete cover.

Reinforced concrete is the most commonly used construction material in the modern construction industry. This is due to its high mechanical strength properties, durability and availability of raw materials sources, possibility of production of structures of any forms, to meet various architectural and technological requirements.

Depending on the method of production, reinforced concrete structures are divided into *monolithic* (placed in the form directly at the construction site), *precast* (assembled from prefabricated elements) and *precast-monolithic* (combine precast reinforced concrete elements and monolithic concrete or reinforced concrete).

Precast reinforced concrete structures by comparison with monolithic have a number of advantages. These include: organization of works is simplified on site, the basic operations of reinforcing, placing and hardening of concrete mixture are performed at the plants, duration of construction is shortened and the labour productivity increases because the casting works are excluded, the output of large size elements with enhanceable manufacture readiness is possible.

However the use of precast reinforced concrete elements requires powerful and specialized lifting-transport equipment, using of cut charts of buildings and considerable financial expenses to achieve the desired joint arrangements, and does not always allow the desired architectural expressiveness of buildings and structures.

Experience of monolithic construction shows that it has technical and economic advantages over the precast ones in a number of cases. It allows reduction with respect to non-permanent charges on creation of production base, expense of steel, cement and energy. Monolithic structures allows a substantial promotion of the performance and reliability parameters of buildings. Monolithic construction plays a special role and effect in seismic areas where the metal saving reaches 20%.

However, the efficiency of monolithic reinforced concrete is reduced by considerable specific gravity, the cost and labour content for construction of form work, low degree of mechanization of reinforcing operations, placing and distributing of concrete mixture, and also cost of transporting the concrete mixture.

The question of whether to apply precast or monolithic method in the construction of reinforced concrete is solved only by deciding which has the possibility of satisfying the project requirements of the structures and buildings with respect to time and cost.

Buildings with large length, in relation to their volumes and sizes (channels, dikes, sluices, underwater parts of hydroelectric power plants, retaining walls, etc.) are expedient to build as precast-monolithic process. The most labour consuming aspects are prefabricated in the plants or grounds whilst less labour intensive component elements and their junction are achieved by the in-situ monolithic constrcution.

14.1. Reinforcing of Concrete

Reinforcing Methods

The use in reinforced concrete of reinforcement like bars, meshes, frameworks and other elements (Figure 14.1) is intended mainly for resisting tensile stresses (Figure 14.2). Apart from the principal reinforcement, embedded fittings for connection of structures during assembling, lifting loops, distributive reinforcement are also applied in elements.

There is ordinary and prestressed reinforcing. Although ordinary reinforcing, increases the load-carrying ability of structures, it has some limited possibilities, predefined by insignificant stretchability of concrete – 0.1-0.15 mm/m. Stresses in the tensioned reinforcement are small at such strains and make up approximately 20-25% of its calculated strength. Increase in concrete strength increases its stretchability insignificantly. That is why cracks exist in the tension regions of concrete structures at comparatively low loadings. On formation of cracks, deflections are increased, moisture and gases get to the cracks and corrosion of the steel reinforcement becomes inevitable.

This shortcoming may be overcome through the use of prestressed structures, first carried out in practice in 1928 by the French engineer E. Freyciet.

The essence of *prestressing* is demonstrated in the concrete compression with the tensioned reinforcement shown in (Figure 14.3). In order to change the sign of tension, which is acting in the concrete of prestressed structure, it is required foremost to neutralize the compression of concrete. At the same time it is required to mean that possible deformation of concrete at a compression exceeds limit tension in 20-25 times.

The most important consequences of prestressing is increasing of crack resistance, reinforcement saving and decline of structures mass or their enlargement. Reinforcement saving is caused by possibility of the use of high-strength steel which can not be rationally used for ordinary reinforcing. In the later case, tension of high-strength steel reinforcement increases with the increase of working stress in comparison with ordinary steel, which results in crack appearance in the tension region of reinforced concrete element and loss of load-carrying ability by it.

Due to prestressing it is possible to produce structures (slabs, beams, trusses) for covering of large spans (more than 9 m), thin-walled spatial structures (doubly curved shells, panels-shells span-sized 12, 18 and 24 m) for structures of differing purposes, etc.

Production of large diameter pipes for penstocks, supports of high-voltage lines of electricity transmissions and a series of other structures are constructed using prestressed concrete.

The application of prestressed concrete is considerably extensive and includes precast elements used for construction of dams, locks, structures of hydroplants, etc.

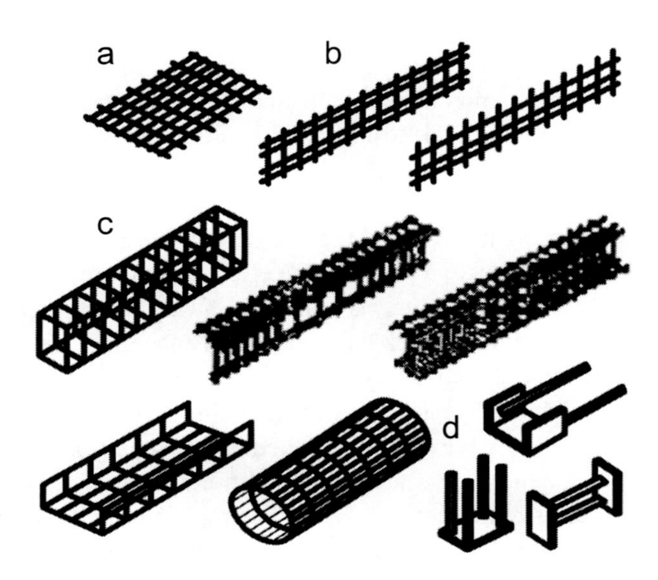

Figure 14.1. Reinforcement of reinforced concrete elements: a– mesh; b– flat frameworks; c– spatial frameworks;
d– embedded elements.

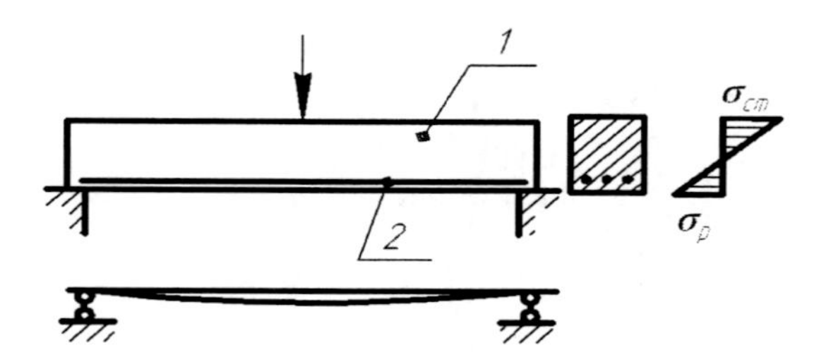

Figure 14.2. Diagram of work reinforcement in a reinforced concrete element:
1- concrete element; 2- reinforcing bar.

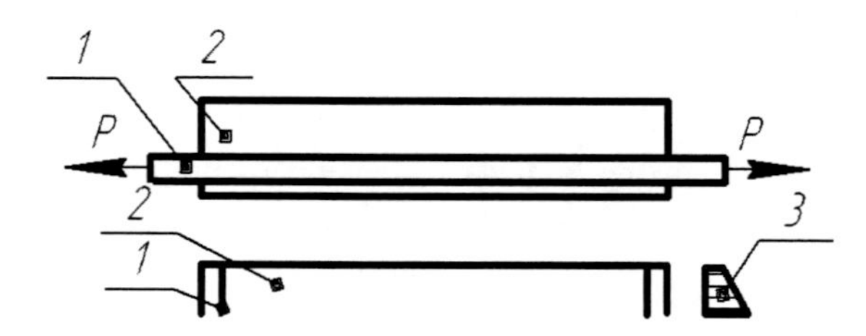

Figure 14.3. Scheme of reinforcement prestressing:
1-reinforcement; 2-concrete element; 3- stress diagram.

During the production of precast reinforced concrete elements prestressing can be conducted either before the concrete hardens or after it has acquired some strength. The first method (pre-tension) is more widespread. The essence of it is that the reinforcement is placed in a form, fixed and then tightened. The form is filled with concrete mixture and the reinforcement freed of tension after the concrete has hardened. For the second method (post-tension), reinforcement is placed in special channels left in concrete and tightened after the concrete hardens. The required adhesion of the tensioned reinforcement with concrete is achieved by injection of cement mortar in the channels. In both cases reinforcement now released of tension, tends to return back to its initial position, shortens and wrings out the reinforced concrete elements.

It is also possible to cause tensioning of reinforcement, due to expansion of cement after the concrete has acquired a strength of 10-20 MPa. The stretched reinforcement has sufficient adhesion with concrete, wrings it out and provides the effect of self-stressing.

A reliable adhesion with concrete is obtained with corrugated bars, twisted reinforcement, and also reinforcement which have additional anchor devices set on their ends.

Stretching of reinforcement is realized by mechanical, electro- thermal, electro- thermo-mechanical and chemical methods. The mechanical stretching of reinforcement is carried out by hydraulic jacks and by other devices. Electro-thermal stretching is based on using the linear expansion of reinforcement when heated by an electric current.

The ends of the extended reinforcement fasten in grip vice (at stretching on stressing abutments) or by anchors (at stretching on a concrete), as a result there is tension in it.

Reinforcement becomes taut by mechanical device and simultaneously heated by an electric current during the electro-thermal stretching.

Chemical tensioning occurs through use of expanding cements which have high energy of expansion.

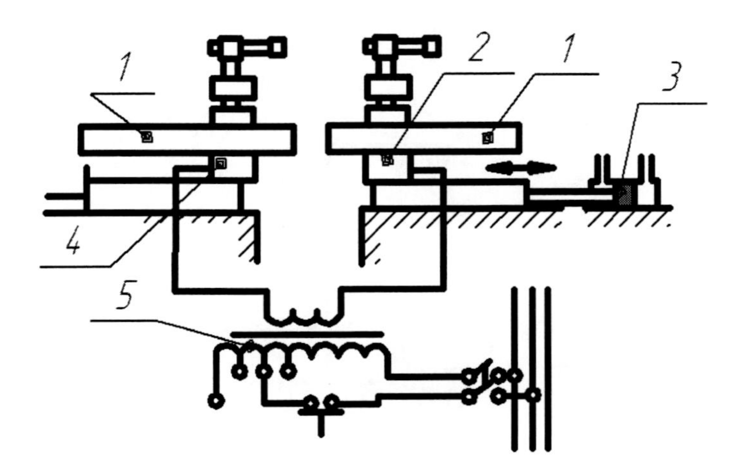

Figure 14.4. Scheme of the butt welding of reinforcement:
1- -bars; 2– mobile clamp; 3-mechanism for rapprochement and clamping of bars; 4– immobile clamp; 5-transformer.

Types of Reinforcement

Reinforcing of structural elements and structures are carried out using steel bars and wire mesh. For ordinary reinforcing, steel bars and corrugated wire profile are widely used as reinforcement. Hot-rolled, thermal fixed bar reinforcement and also high-strength wire and reinforcement ropes are mainly used for tension reinforcement.

The basic types of reinforcement elements for reinforced concrete structures are reinforcing fabrics, flat and spatial cages. *Fabrics* are made from a wire and used as an assembling reinforcement. *Flat cages (mats)* are made of work and distributive bars, using them as bearings elements. *Spatial cages* can have a rectangular, t-shaped and round cross-section. They are used for reinforcing of columns, beams, pipes and others like that.

Prestressed precast reinforced concrete structures are reinforced by separate wires and bars, by wire strands and bundled burs, and also by wire packages with the different number of strings. A choice of reinforcement, which becomes taut, depends on the type of products and equipment for tendon jacking.

Production of reinforcement includes: preparatory operations; welding of bars from separate short small twigs, marking and cut of bars on the elements of required length, bend and grant to them the set project configuration, drafting and welding of reinforcement products.

The most effective process of producing reinforcement are welding operations. The contact butt welding is applied for connection of reinforcement bars at their purveyance for providing of subsequent non-waste cutting (Figure 14.4). The contact point welding is widespread for producing reinforcement products (Figure 14.5). The essence of contact welding is for connection of metallic elements when warming-up by an electric current and application of mechanical effort. The most heat release takes place at the bar joints as a result of passing current through weldable elements.

Mechanized and automated lines, equipped by highly productive equipment are applied for the production of reinforcement elements and structures.

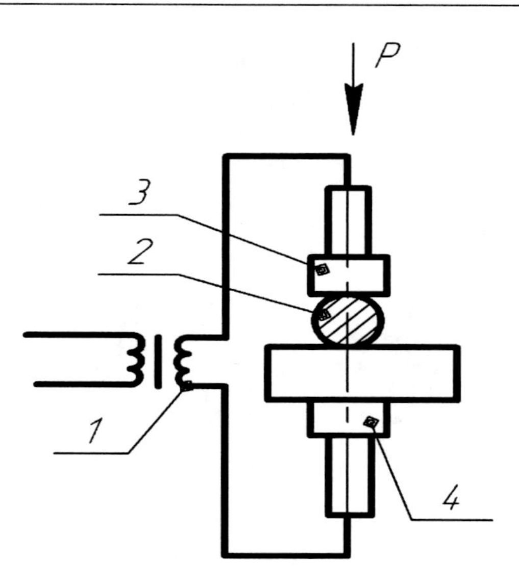

Figure 14.5. Scheme of the contact point welding of rebars:
1-transformer; 2- -bars; 3– mobile clamp; 4 - immobile clamp.

Corrosion of Reinforcement

The alkaline character of the internal environment of concrete is a favorable factor for the protection of the reinforcement and causes surface "passivation" of steel. However for highly porous concrete, passivation of reinforcement steel is violated as a result of penetration of carbon dioxide and other gases from the environment. Corrosion of reinforcement at humidity greater than 80% is the most intensively observed process. Steel reinforcement is affected by highly aggressive environment such as those which exist in the wet and hot climate of marine environments with the air saturated by salts. Corrosion of reinforcement can also be initiated by chloride salts added to the concrete to accelerate the hardening process. They are allowed to be applied only for reinforced concrete structures from non-stress working reinforcement with more than 5 mm diameter, intended for exploitation in non-aggressive gas and water environments.

An effective method of prevention of reinforcement's corrosion in concrete is by use of high density concrete and sufficient thickness of the protective cover layer, which depends on humidity and the aggressiveness of the environment.

14.2. Methods of Production and Types of Reinforced Concrete Products

Methods of Production

The production of reinforced concrete elements consists of the following basic processes: preparation of concrete mixture; production of reinforcing elements; products casting;

hardening of concrete; demolding of products; their facing; completing of construction details for the increase of their factory readiness.

Technological processes are carried out on production lines in moulds, which move, or are immobile. In the first case, the moulds for manufacturing the products move from one specialized post (for example, placing and compacting of concrete mixture, tendon jacking, etc.) to another one. In the second case, the moulds are immobile during all production process but the required technological equipment moves.

Making of products in movable moulds is carried out using aggregate and conveyer technological lines.

Cycle method of production (Figure 14.6) is foreseen for forming of products on hardware devices which include forming machine (for example, vibroground); concrete packer and machine for the establishment of moulds on casting post (mould packer).

Moulded elements with moulds are transferred to the hardening chambers, and then taken out and sent to a warehouse, and disengaged moulds prepared for the next cycle.

This method has a wide application because it is marked by flexibility and possibility of rapid readjustment in transition from one elements' production to another, and also requires comparatively small capital investments. In most cases, aggregate method is the terms used to describe the serial production in the factories of middle and small productivity.

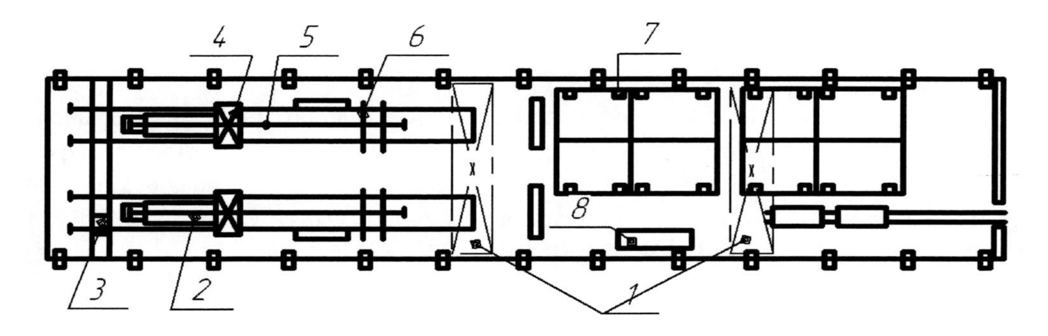

Figure 14.6. Cycle-production line for the production of ceiling slabs 3×12 m:
1- overhead (traveling) crane; 2– vibratory table; 3- distributing hopper; 4– concrete placer; 5- roller conveyer; 6- equipment for reinforcement tension; 7– packetization of forms; 8- table for control and finishing the elements.

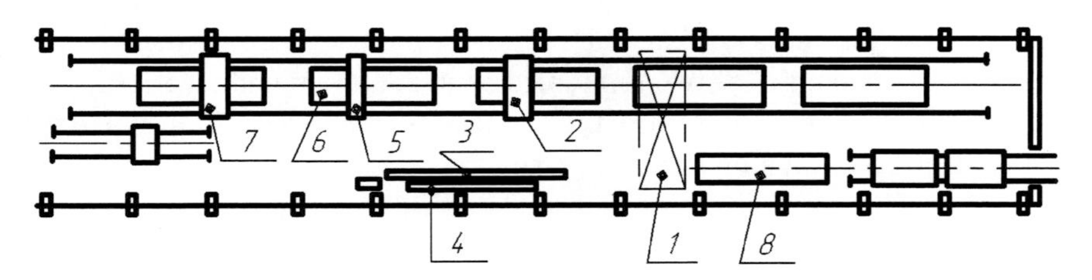

Figure 14.7. Scheme of casting bed line the production of complex slabs:
1- overhead (traveling) crane; 2,7– concrete packers; 3-machine for bars strengthening; 4-line of bars welding; 5-machine for insetting of vapor sealing; 6- power form; 8-post of causing of damp course.

The continuous (conveyer) method of production differs from the cycle method by division of technological process on separate operations and certain rhythm, otherwise speaking, by identical duration of operations. The conveyers are divided into foot-pace and continuous action.

The continuous action allows creating powerful mechanized current process, and it is especially effective for the serial production of the same type products: wall panels, roof slabs, columns and others like that.

Concrete casting bed method (Figure 14.7) products produce in immobile moulds or on the specially equipped stands. Stand technology is expedient especially at producing of large size and prestress products, it is applied also in the conditions of precasting yard, when thermal treatment is carried out in floor chambers or in thermomoulds. For casting bed production complete mechanization and automation of technological processes, enhanceable labour intensiveness is complicated. Duration of technological cycle is usually for 1-2 days.

Cassette method of production is a kind of casting bed process, the feature of which is forming of products upended in stationary sectional group moulds-cassettes. Relatively high exactness of their sizes and good quality of surface is reached by forming of the products in cassettes. At the same time the products are characterized by inhomogeneity of strength by height. Cassette method provides the high productivity of labour, needs less working area and amounts of steam and electric power.

14.3. Basic Types of Reinforced Concrete Products

An industry of precast reinforced concrete provides the products for all branch of construction. It is possible to divide all complex products into linear, flat, block and spatial, depending on their geometrical features. Columns, slabs girders, beams, piles belongs to the linear products; roof slabs and ceilings, panels of building walls, walls of reservoirs and others - to flat products; wall and foundation blocks- to block products; rings of wells, elements of silos, lifts and others - to spatial products.

Wall panels products (Figure 14.8) are widely used in civil buildings.

Large amount of panels for external walls can be made consisting of one layer of lightweight concrete with porous aggregates or cellular concrete. The most widely used is light claydite concrete wall panels for concrete structures with density 900-1100 kg/m^3. As a rule, panels have a texture layer on the facade side measuring 20-30 mm thick. Panels are faced by ceramic, glass tiles, decorative concrete, silicon enamels and other materials.

For reduction of the mass of external walls and improvement of their heat-insulation properties panels which consist of two reinforced concrete layers and layer of heat insulation material between them are manufactured. The thickness of walls from such "sandwich" panels goes down to 250-300 mm, mass diminishes by 50%.

For floors of buildings, hollow and ribbed slabs with ordinary reinforced or prestressed concrete with ultimate compressive strength 20-30 MPa are used. The hollow slabs are produced with round or oval voids, length up to 6-12 m and width up to 1.5-2.4 m. The ribbed slabs have length up to 15 m and width of up to 3m.

Columns and cross-bars are principal elements of building frames. Crane beams and girders are the main elements for industrial buildings.

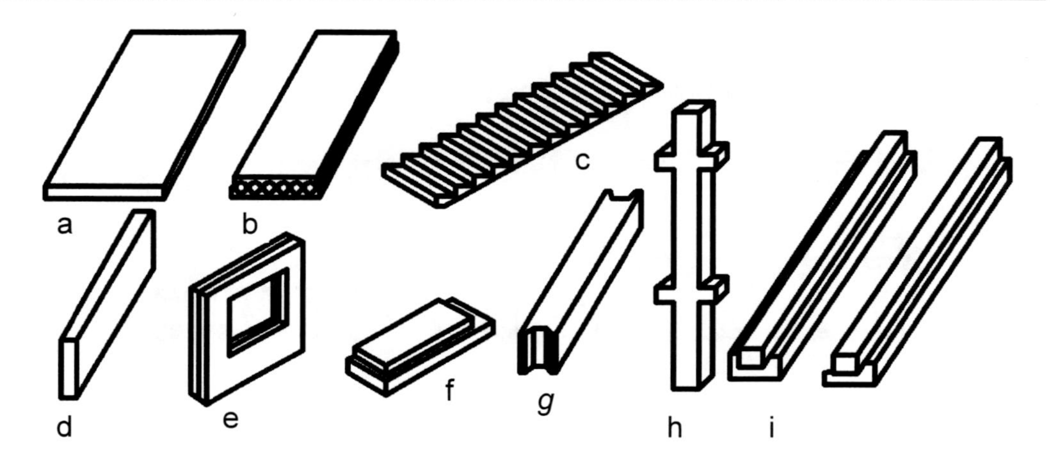

Figure 14.8. Charts of reinforced concrete products for civil and industrial construction:
a– panel void-free; b– panel with round emptinesses; c– flight of stairs; d– beam; e– wall panel; f– resting place; g– girth rail; h– foundation block; i– column.

Columns usually have a rectangular, square or t-shaped cross section. Mainly heavy-weight concrete with ultimate compressive strength 30-40 MPa is applied for columns construction. Considerable reduction in the cost of concrete and steel reinforcement is realized when used as framework elements with ultimate compressive strength of 50-80 MPa.

Reinforced concrete products and structures are widely applied in the construction of irrigational systems, underground pipelines, pumping station buildings, elements of bridges etc. (Figure 14.9). Water-towers, pools, aqueducts and pools, piers and other engineering buildings are economically erected as monolithic constructions.

Reinforced concrete pipes (Figure 14.10) are widely used in the construction of different piping systems. The service life of reinforced concrete pipelines can be as high as 80-100 years, whereas metallic pipes for same use usually last for not more than 30 years.

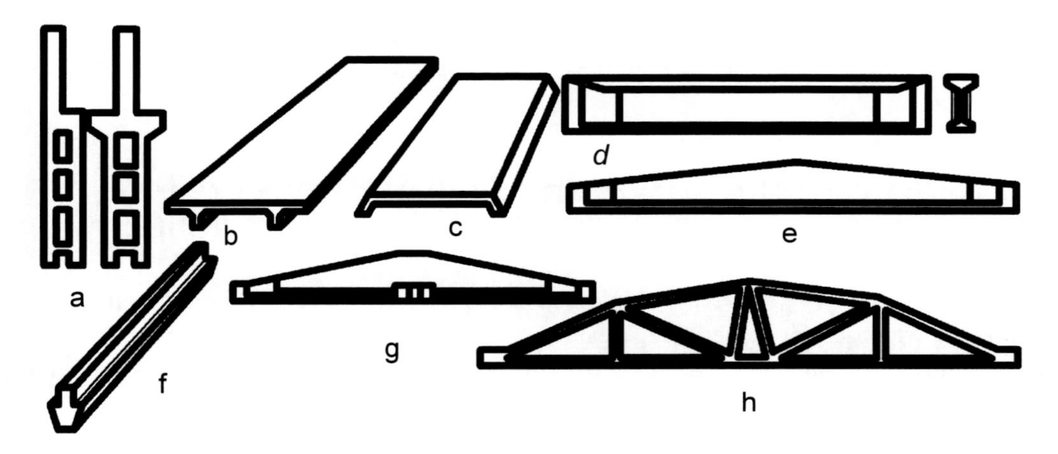

Figure 14.9. Schemes of reinforced concrete elements for industrial construction:
a– columns; b– floor panel; c– roof slab; d– crane beam; e–roof beam; f–collar-beam; g– eaves girder; h- truss.

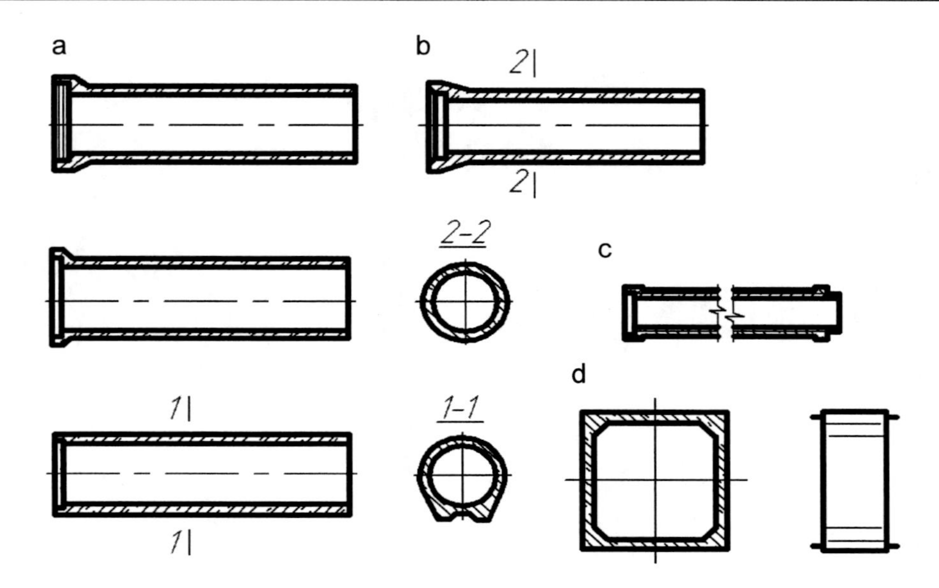

Figure 14.10. Reinforced concrete pipes which are used in an irrigation and drainage construction a – nonpressure; b, c – pressure;d- rectangular.

Reinforced concrete pipes are used for various assignments in irrigational and drainage construction. As a result of their impermeability to water, they may be laid in the embankments of highways and railways; sewage and gully, through which economic domestic, industrial and waste waters are passed. The pipes can have different cross-sections and/or configurations: round, elliptic, rectangular, or various other difficult forms.

Reinforced concrete pipes are usually 3-8 m in length. The minimum diameter of pipes is 300 mm and as a rule, the maximal diameter does not exceed 2500mm.

Concrete with ultimate compressive strength between 30-50 MPa, and water-pressure-resistance 0.4-1.2 MPa is suitable for pipes.

Depending on the external environments, reinforced concrete pipes are divided into pressure (with hydrostatical pressure of 0.5-2 MPa), low pressure (0.3 MPa) and nonpressure.

14.4. Ways of Saving Material Resources during the Production of Reinforced Concrete Products

In the production of precast reinforced concrete, product cost for materials mainly include the expenses for cement, aggregates and reinforcement. On the average, for 1 m^3 of precast reinforced concrete more than 90 kg of equivalent fuel (conditional) are outlaid. Of these, thermal energy constitutes near to 90%, whilst electric energy constitutes about 10% of the energy resources used.

Saving of Cement

The saving of cement in the production of concrete and reinforced concrete is achieved by the removal of its production losses, by improving project decisions, technological processes, forms and methods of production organization, wide utilization of industrial wastes and by-products.

Part of cement's losses take place through the inefficient use of cements of different kinds and strength, their use for the wrong purpose, mixing of different grades of cement, shipping of hot cements with activity lower their quality. Sometimes, a 40 MPa cement is used for making concretes with ultimate compressive strength of 10 to 15 MPa and low strength cement mortars. Part of cement's losses is due to the use of low-grade aggregates in concrete mixtures. Every percent of muddiness of the crushed stone, for example, equals additional amount approximately 1% of cement.

A significant part of cement losses are conditioned by imperfections in the technology of manufacturing of the reinforced concrete products; the use of inappropriate equipment for transporting, unloading and dosage of cement; losses due to preparation and transporting of concrete mixture, the increase of slump and strength of concrete in comparison to specific project indexes; wear of equipment, issue of defective products and others like that. It is has been established that, on average, losses are 10 times less if transportation is in cement carriers, than in simple carriages, and about 40 times less in open transport vehicles.

Reduction in cement content is promoted by use of high-strength concretes and precast reinforced structures which are proven to have reduced cross-section of the elements and their mass. Examples of economic reinforced concrete products include- large size slabs completely factory assembled, which allow reductions of up to 2-3 times the number of elements which are assembled, and centrifugated concrete columns with ultimate compressive strength 50-60 MPa, etc.

Considerable saving on the cost of cement is related to wide use of light-weight concrete. The use of light-weight concrete instead of normal-weight one normally increases the cement content per m^3. However, taking into account the reduction in cross-sectional dimensions, the overall expenses on cement is reduced. With respect to construction mass, the use of light weight concretes for external walls, slabs and ceilings result in up to 30% reduction in the mass of buildings and about 10% on the cost of cement.

Chemical admixtures are a most powerful means for cutting the cement content in concrete. The use of superplasticizers allows the reduction of cement content by 15-20% without worsening the concrete properties. In mortars containing fine sands, the content of cement may be reduced by 5-20% by introducing enlarger admixtures into the concrete mixture. The admixtures may be classified stone sifting out, wastes from ore mining and processing enterprises, or slag produced as waste from power plants.

The use of fly ash in the optimum amount as an admixture in concrete mixture allows cement saving of up to 50-70 kg/m^3, and more. Re-grinding even low-active ash to the specific surface 5000 sm^2/kg allows saving 20-30% of cement without the decline of concrete's class.

A considerable saving of cement can be attained by optimization of composition of concrete mixtures, taking into account growth of concrete strength after thermal treatment, expansion of volume of products' manufacturing with negative tolerances to the sizes and due to other technological means.

Saving of Metal

Reinforcement for reinforced concrete constructions accounts for approximately one-third of metal that is used in the whole of construction. The basic ways of reducing metal consumption of reinforced concrete products are by improvement of the strength indices of reinforcement steel, development of production and use of wire reinforcement, enhancement of reinforcement production and use of effective embedded parts.

The amount of reinforcement steel of any class (T) can be approximately shown to be conditionally equivalent to the strength amount of steel of class (T'):

$$T = T' / K_{tr} \tag{14.1}$$

where K_{tr} - transitional coefficient.

It is possible to save about 40% of metal due to pretensioning of reinforcement. For prestressed structures it is effective to use high-strength reinforcement steel. High-strength thermally fixed reinforcement of enhanced corrosive resistance is widely used in recent years.

A considerable saving of steel can be achieved through production of welding fabrics and cages from reinforcing wires from die-rolled sections in place of the smooth ones. An important reserve of cutback of steel's spending is through production of reinforcement elements on the automated no-waste lines. On average 6-35 kg of steel of metallic forms are outlaid for 1 m^3 of reinforced concretes. The cutting of metal's cost is achieved by improvement in the construction of forms, as well as their rational planning and use.

Saving of Fuel

The principal reason of overrun of fuel at a steaming-out is the insufficient state of steam-curing chambers, and lack of controls for steam flow consumption. The non-productive losses of fuel reduce by the increase of thermal resistance of steam-curing chambers, diminishing of thermal conductivity of chambers walls with the help of different heat-insulation materials and light-weight concrete.

In comparison with pit more profit-proof are vertical, tunnel and low downstream chambers. In the last one, for example, charges of steam are about 30-40 % lower than in pit ones.

For the saving of fuel and energy resources in the production of the precast reinforced concrete the use of energy-efficient technologies is of great importance. For example, application of hot mixtures with a heating temperature of 65-70°C results in the acceleration of thermal treatment for 3.5-4 hours during the production of massive elements. Possible reduction of the treatment duration in the chambers or cassettes does take into account the fact that subsequent curing of products reaches 3 hours and more. Considerable reduction of temperature and duration of thermal treatment is possible through combination of intensive mechanical and thermal influences on concrete, creation of surplus pressure in the chambers, use of more rigid mixtures, use of chemical admixtures to accelerate the hardening of concrete.

The decline of power-consumption of precast reinforced concrete is possible also due to low temperature modes of thermal treatment and reduced heat of hydration of the cement in the process of products warming up, application of non-steam methods of thermal treatment (use of combustion products of natural gas, high temperature organic heat-transfers). In warmer regions energy consumptions for the acceleration of concrete hardening can be reduced considerably by application of sun energy.

14.5. FIBER CONCRETE

Composite materials, which include short segments of different fibers in the cement matrix, are classified as fiber reinforced concretes or dispersed-reinforced concretes. Different types of fibers from steel, glass, synthetic materials, asbestos, carbon and others can be used as fiber.

The disperse reinforcement of cement stone by fibers substantially increase its specific strength in particular with respect to the tension, bending, crack-resistance, impact resistance, vibration influences, abrasive resistance, etc. Application of fiber reinforced concrete allows:

- the realization of effective constructive solutions, for example, thin-walled structures without the bar reinforcement or reticulated distribution rods;
- a decrease in labor expenses for the reinforcing works and to increase the degree of mechanization and automation of the production of the reinforced structures.

It also offers the possibility of applying new, more productive methods of molding the reinforced structures, for example, pneumatic sputtering, roller molding, etc.

Fibers increase the bearing capacity of the concrete matrix. As a result of the difference between the values of the elasticity coefficient of the reinforcing fibers and the matrix, the transfer of load in the contact zone through the matrix to the fiber is accomplished.

Typical stress-strain diagram of fiber in fiber-reinforced concrete includes three zones (Figure 14.11):

I - zone of the elastic strain energy of both the matrix and of the fiber;
II - zone of the alternate displacement of stress on the matrix and the fibers and the formation of microscopic cracks;
III -zone of the perception of load by the fibers.

In the III zone, the complete destruction of material is stopped by the adhesion strength in the contact zone or by the strength of some fibers only.

If the length of the fibers (l) is less or equal to the critical length (l_{cr}), the breaking energy is determined by the energy required for the extraction of fibers. At $l > l_{cr}$ the material is destroyed with the breaking of fibers and the breaking energy is inversely proportional to the length of the fiber. The breaking energy is proportional also to the diameter of the fiber. The increase of the tensile strength of fiber reinforced concrete can be achieved by the application of fiber cables and therefore by increasing their working diameter.

Portland cement based fiber concrete, reinforced by steel fibers (*steel-fiber concrete)* are the most widespread. Steel fibers are usually represented by the sections of wire. Fibers can have different cross sections - round, oval and other sizes from 0.2 to 1.6 mm and length from

10 to 160 mm. The surface of fibers can be shaped; processed by etching and can be smooth. The quantity of fibers introduced into the concrete in the majority of the cases varies from 0.5 to 2% by volume. Introduction of steel fibers into the concrete in quantity of 1-1.5% in volume increases its tensile strength to 100%, bending strength by 150-200%, and compressive strength rises to 10-25%.

Due to the higher crack-resistance of steel-fiber concrete it possesses an increased(1.5-2 times) frost, heat and fire resistance, as well as water impermeability. The valuable qualities of steel-fiber concrete are the increased abrasion resistance, impact and dynamic strength. Thus, the wear-resistance of steel-fibro-concrete is increased by about 50% and the impact strength by about 10-12 times.

Application of steel-fiber concrete is promising both in the precast and in monolithic structures. The application areas include: road and aerodrome paving, sprayed layers for the revetment of the mines of tunnels and fire-retardant stack linings, thin-walled and ribbed floor slabs, the elements of shells, piles, ties and others.

The effectiveness of the application of the steel-fibro-concrete structures is made evident by the observable decrease of labor expenses for the reinforcing works, the reduction of steel and concrete contents and the structures thickness, combination of the technological operations of the preparation of concrete mixture and reinforcement, improvements in the durability of structures and reduction in the expenses for routine repair.

It is expedient to combine fiber and bar reinforcement i.e. the application of the combined reinforcement to resist significant loads in the structure.

The introduction of steel fibers into the concrete mix is characterized by specific technological difficulties connected with the clumping, the formation of "urchins", and the complexity of proper consolidation of material. The fiber is introduced into a preliminarily mixture of cement, water and aggregate or the aggregates and fibers are premixed at first, and then the cement and water are added. The preparation of fibers concrete involves the use of mixers with pulsing action on the mixture and other special mixers.

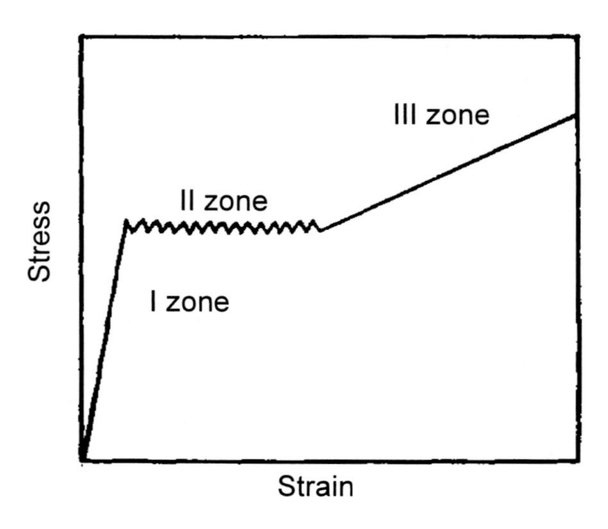

Figure 14.11. Typical curve of dependence stress-strain for the cement compositions, reinforced by the fiber.

There is a positive experience of the application of *glass-fiber concrete* in construction, which makes it possible to reduce significantly the mass of structures additionally. The positive experience is obtained by including cement paste or mortar of alkali-durable fibers (up to 5%) throughout the mass. Tensile and bending strength of concrete reinforced by glass exceeds the strength of the ordinary concrete by about 2-3 times even after 10 years of air storage. The maximum deformation due to the maximum stretching effort in the glass-reinforced concrete,s is 10 times higher than for the non-reinforced concrete.

14.6. Glass-cement Materials and the Glass-fiber Plastics

Glass-cement Materials

This group of materials are obtained by the application of inorganic binding agents and glass (including basaltic) fibers. The application of both hydraulic binders (Portland cement and its varieties, alumina cement) and also air binders is possible (gypsum, caustic magnesite). The cables and clusters, twisted from thin glass fibers, and grids made of the alkali-free fiberglass serve as the reinforcing materials. The fibers, obtained by drawing from basaltic fusion are also used.

The strength-weight ratio of glass-cement is about 1.5 times higher than the strength-weight ratio of steel, and the density is lower than the density of aluminum alloys which is 1.5-2 times. The relative deformations of glass-cement at the moment of the microscopic cracks forming in the matrix are in 30-60 times higher than in reinforced concrete. The use of glass-cement compositions instead of the reinforced concrete makes it possible to decrease the mass of structures by 8-10 times and to reduce the cement content in 2-4 times.

The properties of fiberglass depend on the chemistry of glass and method of obtaining it. According to the chemistry the glass fibers are divided into three groups: alkali-free (content of alkaline oxides not more than 1-2% throughout the mass), low-alkali (from 2 to 10%) and alkaline (more than 10% of alkaline oxides throughout the mass. The tensile strength of fiberglass is considerably higher than the strength of massive glass, which is explained by its lower heterogeneity, presence of the strengthened surface layer. The observed increase in the strength as the diameter of fibers decreased(Figure 14.12) is explained by the smaller thermal gradients, which appear during the glass cooling and with the respectively smaller intensity of crack formation.

The average strength of the elementary fiberglass is 3500 MPa. Atmospheric moisture decreases the strength of glass (Figure 14.13, 14.14). After several months of storage the strength of fiberglass decreases by 10 to 15% in comparison with the initial. To prevent the reduction in the strength of fiber under the action of atmospheric moisture, they are protected by hydrophobic substances during the drawing.

Fiber from alkali-free aluminum-borosilicate glass has the largest strength. Alkaline oxides decrease the strength of fiberglass. Fibers from phosphate and borate glass has the lowest strength.

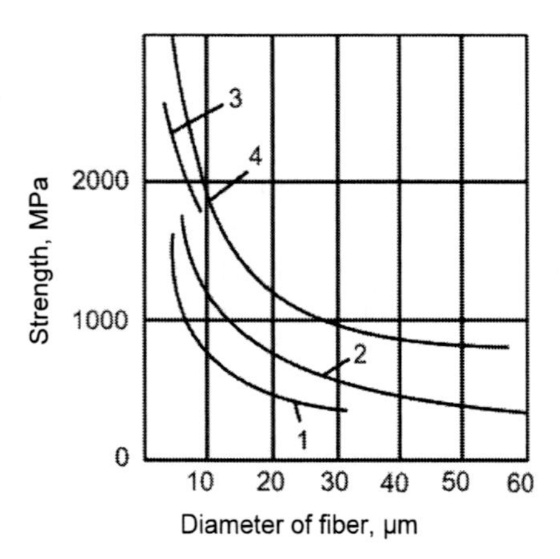

Fig. 14.12. Relationship between ultimate strength and the diameter of fiber, obtained by the different methods:

1- bulge forming; 2- glass rod forming; 3- jet-stretching (compressed air); 4- jet-stretching

Figure 14.12. Relationship between ultimate strength and the diameter of fiber, obtained by the different methods:
1- bulge forming; 2- glass rod forming; 3- jet-stretching (compressed air); 4- jet-stretching (rotary drum)

Glass in the system of Na_2O-MgO-Al_2O_3-SiO_2 in the alkaline environment of the hardening cement stone has relatively high durability if it composition (mol %) is such as Na_2O 0.14-0.3; MgO 10-30; Al_2O_3 0-15; SiO_2 50-70. Industrial fibers are produced in the form of the short (up to 60 cm) models of elementary fibers (staple fiber) and continuous fibers with length of hundred and thousands of meters.

An increase of fiberglass strength is achieved by action on its surface for the purpose of the reducing the quantity of micro-flaws by heat treatment.

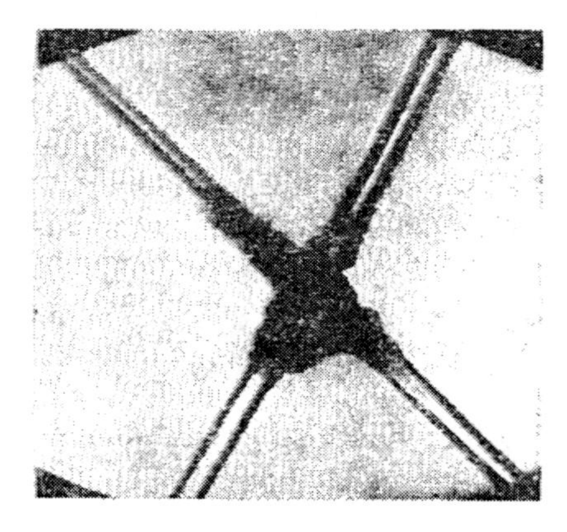

Figure 14.13. Microphotography of glass fiber (Na_2O content is equal to18,9%), 180 days hardening in moist environment.

Staple and continuous fibers are obtained from basaltic fusions just as in the case of glass. Continuous fiber is processed into the roving (plait), woven and non-woven materials. Glass (basaltic) wool, roll materials, mats, linens and others are made of the staple fibers.

Two basic methods of reinforcement of inorganic materials by fibers are used: directed and chaotic. The tentative influence of the method of reinforcement on the strength of fiberglass materials along the fibers is given below:

Distribution of the fibers	Strength, %
Unidirectional	100
Reticulate	45-50
Two-dimensional chaotic	30-37
Three-dimensional chaotic	0-20

The tensile strength of glass-cement grows linearly with an increase of the glass content (Figure 14.15).

On the stress-strain diagram of glass-cement, two sections are observed. The linear section characterizes the joint operation of fibers and matrix. At the point of inflection (a) the matrix undergoes splitting and load is redistributed to the fibers.

There is an experience of the successful application of glass-cement as the decoration layer for different forms of concrete and reinforced concrete products, roof claddings, facing slabs, light-weight facing panels, materials for the construction of oil storages, silos, tubes, pipes, chutes, waterproofing, etc.

A new type of glass-cement is "*transparent concrete*" , that appeared on the construction market in 2004 (LiTraCon - Kight Transparent Concrete). The matrix of this concrete is optical fiberglass, whose volumetric content is 4-5%. Fibers are arranged parallel to each other and transfer light from one front surface of the block to another. Walls from such blocks become luminescent, different shadows and figures can appear on them. With the use of " transparent concrete " a sensation of easiness", is created "the airiness " of structures, the work of optical fibers ensures light transmission almost without losses within a distance of 20 m.

Glass-fiber Plastics

The plastic masses, reinforced by fiberglass materials, belong to the group of glass-fiber materials. A number of the valuable properties is characteristic of this type of materials: high strength, easiness, low thermal conductance and others. Use of light structures on the basis of glass-fiber plastics makes it possible to decrease the mass of buildings by about 16 times in comparison with brick. Glass-fiber plastics are lighter by 1.5 times than products made of aluminum alloys, substantially exceeding the last in terms of mechanical strength. The light transmission of glass-fiber materials can reach 90% with a thickness of 1.5 mm, and up to 30% in the ultraviolet spectrum. Its thermal conductance of 6-10 times is lower than in ceramics and concrete. Industry makes glass-cloth laminates, sheet glass-fiber materials, glass-fiber materials with oriented arrangement of fibers, roll materials.

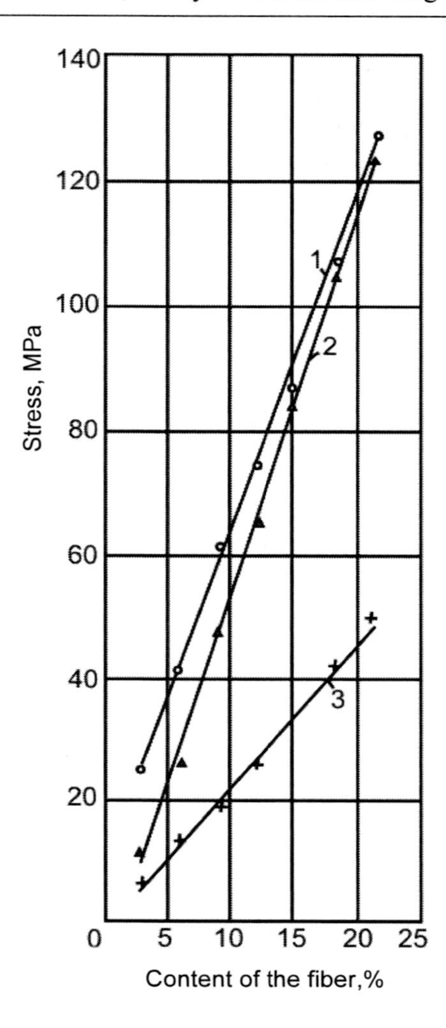

Figure 14.14. Change in the characteristics of the tensile strength of glass-cement along the fibers in the dependence on the glass fibers content: 1 -ultimate strength; 2- stress, which causes crack formation in cement stone-; 3- limit of proportionality.

Glass-cloth laminates – These are laminate sheet materials, produced by the method of the hot pressing of the widths of fiberglass fabric, impregnated with synthetic binding agent and packed by parallel layers. The fiberglass laminate in the form of sheets with a thickness of between 0.5…7 mm is used for the envelope of sandwiches-panels, elements of three-dimensional structures, the device of electrical switchboards and others.

Sheet glass-fiber materials are manufactured as flat and corrugated with the longitudinal and transverse wave. They are made opaque (with the thickness of 1 mm they pass to 50% of incident light), semi-transparent (50-60%) and transparent (60-85%). Basic properties include: density 1200-1300 kg/m^3, water absorption after 24 h, is not more than 1.2-1.5%, ultimate tensile strength is not less than 40-50 MPa. Sheet glass-fiber materials are intended for preparing transparent light enclosing structures of walls and coatings, partitions, enclosures of the small forms of urban construction and others.

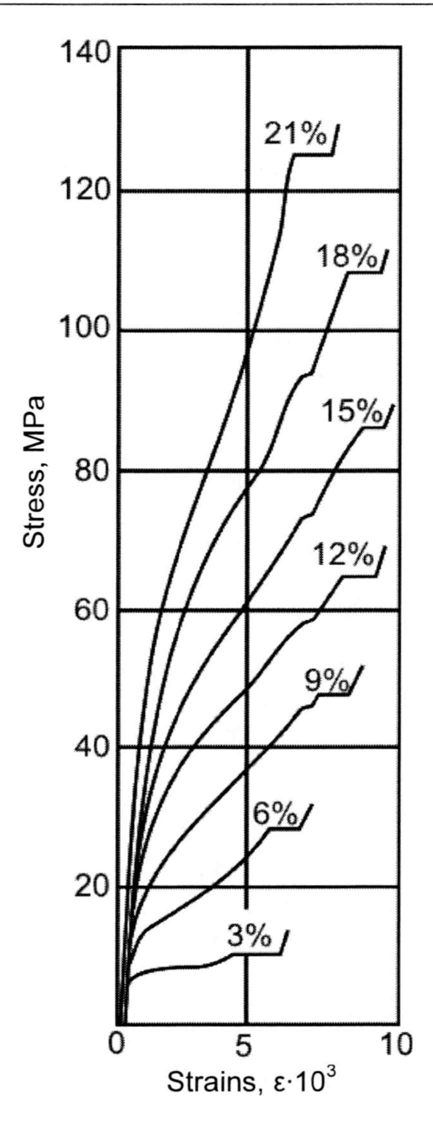

Figure 14.15. Diagrams of tensile strains along the fibers of glass-cement with different glass fibers content.

The basic representatives of glass-fiber materials with the oriented fibers are the glass-fiber anisotropic materials, obtained by placing of elementary glass fibers parallel to each other with the simultaneous putting of binders on them. Obtained by such means glass-fiber veneer is dried and stacked into packages, which are pressed on the hydraulic presses at raised temperatures. The strength of the sheets in the tension achieves 100 MPa. The strength of the anisotropic glass-fiber material in the longitudinal and transverse directions of the sheet depends on the arrangement of glass fibers in the glass-fiber veneer and the method of laying the sheets into the packets and the glassfiber content (Figure 14.16).

Phenol-formaldehyde, polyester, epoxy and silicon polymers are used usually for preparing the glass-fiber materials as the binding agents. The regulation of the properties of binders is provided by the introduction of fillers.

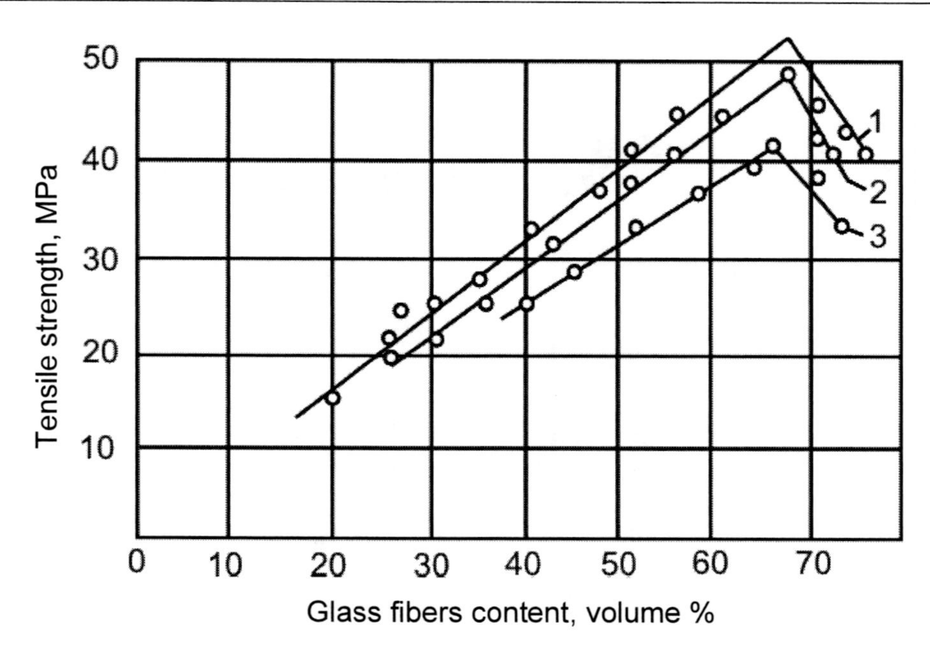

Figure 14.16. Relationship between strength of anisotropic glass fiber material and diameter of the fiberglass: 1-14,5 μm; 2-19,5 μm; 3-9,5 μm;

Self-Assessment Questions

18. Describe the general characteristics of reinforced materials.
19. What reinforcing methods are used?
20. What are the types of reinforcement of concrete structures.
21. What are the methods of prevention of reinforcement corrosion?
22. What are the peculiarities of basic methods of reinforced concrete products manufacture?
23. Explain the casting and thermal treatment of reinforced concrete.
24. Which ways of cement saving are possible at the production of concrete and reinforced concrete?
25. Which ways of metal and fuel saving are possible at the production of reinforced concrete elements?
26. What are the peculiarities of fiber concrete.
27. Distinguish between the properties of glass-cement materials and glass-fiber plastics.

Chapter 15

BITUMEN AND TAR MATERIALS

Materials of this group (asphalt materials) include a great number of road, roofing, waterproofing and other materials, which are produced from bitumen and tar cementing materials. They are used as emulsions, pastes, mastics, mortars and concretes, roll and piece products with such characteristic features as high waterproofing and chemical resistance and adhesiveness.

An overwhelming majority of bitumen and tar materials is applied in asphalt concrete manufacture. Asphalt concrete coatings are used on highways of any load rating. They are characterized by sufficient mechanical strength, required elastic and plastic deformations, and good adhesion with tires and easy in mending. Asphalt carpets are widely used in aerodrome construction. Asphalt materials are also applied for waterproofing, anticorrosive protection of pits, pipes and water storages.

Bitumen and tar varnishes and paints are used for waterproofing concrete, wooden and metal structures, hot- and cold- asphalt mastics and mortars are applied for plastering damp-course, and roll materials - for membrane waterproofing.

Considerable part of civil and industrial roofs of buildings are constructed with use of bitumen, roll and mastic roofing materials. Effective and endurable roofing materials include glassy bitumen felt, overlay bitumen felt, bitumen felt with colored grit and other. Asphalt concrete properties are improved fundamentally by adding polymer additions.

15.1. BITUMEN AND TAR BINDERS

Common Specifications

Bitumen and tar along with synthetic polymers are organic binders and consist of high-molecular hydrocarbons and their non-metal derivative mixes.

Bitumen and tar are produced from different materials and in different ways, but they are united by common properties, which is caused because of similarity of structure and composition. The most important components as for bitumen, so for tar are oils and resins, which are flow medium with dispersed solid phases in it: asphaltenes for bitumen, free carbon for tar.

Asphaltenes are high-molecular compounds with molecular weight between 900–6000, which contains mainly compounds of carbon with oxygen, sulphur and nitrogen. Modifications of asphaltenes, which appear at more deep oxidization, are carbenes and carboides.

Resins have molecular weight from 500 to 2000, their consistency changes from viscous and sticky to solid amorphous mass.

Bitumen and tars are characterized by a micellar structure. The kernels of Micelle are asphaltenes in bitumen - particles of colloid sizes (<20 micrometers), and in tars is a free carbon. The shell of micelle in both cases consists of oils and resins. Aggregate state of bitumen and tars depends on correlation in them of oils and solid components, and also from a temperature. When the oil content of micelles is excessive, there is no contact and system is in a liquid state. Reduction of viscidity and transferring into the liquid takes place also on heating, and results in destruction of micelle. At lower temperatures, the concentration of micelles is increased, come in a direct contact and consequently, the system passes into a viscid or solid state.

The main qualitative parameters of bitumen and tars are viscidity, deformability and heat capacity. These properties depend mainly on their composition. While increasing content of oils, the viscidity reduces, deformation ability grows and the softening temperature decreases. Resins stipulate the cementing properties of bitumen and tars, give them plasticity, and increase tack (adhesion ability). Asphaltenes in bitumen and the free carbon in tars improve the softening temperature and solidness.

Viscidity of tars and liquid bitumen is determined by viscosimeters which measure time of flow through an opening with diameter 5 or 10 mm at a constant temperature.

Viscidity of the semisolid and solid bitumen is relatively estimated from the depth of penetration of certain diameter needle at a temperature of 25°C into the tenth part of millimeter (degrees of penetration). With the increase of temperature viscidity of tars falls quicker than that of bitumen, due to insignificant tar micelles firmness.

Plastic flow capacity is determined for bitumen. It is insignificant for a solid tar product (pitch). Deformability of bitumen is characterized by *extension* (inklength) - ability to stretch in a filament. The greater the extension, the higher plasticity will be. Solid bitumen and tar materials, as well as other materials of amorphous structure, don't have a specific temperature of melting, but are characterized by the *softening temperature*, which is determined on a standard tester «ring and ball». A softening temperature is one at which bitumen or solid residue after the distillation of tar (pitch), embedded in a ring, under the action of steel ball is pressed out at a certain depth.

It is also usual for the flash temperature, fragility, solubility and other properties of bitumen and tar binders to be determined if it is required.

In the final analysis, the basic properties of bitumen and tars are determined by their chemical composition with which they are interconnected. So, increasing penetration and stretching will be accompanied by diminishing softening temperature and vice versa.

Bitumen and tar binders are hydrophobic materials which do not moisten in water and are water-insoluble. Values of their real and average density range between 0.9-1.3 g/cm^3 and are practically equal. Due to high water resistance and insignificant porosity, they are successfully used as basic components of waterproofing materials which have high water impermeability.

Bitumen and tars are chemically inert to water solutions of mineral salts and acids, which allow them to be widely applied for corrosive protection. However, concentrated alkalis and strong concentrated acids which have oxidizing ability would have aggressive influence on these materials (e.g, 10% solution of alkali causes the saponification of bitumen).

Bitumen and tars are soluble in organic solvents, especially the non-polar type (dichloroethane, benzene, chloroform, lacquer petroleum and others like that).

In the conditions of moist environment or foul with microorganisms, bitumen do not protect organic materials from rotting. Unlike bitumen, tars are fully biostable. That is why tar materials can be advantageously applied in conditions of concealed flashing. Furthermore, tars have higher adhesiveness than bitumen on account of the enhanced content of surfactants (phenols, carbazole and others).

Both bitumen and tars are susceptible to deterioration (with respect to worsening of quality parameters, decline of longevity) through the action of atmospheric factors. This is the essence of the senescence process which involves the successive oxidization of components, which slowly flows at ordinary temperatures. A time dependent decline of the oils and resins in bitumen and tar materials results in the reduction of their plasticity and increases their fragility.

Types of Bitumen and Tar

Bitumen are divided into natural and artificial –oil-firing.

Natural bitumen are products of natural oxidization of oil which flows in the earth's crust for ages. Per se they meet rarely, mainly impregnate the porous mountain rocks - sandstones, limestones, dolomites. These bitumen rocks are called asphalt. Bitumen are extracted by organic solvents from them, at a content of not less than 10–15 %. Natural bitumen contain a negligible quantity of harmful admixtures - paraffin and are characterized by enhanced longevity. Frequently, asphalt rocks are ground down and are used as fillers of mastics and asphalt concrete.

In construction practice mainly, artificial *oil bitumen* are appied. Oil bitumen are produced as the finished products of direct distillation or oil cracking.

They are divided into two basic types of oil bitumen according to the production method use as - residual and oxidized (Figure 15.1). Residual bitumen appears as the leftover after distillation of oils from black oils and tar oils. The output of residual bitumen from distillation of aromatic oils can be as high as 15%. Residual bitumen contains a little amount of asphaltenes. That is why they have viscid consistency and belong to the soft types.

The oxidized bitumen is produced by oxidization of fluxing oils or residual bitumen at the temperature range between 180 - 280°C. Under the action of oxygen condensation reactions take place, which results in concentration of bitumen asphaltenes due to the decreasing of oils and resins part, at such temperatures. Increasing oxidation of part of the asphaltenes in the bitumen causes a marked change of physical-mechanical properties, such as - increase the softening and flash temperature, or diminishing the penetration and stretch.

Depending on the function, bitumen may be distinguished as - road, structural, roofing and special. The basic standard requirements for the road and structural oil bitumen are shown in Table 15.1.

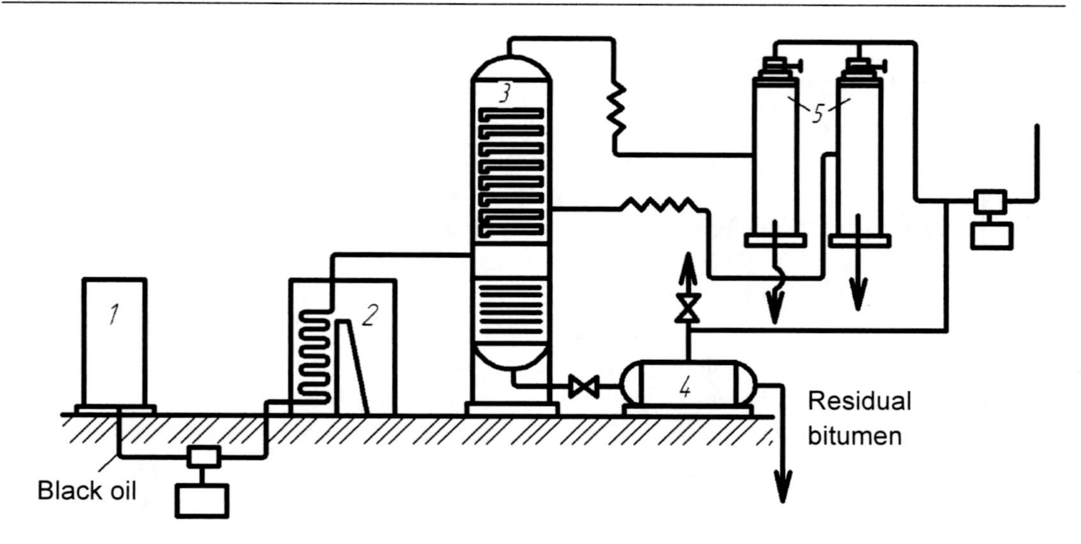

Figure 15.1. Scheme of residual bitumen manufacture:
1– stock measuring device; 2– tube furnace; 3– vacuum column; 4– bitumen tank; 5– condensers.

Three classes of liquid bitumen exist and are applied for using asphalt materials along with viscid bitumen These include:

- Bitumens which thicken quickly;
- Bitumens that thicken with moderate speed;
- Bitumens that thicken slowly.

Liquid bitumens can be produced from viscid bitumen with solvents.

Tars are used as construction binders in considerably less measure, than bitumen. This due to a series of reasons, which include; a narrow interval of thermal stability, enhanced capacity for senescence under the action of atmospheric factors, deficit of tar products, which are used as valuable chemical raw material, and their toxicity.

Table 15.1. Common properties of oil bitumens

Bitumen grade	Softening temperature not below, °C	Penetration in 0.1 mm		Stretch at 25°C not lesser than, cm
		at 25°C	at °C	
Road viscid				
200/300	35	201-300	45	-
130/200	40	131-200	35	65
90/130	45	91-130	28	60
60/90	48	61-90	20	50
40/60	52	40-60	13	40
Construction				
50/50	50	41-60	-	40
70/30	70	21-40	-	3
90/10	90	5-20	-	1

The condensed products of destructive (without access of air) distillation of solid fuel and oil produce tars. Coal tars are the most widespread. The processing of anthracite coal produces coke, coke gas, benzene, ammonia and coal resin - "crude tar". Ammonia water oils and solid remains – *pitch* are manufactured by processing of raw tar. In construction, refined or composite tars are mainly applied. Refined tars are produced by rectification of raw tars from water and other volatile substances, whilst the composite tars are produced from composition of pitch and tar oils.

15.2. Emulsions, Paste. Composite Binders

Along with bitumens and tars, water *emulsions*, which are the dispersible systems in which the shallow particles of organic binders (lesser than 1 μm) are evenly diffused in water, are also used in construction. The manufacture of stable emulsions is possible due to application of surfactant additions – emulsifiers. Water-soluble high molecular compounds and soaps of alkaline metals (saponated rosin, fatty acids, lignosulphonates and other) are used as emulsifiers. The mechanism of emulsifier action consists in adsorption on the particles of bitumen or tar and formation of protective shields which are instrumental in maintenance of dispersed particles in the hanging up state.

The emulsion variety is the so-called *emulsion pastes*, emulsifier of which is superfine hydrophilic mineral powders (lime, clay, tripoli powder and other like that).

In Table 15.2 the characteristic compositions of bitumen emulsive pastes, applied for cold asphalt mastics are presented.

Emulsions and pastes do not contain volatile solvents; they can be used to coat both dry and wet surfaces. Upon evaporation of water the dispersible phase of emulsions coagulates and homogeneous coverage appears.

Bitumen and tar emulsions and pastes are widely used in road and hydraulic engineering construction. Pastes are used as a binder for cold asphalt mastics.

Table 15.2. Recommended compositions of bitumen pastes

Emulsifier	Composition of components by mass %		
	Bitumen	Water	Emulsifier
Clay:			
high-plastic	55-65	25-87	8-10
plastic	45-50	31-45	10-14
Loam:			
heavy-weight	40-50	30-45	15-20
light-weight	35-40	25-30	30-35
Slaked lime:			
1 grade	50-55	33-42	8-12
2 grade	45-50	34-43	12-16
Tripoli, loess, diatomite	50-55	30-40	10-15

For making emulsions and pastes the viscid oil bitumens are used. The technological process consists in heating and mixing of the initial components in the special mixers.

The improvement of the properties of bitumen and tar materials can be attained by application in the form of composite binders: tar-bitumen, rubber-bitumen, bitumen- and tar polymeric.

Binders which are formed by the combination of bitumen with tar or pitch, have the advantages of both components: enhanced biostability and atmospheric resistance, high adhesive capacity, less sensitivity to temperature drops and etc. The most widespread are composites which have a 65-70% bitumen and 25-30% tar.

Rubber matters are effective additions for increasing the elasticity, softening temperature, durability and longevity of the bitumen. Hence, by combining rubber and bitumen in the ratio of 1:1, the relative lengthening improved by more than 200%, elasticity – by almost 3 times, resistance to breaking by 0.8 MPa, the softening temperature to 120°C, and the brittle temperature reduces down to - 20°C.

Content of rubber in *rubber-bitumen binders*, as a rule, is in the limits from 5 to 30%, thus its optimum concentration grows to the extent of decline of viscidity of bitumen. The ordinary ground up rubber from old motor-car overlays, and also wastes of production are used for making the binder. Production of rubber-bitumen binder consists in melting of the ground rubber with bitumen at a temperature 160–200°C and continuous interfusion within 1-2 hours to complete homogenization. With respect to mechanical influence, molten bitumen rubber is devulcanized and acquires high plasticity.

A number of bitumen composites with additives of latex, polyethylene, epoxies have been developed. *Bitumen-polymeric compositions* have higher heat-resistance, mechanical strength and deformative ability, especially at subzero temperatures. They both are produced by mechanical method - by treatment of the mix on rollers and method of alloying at temperatures ranging from 160-190°C.

15.3. Asphalt Mastics, Mortars and Concrete

Mastics

Plastic substances which are produced using combination of organic binder with fine powder-like or fibred fillers are categorized as mastics. Fine powdered mineral matters – limestone, dolomite, chalk, talc, tripoli powder, ash, cement, etc. are used as powder-like fillers; low sorts of asbestos, mineral wool, etc - as fibred fillers. Mastics can consist of 10 – 70% fillers, depending on the function and peculiarities of the components.

The name of mastics usually specifies the kind of binder, applied: bitumen, tar, rubber-bitumen etc. Depending on the method of application mastics may be categorized as hot or cold.

Hot mastics are applied with previous warming-up to 130-180°C. They harden quickly as a result of diminishing viscidity of binder at cooling and have high water resistance and adhesive capacity.

At the same time its application involves a number of difficulties: by the danger of working with hot materials, enhanced adhesiveness, by the necessity of heating of transport devices and etc. These difficulties are absent with the application of *cold mastics* which allows waterproofing work to be mechanized, increase their productivity and improve the labour time. Furthermore, cold mastics allows getting a more lubricating skim and more economy with respect to the cost of organic binders. Cold mastics are mixtures of organic binders, thinned with solvents, or water-emulsion pastes with mineral fillers. Petrol, ligroin, white spirit, kerosene, oil butter, etc. are applied as solvents. At the normal ambient temperature, cold mastics are applied without heating, their hardening progresses due to evaporation of solvent or water evaporation.

Depending on the required function, mastics are divided into roofing, waterproofing, pressurizing, spackling, lining, etc.

For roofing mastics are used for agglutination of roll materials and waterproofing of roofs, the basic indexes are heat-resistance and stretchability.

Bitumen compositions have got the widest application among roofing mastics.

Tar mastics are applied for gluing on tar paper materials. Tars and their mixtures with polymers serve as binders in it.

Waterproofing mastics are applied for waterproofing plaster creation and filling of deformation joints. Major requirements for waterproofing mastics are small values of water saturation and swellings, sufficient deformability.

The basic requirements for the hot waterproofing asphalt mastics are presented in Table 15.3.

For waterproofing of plasters and arrangement of unroll roofs, cold asphalt mastics are widely used. They are produced by mixing bitumen pastes with mineral fillers. They have enhanced water saturation in comparison with hot asphalt mastics; however, the process of water absorption is halted with the lapse of time. The advantages of cold asphalt mastics are: enhanced durability and heat-resistance, possibility of high mechanization of works. Among bitumen-polymeric mastics the most widespread are: bitumen-rubber, bitumen-latex and bitumen-epoxy composites.

Asphalt Concrete

Asphalt concrete is the name for material which is produced as a result of consolidation of compacted and rationally composed mixture of asphalt binder and mineral aggregates. Asphalt binder is the mixture of bitumen with mineral filler – micronized powder. Sand and crushed stone or gravel is used as mineral aggregate.

Table 15.3. Physical-mechanical indexes of waterproofing asphalt mastics

Index	Category of heat-resistance		
	I	II	III
Softening temperature, °C	90-105	75-90	60-75
Depth of penetration of needle at 25°C is not less than, 0.1 mm	5	10	15
Tensility at 25°C not less than, cm	0.5	1.5	3
Water saturation under a vacuum by mass no more than, %	0.5	0.3	0.1
Swelling under a vacuum by volume no more than, %	0.05	0.05	0.05

Asphalt concretes are classified according to structural features – grain and porosity, the temperature at a packing, physical state and function. Depending on the maximal fineness of grains of aggregates in the mixture, asphalt concretes may be distinguished as: coarse-grained (to 40 mm), medium-grained (to 25 mm) and fine-grained (to 15 mm). Maximal fineness of aggregates should not exceed 0.6–0.7 thickness of asphalt coverage. In the absence of in-bulk grains of coarse aggregates (more than 3–5 mm), the material is called sand asphalt concrete or asphalt mortar. For porosity less than 5% asphalt concretes are considered as dense and more than 5% - porous asphalt. Depending on the temperature of mass at packing asphalt concrete may be distinguished as hot, warm and cold asphalt concretes. The first are packed at a temperature of the masses not below 120°C, the second - not below 60°C. Cold asphalt concrete is packed at the temperature masses 25-30°C and ambient air not below 10°C. The positive feature of hot asphalt mixtures is high-rate of hardening, and cold is the protracted storage.

Asphalt concrete mixtures depending on workability may be stiff, plastic and poured. Diminishing the stiffness of the mixtures leads to difficulty in the placement of mass, its compacting is however facilitated.

Depending on the functional requirements asphalt concretes are classified as: road, aerodrome, decorative and hydraulic. H*ydraulic asphalt concrete* is intended for permanent work in water. That is why it should have enhanced water impermeability, water resistance, elasticity and heat-resistance.

Viscid road bitumen are used for making hot asphalt concrete. Thus more viscid asphalts are applied if it is required to provide enhanced heat-resistance and water resistance of coverage. In warm mixtures, bitumen of reduced viscidity are used and also liquid bitumen which thicken quickly and those, which thicken with medium speed. Cold asphalt concrete is prepared on the basis of liquid bitumen which thicken with middle speed or slowly. For asphalt concrete, used in damp-proofing, it is more effective to use bitumen-polymeric binders.

An important component of asphalt concretes is the filler, which improves the basic properties of material, connects the bitumen on the surface and structures it. Mineral filler is instrumental in increasing the density of asphalt concrete (Figure 15.2) and the reduction of bitumen consumption. The fine powders from such materials - limestones, dolomites and asphalt rocks are applied as fillers. Sometimes pulverized industrial wastes are also used. The content of clay particles in the filler must be no more than 5%. The optimum dispersion of powder, is that at which its specific surface measures 4,000-5,000 cm^2/g. Mineral powder, produced from rocks, which contain acidic compounds, does not have the necessary positive influence on bitumen. An effective method of improvement of the filler quality is its activation by treatment with bitumen and surfactant matters in the process of grinding.

Heat and crack resistance of asphalt concrete is considerably improved by addition of 2-3% fibred filler(– low-grade asbestos). Surplus asbestos results in a negative effect such as: increased porosity and reduced water resistance of concrete.

Sand and crushed stone or gravel used for asphalt concrete is mainly the same, as those used for cement concretes. It is thus important to create such grain-size distribution of the mineral part of asphalt concrete which would provide minimum voidage of the mixtures.

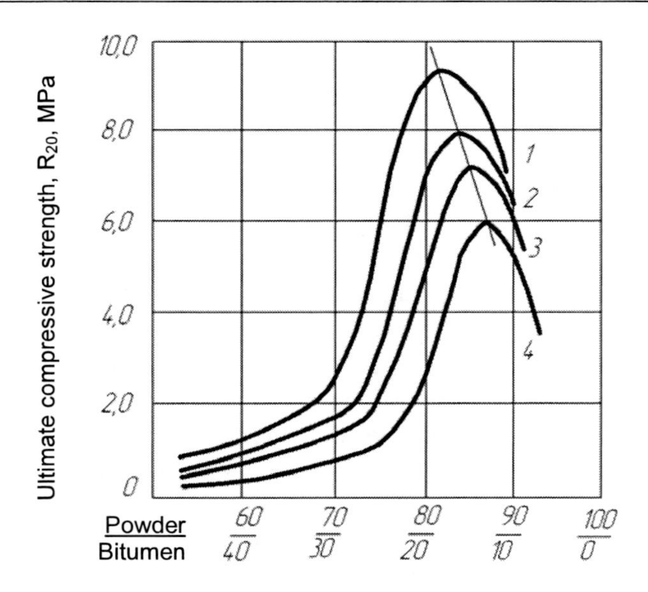

Figure 15.2. Relationship between content of mineral filler and concrete strength:
1– bitumen 40/60; 2–90/130; 3–130/200; 4–200/300.

Among the methods of proportioning asphalt concrete mixtures the most widespread is the method of that involves the use of dense mixtures curves, which includes determination of grain-size distribution of the crushed stone, sand and filler; the choice of the proportion of mineral materials with condition of 100 % total content of all fractions: setting of approximate amount of binder with next correction by the trial batches and researching of specimens.

In Table 15.4 approximate compositions of sand and fine-grained mixtures, recommended for hydraulic asphalt concrete are presented.

The basic properties of asphalt concrete are closely related to its structure, which is determined by the structure of mineral part, by the features of mineral materials and their interaction with bitumen.

Table 15.4. Composition of hydraulic asphalt concretes by mass %

Component	Sand asphalt		Fine-grained asphalt concrete		
	compacted	poured	compacted	poured	porous
Crushed stone or gravel of fineness to 15 mm	-	-	10-25	10-25	40-65
Stone chips of fineness to 5 mm	20-35	20-35	40-65	40-65	20-35
Medium-grained sand	40-65	40-65	15-25	15-25	15-25
Mineral powder	15-25	15-25	15-25	15-25	0-5
Shortly fibred earth-flax	1-3	-	1-3	-	-
Bitumen	7-12	12-18	6-10	10-15	4-7

The physical-mechanical properties of asphalt concrete are related, first of all, with its density. So the maximum density correlates to the maximum strength of material. Influence of moisture on asphalt concrete increases with the increase of the amount of connected pores. The volume of these pores can be correlated with the amount of water absorption.

Bitumen has the most high viscidity and strength in the saturated state, i.e., in the area of contact with the surface of mineral components. Increasing the unabsorbed bitumen-free amount results in structure weakening and strength decreasing (Figure 15.3) of asphalt concrete. However a certain amount of free bitumen is needed for a granting of plasticity and corrosive resistance to the asphalt concrete.

Ultimate compressive strength of asphalt concrete is determined by testing cylindrical specimens at temperatures of 0; 20 and 50°C.

The ratio between strength at the normal temperature and strength at a temperature 50°C is called the *heat-resistance coefficient whilst the* ratio between the indexes of strength at 20 and 0°C is called the *elasticity coefficient*. Shear deformations take place due to insufficient heat-resistance, which cause dulling of asphalt concrete on the slopes. Deformability and crack-resistance are reduced due to insufficient elasticity at low temperatures. Additions of some polymers and special rubbers improve structural and mechanical property of asphalt concretes. Rubbery polymers in the amount 2-3% introduced to the mass of bitumen increase the interval of plasticity of asphalt concrete mixtures up to 100-120°C, decrease the temperature of fragility by 10-20°C, considerably improve their elasticity over a wide range of temperatures, and improve a range of other properties. To ensure good resistance to temperature of asphalt concrete ,it is important also to provide optimum content of mineral filler in it.

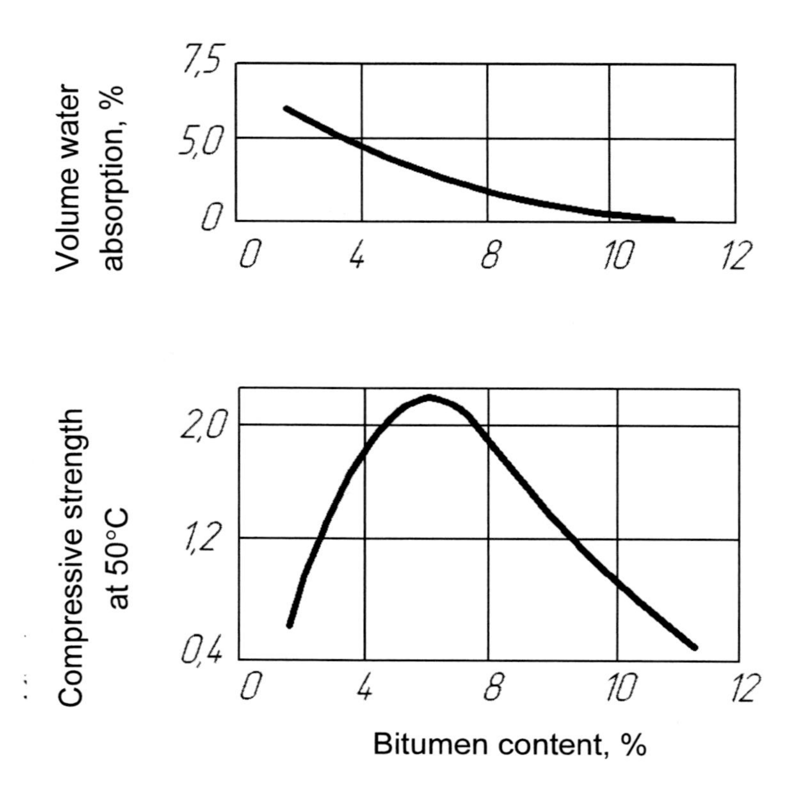

Figure 15.3. Relationship between bitumen content and concrete properties.

Table 15.5. Properties of hydraulic asphalt concrete

Indexes	Compacted asphalt concrete		Porous asphalt concrete	
	normal	improved	normal	improved
Ultimate compression strength not less than, MPa: at 20°C at 50°C	2.5 1.2	3 1.5	1.6 0.6	2 0.8
Coefficient of thermostability not less than	3	2.5	4	3.5
The water resistance coefficient is not less	0.85	0.90	0.75	0.80
Elasticity coefficient	2.0-3.0	2-2.8	2-4	2-4
Remaining porosity %	1-3	1-2.5	6-20	4-8
Water absorption in vacuum by a volume no more %	2	1.5	5-18	3.5-7.5
Swelling under a vacuum no more %	1	0.5	-	-

Water resistance of asphalt concrete is characterized by the ratio between ultimate compression strength of water-saturated and dry specimens at 20°C, and also water absorption and swelling under vacuum. Increasing the density, improvement of bitumen coupled with mineral components, positively influence the water resistance. Water resistance also depends on the amount and chemical composition of bitumen and the mineral powder. In hydraulic asphalt concrete the content of bitumen and powder is recommended to be 1-2% higher, than for road asphalt concrete. The basic technical requirements for the properties of hydraulic asphalt concretes are presented in Table. 15.5.

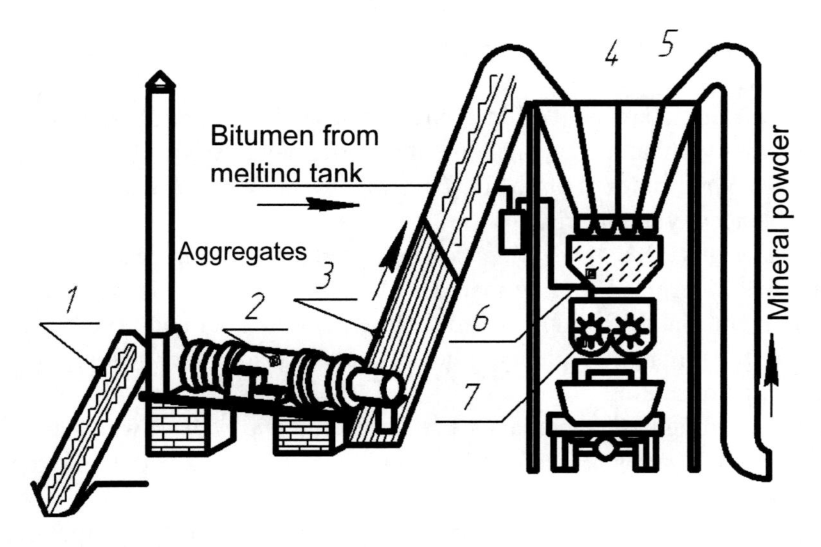

Figure 15.4. Scheme of asphalt concrete manufacture:
1– elevator for sand and crushed stone; 2– drum dryer; 3– "hot elevator"; 4– batch bin with vibrotrunk; 5– mineral powder elevator; 6– batcher; 7– blade mixer.

Asphalt mixtures are made on the specialized enterprises of stationary or temporal movable type (Figure 15.4). Such basic operations are included in the technological cycle of hot asphalt mixtures production: drying, heating of aggregates to 160-200°C, separating them into fractions; preparation of mineral filler in the mill unit; melting of bitumen in melting tank at temperatures between 150-190°C, dosage and careful mixing of all components.

Readymade road concrete mixtures with the temperature 140-160°C is transported on industrial site, where it is specially placed with a mechanical paver and is compacted by static and vibrating rollers, surface vibrators and other special vibratory compactors. Bases under an asphalt concrete should be compacted, dry and plane, to provide good coupling and take up the operating loadings without considerable deformations. If it is required, the base is treated by insecticides.

For asphalt concrete revetments of hydraulic structures for the purpose of improving water impermeability, the surface treatment with hot bitumen, emulsion or asphalt mastics is conducted and it is sprinkled with coarse-grained sand.

15.4. Roll Materials

Roll materials are supplied in rolls of 10–30 m long. They are divided into basic roll materials, which are produced by impregnation of bases (the cardboard, glass fabric, etc) by organic binder (Figure 15.5) and unsupported, which are made by rolling of mixtures binder with fillers. Basic and unsupported roll materials are applied for the installation of roofs and damp-proofing. Specific requirements, which are advanced for roofing materials are in respect to its atmospheric- and thermal resistance, to waterproofing, water-impermeability, deformability, and biostability. A roof carpet from roll materials for industrial and civil buildings is used in three - five layers as a roofing carpet. For arranging of overlays one material with improved physical-mechanical properties are used. For lower layers one - sarking felt can be used.

Surface waterproofing from roll materials is also used as multi-layered coverage and is the most widespread.

Basic roll materials are divided into uncovering and covering. The first are produced from dipping bases without next application of covering layer; second - by application on the dipping basis from both sides of protective covering layers, which provide enhanced technical properties and longevity of materials.

Asphalt paper and sarking felt which are produced by the impregnation of paper with oil bitumens are the widespread roofing material. Asphalt paper basis is a roofing carton with mass 300-350 g/cm^2, made from the mixture of rag, waste-paper and fibred wood-pulp. It is used mainly for the roofs of temporary structures, and also for vapor sealing. Asbestos paper serves as the basis for sarking.

Widespread roofing basic materials are *ruberoid* and *pitch paper*, which are produced by the dipping of roofing carton respectively with oil bitumens or tars with the next covering the layer of organic binder with fillers and grit.

Ruberoid is divided into roofing and lining depending on setting. The first one serves for arranging of overlay of roofing carpet, second one is for lower layers of roofing carpet and waterproofing of structures.

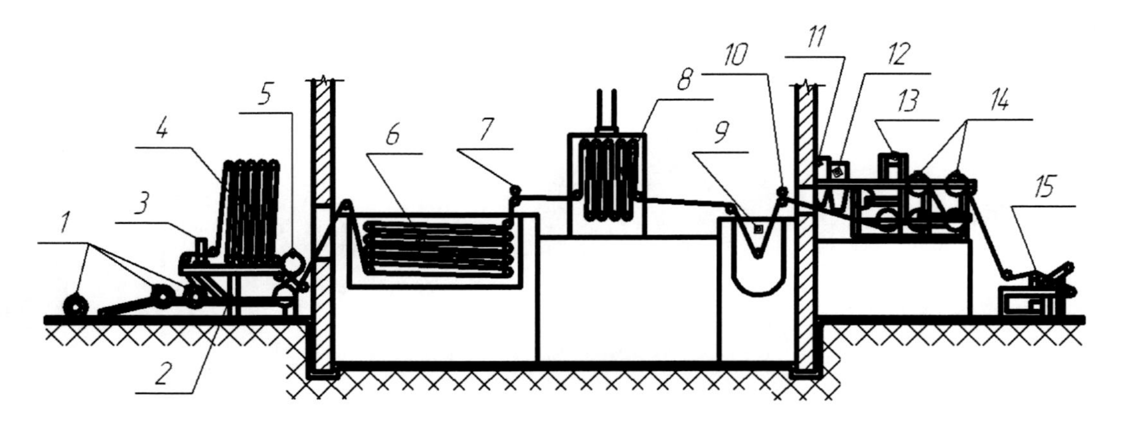

Figure 15.5. Scheme of ruberoid manufacture:
1–carton reel; 2– uncoiler machine; 3– stitcher; 4– carton stock box; 5– drier roll; 6– dipping tank; 7,10– squeeze rollers; 8– additional dipping cell; 11,12,13– grit bin; 14– cooling roller; 15– reeler.

Waterproofing properties of the roll materials are improved considerably by the replacement of roofing cardboard with fiberglass materials and metallic foil.

The nonwoven fabric made of chaotically placed fiberglass, and agglutinated by polymers is used as the basis for glass-ruberoid. Alfol and bitumen-polymeric binders are used for the manufacture of special roll materials, intended for roofs making and waterproofing of structures.

The pitch paper is applied mainly for arranging of the roofs of temporary buildings, for vapor sealing, as lining material and for damp-proofing of foundations.

Rubber-bitumen materials are high-quality roll materials. They are produced by the use of bitumen-rubber binder, fillers and admixtures. By comparison to an ordinary ruberoid, rubber-bitumen materials differ in terms higher extensibility, lesser water saturation, rot-resistance and flexibility.

Self-Assessment Questions

1. Distinguish between the general characteristic of bitumen and tar binders.
2. What is the value of bitumen and tar binders for construction?
3. What are the requirements for bitumen?
4. Discuss the characteristics of the bitumen and tar emulsions and pastes.
5. What are the features of asphalt mastics?
6. Explain technology and properties of asphalt concrete?
7. Provide a short description of the properties of roll bitumen and tar materials.

Chapter 16

Polymer Materials

Polymer materials are obtained on the basis of macromolecular substances - polymers. Synthetic organic polymers, obtained by syntheses of the simplest substances — monomers, are mainly used in construction, Polymers molecular weight exceeds more than 5000 and culminates hundred thousand units, while the weight of common low-molecular substances' molecules changes from units to few hundreds (usually lower 500). That truthful difference in molecular weight explains the big difference of polymers physical properties from the low-molecular substances properties.

The row of general properties is distinctive to polymer materials, as it determines their using in construction: lightness in combination with high strength, water and chemical resistance, high wearing resistance, processability, colouring ability, low thermal conduction. Low heat-resistance, considerable linear increasing, creep, degradation ability– physical-chemical deterioration because of environmental factors are general defects of the polymer materials

The most of the polymer materials are used as plastics, including polymer binder, fillers, plasticizers, stabilizers and other components.

Plastics belong to the most modern construction materials, as they prevail over traditional materials by many parameters. For example, the construction quality coefficient, which means ratio between the compressive strength and the average density achieves for plastics usually 1-2, as for light-weight metal alloys, at the same time for bricks it achieves 0.02, for heavy-weight concrete with strength 20 MPa — 0.08; for pinewood – 0.7.

At replacement of metal, concrete, reinforced concrete, wood by plastics in construction in many cases one gains high technical economic effect. Plastics manufacturing allows providing high level of the comprehensive mechanization and automation of technological processes, their using leads to the high level of construction industrialization and its quality.

Depending on plastic assignment it divides on constructing, finishing, insulating and sealing, piping materials, hygiene and sanitary products, etc.

16.1. Composition and prOperties of Plastics

Plastic masses are materials, plastic on some manufacturing stages, when the polymers are binder. Many of plastic masses beside the polymer binder include the fillers. Such plastics are called filled. At the same time, in some cases, transparent plastics for example where are no fillers (unfilled plastic) are manufactured.

Plastics Components

Synthetic polymers are classified according to different signs: process of manufacture, peculiarity of atomic location in molecule and length of main chain, relation to temperature, special physical-mechanical properties and chemical composition.

Initial materials that are used for synthetic polymers manufacture can be natural gas, coal and petroleum.

There are polymerization and polycondensation polymers according to process of manufacture.

High molecular substances at *polymerization* are obtained because tearing of multiple bonds or ring in cyclic substances and forming of the macromolecule like a chain (Figure 16.1) under the influence of different factors: temperature, light, substances- initiators, catalysts, etc. Accordingly, there are thermal, photochemical, initiative and other kinds of polymerization according to stimulating factor character.

The simplest example of polymerization is reaction of polyethylene formation $(-CH_2-CH_2-)_n$ from ethylene $CH_2=CH_2$ – inflammable colorless gas, obtained from oil refine products or coal.

There are five main ways of polymerization: block – polymerization, polymerization in solution, suspension, emulsion and in gas state. At the block polymerization polymer is obtained already as products of some shape – blocs (Figure 16.2). Representative of such polymers is polymethylmethacrylate (plexiglass), obtained in the form of colorless plates. At the bloc polymerization monomer with additive of initiator or catalyst is inundated in a form and heated. Solution polymerization is used in products obtaining with short chains, used in varnishes and glue manufacturing (varnishes polymerization), etc. In such case monomer previously is dissolved with the help of the solvent, later mixed up with initiator. At the suspension and emulsion polymerization monomer and initiator are dispergated in the water to the smallest globules. The protective colloids (gelatin and others) are brought in to suspension to guarantee the stableness of globules, to emulsion - surfactant species – emulsifiers. Monomers are in gaseous state at the gas polymerization.

Polycondensates have the shorter chains and the consistently lower molecular mass, than polymerization polymers.

Examples of polymerization and polycondensational polymers are represented in Table 16.1.

Polymerization the same as polycondensation polymers may be characterized by lineal, branched and spatial molecular structure (Figure 16.3.) Chain macromolecules have profile forks at branched structure and at spatial structure they are bonded with each other in three-dimensional net by transversal chemical bonds.

Polymers, that are iteratively capable to softening and acquire plasticity at heating, but to hardening at cooling, call thermoplastic. *Thermoplastic polymers* have lineal or branched structure and appear mainly after polymerization reaction.

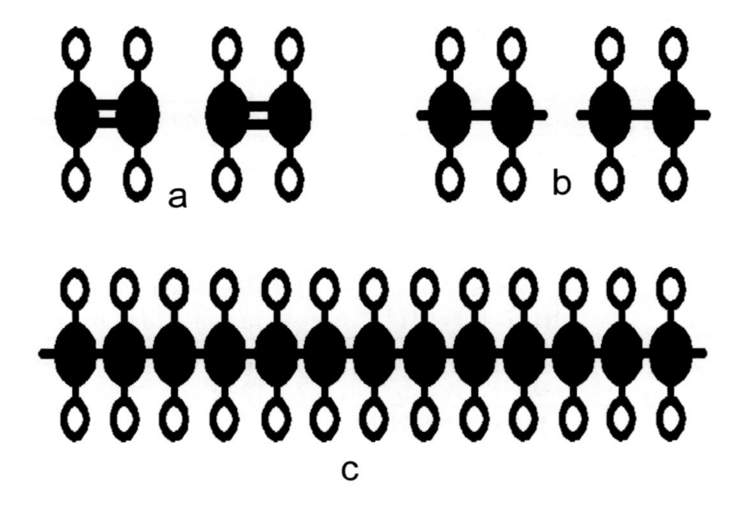

Figure 16.1. Structural scheme of polyethylene molecule formation:
a- ethylene molecule; b- ethylene monomer; c - polyethylene molecule.

Polymers with macromolecule spatial structure after hardening cannot become plastic at heating again. They are called *thermoreactive*. The most of polycondensates belong to them. The more transversal bonds in such polymer's macromolecules (compact net) are, the higher strength is, the fluidity lower is, the higher resilience is, etc. Characteristic properties of thermoplastic and thermoreactive polymers are given in Table 16.2.

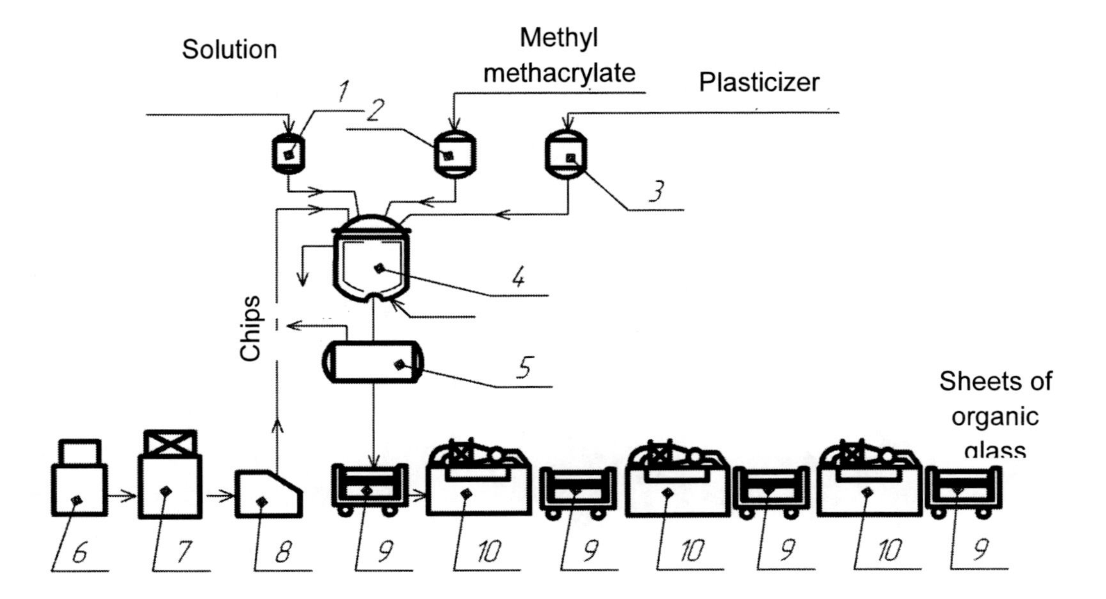

Figure 16.2. Scheme of polymethyl methacrylate manufacture:
1– apparatus for dilution ; 2– chips making machine; 3– heat- treatment box; 4– chips box; 5,6,7– initiator solution batchers; 8– vacuum maker; 9– molds; 10– polymerization boxes.

Table 16.1. Examples of polymerization and polycondensation polymers

Polymers	Monomers	Method of production
Polymerization:		
Polyethylene $(-CH_2-CH_2-)_n$	Ethylene $CH_2=CH_2$	Polymerization in gas phase at the high (120-250 MPa) or low (0.1-0.5 MPa) pressure
Polystyrene $(-C_6H_5-CH-CH_2-)_n$	Styrene $C_6H_5-CH=CH_2$	Block, emulsion or suspension polymerization
Polyvinylchloride $(-CH_2-CHC1-)_n$	Vinyl chloride $CH_2=CHC1$	Suspension or emulsion polymerization
Polytetrafluoroethylene $(-CF_2-CF_2-)_n$	Tetrafluorethylene $CF_2=CF_2$	The same
Polycondensation:		
Phenol-formaldehyde	Phenol C_6H_5OH Formaldehyde CH_2O	Phenol polycondensation with formaldehyde
Carbamide	Carbamide $CO(N_2H)_2$ Formaldehyde CH_2O	Carbamide polycondensation
Polyether	Ethylene glycol $HOCH_2-CH_2OH$ Glycerol and other	Acid polycondensation with polyatomic alcohol
Epoxy	Epichlorhydrin Diphenylpropane	Epichlorhydrin polycondensation with diphenylpropane

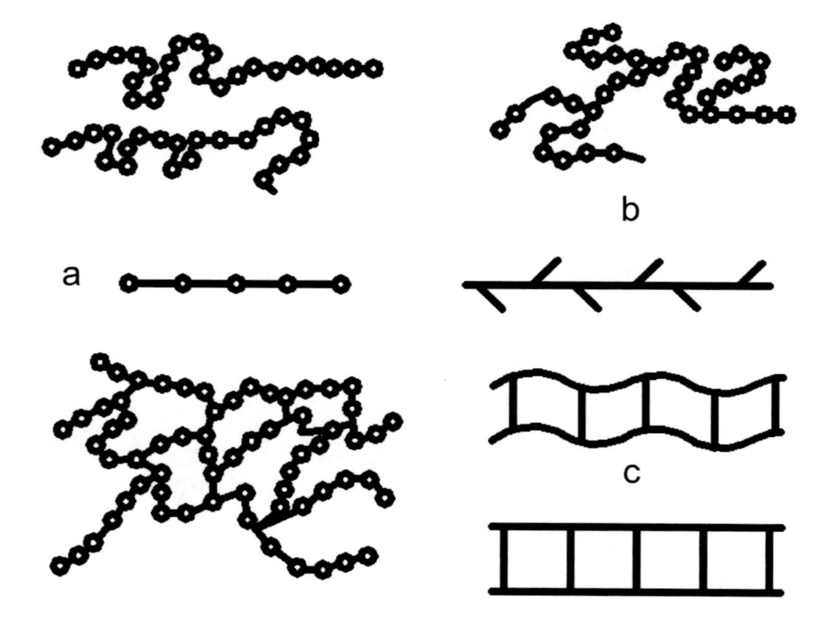

Figure 16.3. Scheme of different polymer types:
a– lineal; b– branched; c– spatial.

Table 16.2. Physical-mechanical properties of synthetic polymers

Polymers	Density, kg/m^3	Ultimate strength at stretching, MPa	Thermal resistance (according to Martens), °C	Hardness according to Brinnel, MPa	Elongation, %
Thermoplastic:					
Polyethylene of low pressure	940-960	22-32	75	45-58	400-1000
Polyethylene of high pressure	920-930	12-16	60	150-250	150-600
Polystyrene	1050-1100	30-50	75-80	200-300	1-5
Polyvinylchloride unplasticized	1380-1400	50-70	65-80	130-160	0-50
Polyvinylchloride plasticized	1200-1400	7-14	30-60	40-80	100-300
Polytetrafluoroethy-lene	2120-2250	15-30	250	250-500	3-4
Thermoreactive :					
Phenol-formaldehyde	1250-1300	25-55	80-120	8-15	
Polyethers	1310-1420	25-55	45-80	50-100	10-26
Epoxy	1150-1240	40-100	50-180	25-80	18-35
Organic silicon	1600-2000	10-50	250-350	1-4	25-30

Structure of macromolecule chains can be divided according to chemical structure into the *carbochained* and *heterochained.* Only the carbon atoms enter to the chains of carbochained polymers, and other atoms can enter to the chains of heterochained polymers. *Element-organic polymers* are the variety of heterochained polymers, that along with elements, enter into the common organic compounds (carbon, hydrogen, nitrogen and oxygen) contain and other elements – silicon, phosphorus, aluminum, titan, tin and so on. Organic silicon compounds (silicones) refer to representatives of elementorganic polymers, main silicones chain contains siloxane bonds (-Si-O-Si). Organic silicon and other compounds due to the perculiarities of chemical structure have a series of positive properties of materials, both organic and inorganic by origin, such as thermostability, hydrophobicity, elasticity etc.

Elasticity and deformability are the characteristic physical – mechanical properties, by which polymer materials are classified. High molecular compounds capable to reverse deformation under the action of external forces are called *elastics (elastomers)*, capable to plastic deformations which are irreversible - *plastics (plastomers)*. Different rubbers, for example, belong to the elastics, the most of polymers form plastics.

Fillers can improve mechanical and dielectric properties, increase thermal stability and atmosphere resistance, diminish contraction, etc. The prime price of plastics with entering filling agents falls down considerably.

Fillers are divided by origin into organic and mineral, powdery, by form – into fiber and leaf.

The sawdust, wood, quartz and mica powders, talc, soot, graphite, kaolin, asbestos dust and others are widespread powdery fillers. Application of powdery filling agents along with the polymers of mainly phenol-formaldehyde type allows obtaining molding powders, which are widely used for making of various technical, domestic and insulating goods, and also products of the special assignment, which have enhanceable shock resistance, chemical resistance, waterproofness and heat resistance.

Especially high mechanical strength of plastics is achieved at application of fibrous (fiberglass, earth-flax, cotton, synthetic filament and other) and sheet (paper, wood leads, foil and fabric) fillers. Fibreglasses are especially effective among fibred fillers. Plastics under the general name of *glass fiber plastics* can be manufactured on their basis with application of the various synthetic polymers.

Along with fillers content of which changes in wide limits, plasticizers, stabilizers and pigments can be added in plastic if it is required.

Plasticizers can be added in an amount of 10-100 % by polimer weight to increase the elasticity, to improve fire-retarding ability and frost-resistance, to increase resistance to the ultraviolet rays and to improve the terms of processing. Essence of plasticizers action consists in penetration in the polymers macromolecules and diminishing of intermolecular forces of tripping.

Stabilizers are applied for deceleration of the plastics senescence processes during their exploitation and processing. Depending on the nature of plastics senescence stabilizers divide into two groups - thermostabilizers and lightstabilizers.

Plastics are processed into constructive products by different ways, the selection of which depends on components properties and structural features of products. So, products based on thermoplastic polymers are frequently obtained by casting under pressure, which consists in the periodic injection of molten mass portions in a form by moulded machines. An extrude-squeezing out of mass through the cannon-bit of screw extrusion machines are also applied, roll-forming - forming in a gap between the revolving rollers, thermoforming, pressing and another ways.

General Properties of Plastics

The average density of plastics changes in a wide range of 15-2200 kg/m^3. The porous plastics have the lowest density. Fillers have a considerable influence on density. Plastics can be in 6 times lighter than steel and in 2.5 lighter than aluminum.

Plastics, usually, have high strength both at a compression and at tension and bend. Ultimate compression and tensile strength of the most high-strength plastics (glass-fiber material, wood laminated plastic and other) achieves to 300 MPa and more.

Hardness of plastics does not correlate directly with their strength unlike the metals and number of other materials. Even for such the hardest plastics, as textolites (filling agent is a cotton fabric), hardness approximately is in 10 times lower than steel's. In spite of low hardness, plastics (especially elastic) have low abrasiveness that allows their wide using in floor coatings. Abrasiveness, for example, of one-sheeted polyvinylchloride linoleum equals 0.06, multi-layered – 0.035 g/cm^2, which approximately is equal to the granite.

Plastics resistance to percussive action which is determined by the proportion of shock energy for destruction to the cut transversal area of a standard arrive the high levels for dense plastics (50-150 kJ/m^2) and can sharply go down according as their porosity increasing.

Lot of plastics that are stretched out may be characterized the deformability. Relative lengthening, that means increasing of materials length in the moment of breaking up to its initial length, for polyethylene tapes, achieves the level of 300, to polyvinylchloride - 150, butyl rubber - 10%.

Module of elasticity is the description of resilient materials properties. This parameter for plastics is considerably lower than others construction materials. So, it is equals for steel (2-2.2)$\cdot 10^5$, woods (0.063-0.14)$\cdot 10^5$, paper-stratified plastic (0.021-0.028)$\cdot 10^5$, polyester glass-fiber material (0.022-0.028)$\cdot 10^5$ MPa.

There are rigid, semi-rigid, soft and elastic plastics depending on the module of elasticity. Phenol-formaldehyde and alkyd (polyester) plastics are examples of rigid plastics, which fragily collapse with the insignificant lengthening at breaking up; their module of elasticity mounts to more than 1000 Pa. Soft plastics (polyethylene and other) have the module of elasticity 20-100 MPa; the high relative lengthening characteristic is for them. Semi-rigid plastics (polypropylene and other) have intermediate values of the module of elasticity 400-1000 MPa. The module of elasticity does not exceed 20 MPa for elastic plastics (rubber and similar materials). Their deformations are mainly reversible at the normal temperature.

The low values of the plastics elasticity module promote the gradual increase of irreversible deformations at permanent load - creep. The creep of plastics can be largely explained by macromolecules sliding of polymer binder. It considerably grows even at the insignificant increase of temperature. Creep is considerably less for plastics, based on spatial polymers, the molecules of which are "sewn" together by transversal copulas. Enhanceable creep limits application of plastics in structural parts which work under the large loads.

Thermal conductivity of dense plastics without any fillers is 0.116-0.348 W/(m°K). Introduction of mineral fillers increases the thermal conductivity of plastics. Heat-reflecting properties of plastics discover the possibility of wide application in the house non structural parts.

Along with low thermal conductivity plastics can be characterized by large thermal expansion. Coefficient of linear thermal expansion of polyethylene (160-230)10^{-6}, of polyvinylchloride (80-90)$\cdot 10^{-6}$, phenol-formaldehyde polymers (10-30)10^{-6}, of steel $12\cdot 10^{-6}$°C^{-1}. Thermal expansion of plastics should be taken into account at constructions planning and exploitation to shut out deformations and cracks formation.

Destruction of plastics develops to the extent of temperature increasing that means the polymers destruction, or beginning of their melting. Initial melting temperature of most thermoplastic polymers is 105-165°C. Plastics thermostability, which is characterized by the temperature of the maximum possible deformation, is mainly in a range of 60-180°C. The minimum possible performance temperature, when plastics become fragile changes in a wide range: from -10°C for vinyl plastic to -270°C for polytetrafluoroethylene materials.

Most of the plastics are highly inflammable and may be burned; they burn with the opened flame both in the area of fire and outside of it. Plastics based on polyvinylchloride, phenol-formaldehyde, carbamide, organic silicon polymers belong to hardly burned plastic materials. Introduction of the special additives – fire-retardants to the burned polymers also transfers plastics in the group of hardly burned. Fluoroplastic and perchlorovinyl plastics do not burn and do not smolder under the action of the fire.

Dense plastics are waterproof and steam-tight. Hydrophobic polymer materials (polyethylene and polyvinylchloride tapes, glass fiber plastics, vinyl plastic and other) are characterized with the least water absorption (0.1-0.5%).

Content of the large amount of hydrophilic fillers for example wood shavings, sharply increases the water absorption.

Number of non-filled plastics has high transparency. It allows production of organic glasses on their basis, used for panning of greenhouse, hothouses and buildings of the medical assignment. Organic glass – acrylic resin is the most widespread – skips up to 94% of radiated visible particle of spectrum and 73.5% of ultraviolet rays, in that time as ordinary silicate glass approximately 84-87% and 0.3-0.6%.

Usually, polymer materials are good dielectrics. It is necessary to take into account the possibility of accumulation of electrostatic charge on their surface which appears under the action of friction forces. The degree of electrization of such roll materials, as polyvinylchloride linoleum, can achieve 65 V/cm^2. It is necessary to take into account at prevention of the fire especially in apartments, where the vapors of inflammable liquids concentrate. Ability to the electrification of plastics diminishes by introduction to their composition of the special matters - antistatik and filling agents which conducts a current.

Synthetic polymers and plastics based on them have high resistance to the aggressive environments. Carbon-chain polymers are the most proof to the action of acids, alkalies, salts and different oxidants. Heterochain polymers are easier affected the action of chemical reagents. Presence in the macromolecules of hydroxyl-group, for example, in the polivinil alcohol molecules, reduces resistance of polymer to the action of water and acids. Opposite, replacement of hydrogen atoms by the fluorine micromoles in composition increases chemical resistance of polymers. Fluorocarbon polymer on the chemical resistance prevail noble metals, special alloys, rust ceramics.

The most of plastics are corrosion-proof not only to the action of chemical reagents but also to the action of fungus, bacterium, insects and rodents. Plastics that consist of wood fillers (wood chipboards and fibreboards), some high-porous plastics (microporous rubber), and the polyethylene products are biologically unstable. Elements of buildings with application of wood chipboards and fibreboards, and also microporous rubber can be deteriorated by fungus and bacterium at the enhanced humidity and temperature. Pipes, tapes and other products, based on a polyethylene may be strucked by rodents. The plastics biostability is improved with wood preservative additives. Tar pitch and some other matters are added in polyethylene products to prevent the rodent damages.

It is important to take into account the sanitary-hygienic properties of floors plastics, internal revetment of walls at their application. Row of plastics, especially on the basis of phenol-formaldehyde, polyester, epoxy polymers, at incomplete processes of polymerization or polycondensation, containing toxic plasticizers, hardening agents, solvents can secrete matters to the environment, insalubrious for people and animals. Static electricity which accumulates on plastics can have a stimulant influence on a microflora.

The change of operating properties - ageing - passes in plastics in a certain degree under the heat, light and air oxygen action. The process of ageing is accelerated due to the action of the mechanical loadings. The ageing of plastics is sharply retarded by introduction of stabilizing additives.

16.2. BASIC TYPES OF CONSTRUCTION PLASTICS

Structural Plastics

Wood laminated plastics, glass fiber plastics, sheet vinyl plastics, plexiglas and polymer concrete are widely used for framing and non- framing structures.

Wood -laminated plastics — are the variety of plastics with filler - wood veneer - thin sheets of wood of thickness 0.3-2.1 mm, obtained from the steamed out blocks of birch, alder and beech. The veneer is satured by polymer solutions and collects in packages which are subjected to the hot pressing on hydraulic press. An average density of sheets is 1,250-1,330 kg/m^3, the ultimate tensile strength along fibers – 140-260 MPa, water absorption after 24 hours is no more than 2-3%.

Usually, physical - mechanical properties of wood-laminated plastics are better than initial wood. They can be applied for making of different framing structures - beams, farms, archs, etc.

Glass-fiber materials are widely used in construction due to the high longevity, lightness, resistance in different environments and transparency.

The fiberglass as filaments, plaits and fabrics is a filler of glass-fiber materials. Fiberglass has tensile strength 300-500 MPa at diameter approximately 50-3 μm. Formaldehyde, epoxy, polyester, polyamide and silicon organic resins are applied as binding agents.

Fiberglass properties depend on the type of glass-fiber filling material, resin and on their correlation. The oriented fiberglass, besides glass-fiber anisotropic material which has especially high strength belongs to it, is manufactured at the parallel placing of fibers or plaits. strength of material at longitudinal or transversal stretching is not lower than 350-450 MPa, and at stretching under the corner of 45° - lower almost in 2 times. The strength indexes of fibreglasses reduce under the action of enhanceable temperatures and water. Fibreglasses can be treated by all the types of mechanical operations.

Light-weight structures, based on the glass fiber plastics, allow constructing structures in 8 times lighter than from heavy-weight reinforced concrete panels.

Polyester fibreglasses and glass-cloth laminate can be used for facing of the walls and ceilings panels, window and door blocks and sanitary products. High chemical resistance allows applying glass-fiber material in structures, exploited in various aggressive environments.

Vinyl plastic is a rigid material, which is obtained by the hot pressing of the polyvinylchloride mixtures. Vinyl plastics are produced as the transparent and opaque painted and achroous sheets with the average density of about 1400 kg/m^3. Vinyl plastic is non-heatproof, and its strength decrease approximately twice at 40°C. Temperature interval of the material function is from -50 to + 60°C. Vinyl plastic is easily processed and welds, but sticks together only with some glue (for example, perchlorovinyl).

Vinyl plastic can be used in translucent non-framing structures, for facing of panels, partitions and ceilings and also for chemically-proof structures. Vinyl plastic can be also used for manufacture of ventilation boxes, pipes, reservoirs, etc.

Facing Plastics

The positive features of plastics are: decorativeness, low wearability, elasticity, high thermo- and sound-insulating properties that caused their wide application in floors arrangement and wall finishing.

Floor plastics are divided into roll, tiles and mastic.

Roll materials - *linoleum*, are made from polyvinylchloride, alkide and rubber polymers without the underlying layer or on textile, wool and other types of the underlying layer. The most mass linoleum without the underlying layer is formed by calander and extrusion methods (Figure 16.4).

Polyvinylchloride linoleum is the most widely used roll floor material. Polyvinylchloride linoleums can be manufactured single- and multi-layered, both with the decorative finishing and without it. It is not recommended to use polyvinylchloride linoleum in humid operational conditions, at action of fats, oils and abrasive materials. Water absorption of linoleum without the underlying layer equals not more than 1.5%, wearability – 0.05 g/cm^2.

Rubber linoleum can be successfully used in humid operational conditions. It has high waterproofness, chemical resistance and sound absorption.

Linoleum coatings are widely applied in dwelling houses and industrial buildings. They allow decreasing the duration of floor arrangement works in few times comparatively with application of other materials.

Linoleums are glued by bituminous, bituminous-rubber and other mastics.

Synthetic carpet materials are widespread in civil construction along with linoleum materials. They have high acoustic and decorative properties. For example, synthetic carpet material is manufactured with emulsive polyvinylchloride film as underlying layer and polyamide or polypropylene fibers as top layer.

Polyvinylchloride and coumarone tiles are the most widespread *polymer tiles* for floor arrangement. Tiles can be manufactured by pressing or other methods. Tiles can have different dimensions (for example, 300×300, 200×200, 300×150, 200×100 mm), thickness in the range of 1.5-3 mm.

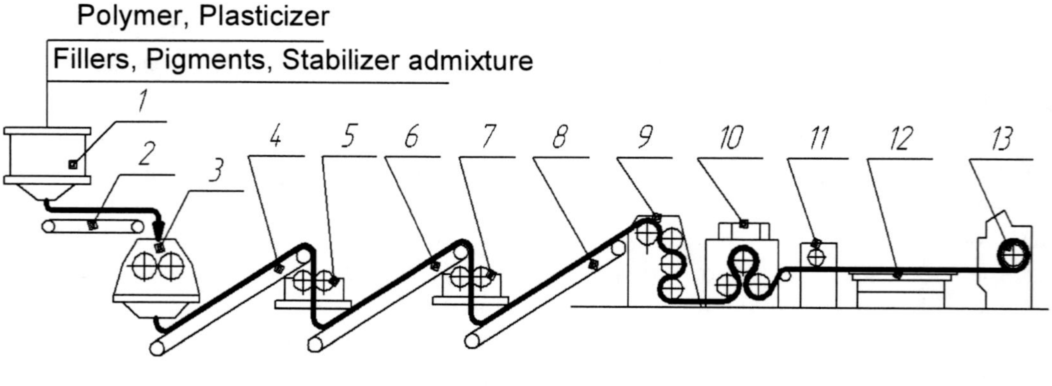

Figure 16.4. Scheme of single-layer polyvinylchloride linoleum manufacture by calander method: 1, 3- mixers; 2,4,6,8- conveyers; 5,7- rollers; 9- calander; 10- cooler; 11- trimming machine; 12- table; 13- roll.

Floor mastics are fluid polymer (mainly polyvinylacetate or latex) mixtures which coated by pouring, sprinkling or spraying on different bases with their subsequent hardening. Floor coating based on latex mastic differs by higher water-resistance than polyvinylacetate. Application of polyvinylacetate mastic is not recommended in apartments with the humid operational conditions, at the shock loadings, acid or alkali action.

Styrene, polyvinylchloride, phenol-formaldehyde and other tiles, sheets and roll plastics can be used for wall facing. Styrene tiles and sheets are widely used for facing of internal walls and partitions from noncombustible materials.

Decorative *paper - laminated plastic sheets* can be manufactured by the hot pressing of paper, saturated with thermoreactive polymers. Surface of sheets can imitate valuable stone or wood surfaces. Paper - laminated plastic is thermo- and wear-resistant, can be easy processed. Sheets from this plastics are hygienic, lightfast and heatproof. Due to sufficient chemical resistance, it is not destroyed under the action of different cleansers, soluble acids and alkalis, organic solvents and mineral oils. Also, they are stable at heating up to 130°C.

Polyvinylchloride decorative tapes produced of different colors and patterns, transparent and opaque, smooth and embossed are widespread roll materials for finishing.

Transparent tapes can be used for arrangement of waterproof and decorative curtains, coverages of hothouses, etc. The opaque tapes are applied for finishing of walls, partitions and ceilings to which enlarged hygienic requirements are specified.

Wallpapers are the roll material based on paper. Moisture resistant wallpapers have protective tape on a front-face area from polymer emulsions or varnishes. They can be used for kitchens and bathrooms facing.

Waterproofed and Sealing Plastics. Pipes

Application of waterproofed and sealing plastics is the most effective way for waterproofing and sealing of structures.

Polymer films can be applied for roof, reservoirs and channels insulation. Film materials are up to 1 mm thick and made of polymer binder by different ways (Figure 16.5). Polyvinylchloride and polyethylene films are widely used. Polyethylene film is more durable, than polyvinylchloride one, does not lose elasticity, more resistant to the low temperatures, but it should be protected from the destroying sun action. Polyvinylchloride films have lower than polyethylene films resistance to the heat and light action, frost-resistance. Industry also manufactures polisobutylene, polypropylene, polyamide and other films. Polymer films have the low weight, chemical resistance, strength and watertightness. Films can be placed on moist basis.

Films allow improving the conditions of work at insulation works and promote their economic efficiency.

Basic properties of polyethylene and polyvinylchloride films are shown in Table 16.3.

Films can be glued to concrete, stone, metal and wood. They protect these materials against aggressive environments.

The production of the *reinforced films* consists in placing reinforcement - nylon, cotton or flax filaments and glass-fiber fabric between two films. The reinforced films have a higher elasticity and strength, than ordinary one. Welding and agglutination can be used to connect films between themselves.

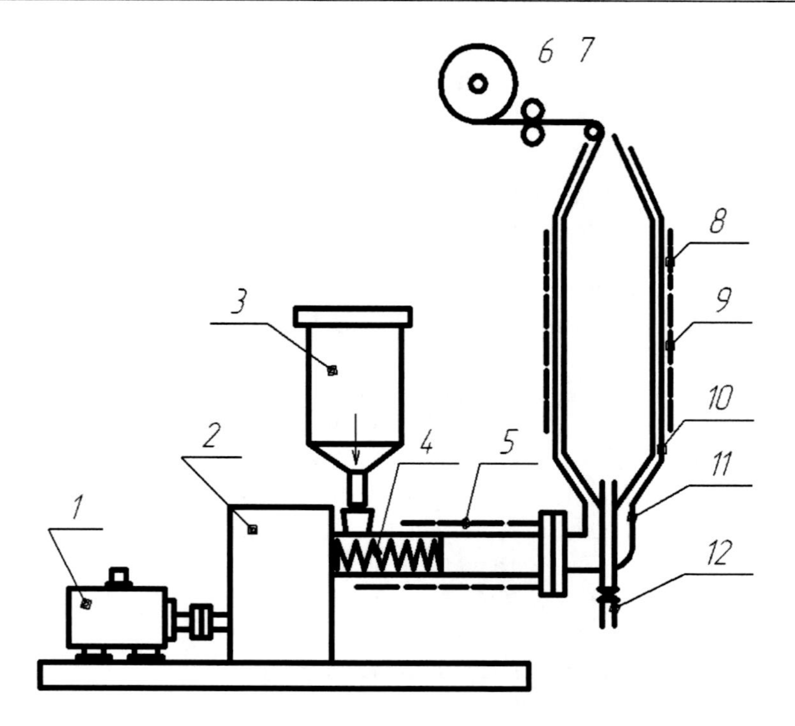

Figure 16.5. Scheme of polyethylene film manufacture by inflate method:
1- electric motor of extruder; 2- reduction gear of extruder; 3- bunker of granular polyethylene of low-density; 4- worm of extruder; 5- device for extruder heating; 6- roll of polyethylene film; 7- rollers; 8- area of film cooling; 9- device for thermal chamber heating; 10- thermal chamber; 11- head of extruder; 12- air introduction for film inflate.

The large group of polymer materials is *glues*. Their gluing ability is determined by the forces of adhesion interaction. Strength of the glue film is depended on the cohesive forces.

Glues based on the linear polymers solutions harden at normal temperature. Most of glues based on the spatial polymers harden at high temperature. Selection of optimal temperature and hardening duration of polymer considerably improves the quality of glue film.

Application of polymer melts as glues is more effective than solutions, as allows getting more dense glue film.

Agglutination quality substantially depends on character of adhered materials surface; content of solvents, fillers, plasticizers; thickness of the glue film. At the estimation of adhered materials surface in most cases the next statement is correct: Polar materials are glued together with polar glues and vice versa. Quality of agglutination is improved at surfaces cleaning and degreasing, and also at chemical or mechanical treatment with the purpose of impaired layers removal. Reduction of glue layer thickness and selection of necessary agglutination pressure, and also introduction of plasticizers and filling agents positively influences on the glue line properties.

High-strength and heatproof lines form glues based on the thermosetting polymers: phenolpolyvinylchloride, epoxy, carbamide, etc. Usually, they consist of two components: polymer and hardener. Components must be mixed before the glue using.

Table 16.3.Physical-mechanical properties of film materials

Property	Polyethylene film	Polyvinylchloride film
Average density, kg/m3	920-940	1230-1400
Liquid limit, MPa	9-10	9-10
Ultimate tensile strength, MPa	12-16	10-25
Percent elongation, %	150-600	140-400
Module of elasticity, MPa	50-250	2-8
Heat-resistance, °C	80-115	60-80
Fragility temperature, °C	-60 до -75	від -25 до -50
Water absorption, %	0.035-0.22	0.5
Coefficient of linear expansion, °C-1	$1.2\cdot10^{-4}...9.4\cdot10^{-4}$	$1\cdot10^{-4}...2.5\cdot10^{-4}$

Mastics by their properties are similar to glues and differ from them with increased viscidity. They can connect different materials. Also, they can be used for formation of protective impermeable layer on surfaces of materials Mastics based on the indene-coumarone resins, epoxy, polyester and other polymers are used as gluing. Epoxy-polyester, epoxy- rubber and other mastics, which have high adhesion to dry and wett materials, high water- and frost-resistance, can be also used.

Sealing materials (*sealants*) can be used for the joints insulation of prefabricated reinforced concrete elements in buildings and structures (Figure 16.6), and also for sealing of glazing, reservoirs, etc. Quality of sealants is estimated by the adhesion ability, impenetrability, resistance to the environmental aggressive factors, relative elongation.

There are two types of mastic sealants: non-hardening mastics based on the polyisobutylene; cold hardening mastics - thiokols (based on liquid polysulphide rubbers), butyl-rubber, silicone, bituminous-polymer.

The widespread sealants consist of polyisobutylene, plasticizer (mineral oil) and fillers (chalk, milled limestone, etc.).

Two - component mastics based on the thiokols consist of basic and hardener pastes. The irreversible process of vulcanization and transformation of paste-like mass into rubber-like material passes after components mixing in a certain proportion. Along with two-component, one-component thiokol sealants are also used. The process of their vulcanization is carried out on the air at 10-25°C under the moisture action and proceeds 7-10 days at humidity 95-100%, and 4-5 weeks at humidity 50-60%.

Mastics which are vulcanized and mastics which are not vulcanized are based on butyl-rubber. Two-component mastics, which are vulcanized, based on butyl-rubber solutions or butyl-rubber mixtures are widely applied.

Foamed polyurethane sealant is obtained by the method of chemical interaction of basic components directly in structure.

Polymer pipes are applied for arrangement of the water-supply and sewage systems, drainage and spray systems. They have high corrosive resistance, sufficient resistance to the destroying actions of freezing water, small mass. Pipes are made from plastics on the basis of thermoplastic and thermoreactive synthetic resins. Pipes from polyethylene, polyvinylchloride and polypropylene are mainly used.

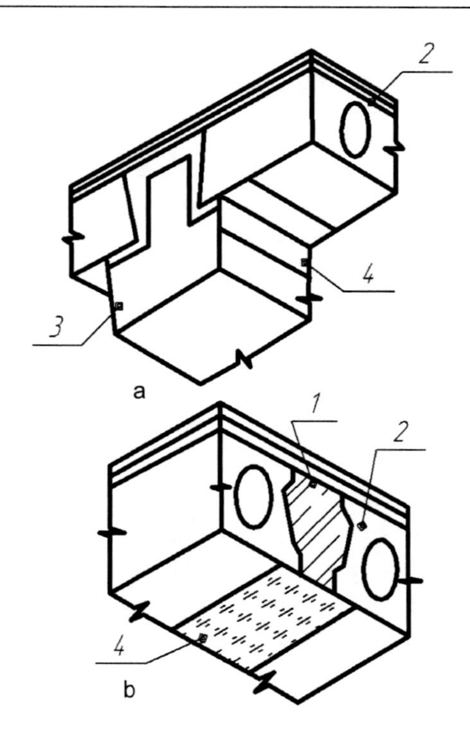

Figure 16.6. Joints insulation of ceiling elements:
a- insulation of places of ceiling panels based on crossbar; b- insulation of ceiling panels joints; 1- cement-sand mortar; 2- ceiling panel; 3- crossbar; 4- mastic.

Polyethylene pipes are mostly used. They are made from the polyethylene of high (HP) and low (LP) pressure with the help of extrusion machines The LP polyethylene has higher strength. Ultimate breaking strength of the LP polyethylene is in the range of 20-25 MPa, and HP polyethylene – 8-25 MPa.

Table 16.4. Comparative properties of plastic and steel pipes

Property	Pipes			
	Polyvinyl-chloride	Polyethylene	Polypropylene	Steel
Density, kg/m^3	1400	950	850-900	7800
Application temperature, °C	60	90	120	150
Frost-resistance, °C	-15	-60	-50	—
Ultimate tensile strength, MPa	50	14	35	200
Resistance: in 60% sulphuric acid	+	+	+	
in 20% muriatic acid	+	+	+	-
in caustic soda	+	+	+	-
in salt water	+	+	+	-

Note. Sign "+" means proof pipes, "-" non-proof pipes.

Polyethylene pipes are manufactured usually with diameter 10-640, polyvinylchloride 10-400 and polypropylene 15-80 mm. Hydrostatical pressure in the polyethylene stand-pipes can achieve up to 2.5 MPa.

Vinyl plast (polyvinylchloride) pipes can be applied in the plumbing and sewage systems. They have high chemical resistance and low thermal conductivity, but are heavier and less frost-resistant than polyethylene pipes.

Comparative description of plastic and steel pipes is resulted in Table 16.4.

Main defects of polymer pipes are low thermal stability and considerable linear expansion.

16.3. Polymer Concrete

Polymer concrete is the special type of concrete with the application of synthetic polymers as the binders. According to their properties polymer concretes occupy intermediate place between plastics and cement concrete. Polymer concrete can be classified according to the density, the peculiarities of structure and the range of application. Furthermore, polymer concrete can be classified according to the type of binders. Polymer concrete differs from other plastics by the high content of the mineral fillers and aggregates. Thermosetting polymers - carbamide, phenol, polyester, furan, polyurethane and epoxy can be used as the binders for polymer concrete. Thermoplastic polymers such as coumarone-indene, methylacrylate, perchlorovinyl are used in the considerably smaller degree. Furan, polyester and carbamide resins are the mostly used.

The hardening of polymer concrete on the basis of thermosetting resins is achieved at a normal temperature (sometimes with the preheating) and usually after introduction of hardeners.

Furan resins are condensation products of furfurol and furfuryl alcohol with phenols and ketones. Furan resins differ from others by considerably smaller cost.

Along with the aggregates, dispersed mineral fillers have essential influence on the properties of polymer concrete. Fillers can be adsorptive and adhesive interacted with the synthetic polymer that is the way to regulate the properties of polymer concrete. Thus, the introduction of the carbon-containing fillers (to 5%) in the mixtures based on furfurol-acetone monomer substantially increases their water and frost resistance.

Compressive strength of polymer concrete based on the furan, polyester and epoxy resin is in the range of 50-125; bending strength – 15-40; tensile strength – 8-16 MPa.

The high wearing resistance and impact strength are the special positive features of polymer concrete. The resistance to abrasion of polymer concrete depending on the type of polymer binder is in the range of 0.001-0.04 g/cm^2. The impact strength of polymer concrete is 5-10 times higher than of cement concrete. One of the important special features of polymer concrete is the high chemical stability.

Polymer concrete can be used as construction material at the influence of chemically aggressive environment due to the combination of high chemical stability with the strength and the longevity.

Polymer concrete can be considered as quick hardening material. It achieves the high strength already in a few first days at the normal conditions of hardening. Strength growing is insignificant in a next period of hardening.

Brief heat treatment, which accelerates binder polymerization, has the positive influences on strength growing of polymer concretes. The main defects of polymer concretes which limit application of these materials in framing structures are: enhanced contraction, creep and low heat-resistance.

Considerable economic effect is achieved at the using of chemically proof steel-polymer concrete structures at aggressive influence of different technological environments.

Self-Assessment Questions

1. What is the value of polymer materials for construction?
2. Tell about the synthetic polymers and describe the basic processes of their manufacturing.
3. What are the advantages of plastics in comparison with other materials?
4. What are the typical defects of plastics?
5. Tell about the compositions and methods of plastics manufacture.
6. What are the features of polymer glues?
7. Tell about the film polymer materials.
8. What are the peculiarities of polymer pipes?
9. Tell about the construction plastics.
10. Tell about the facing plastics.
11. Tell about the roll polymer materials.
12. Tell about the polymer mastics and glues.
13. Tell about the waterproofed and sealing plastics.
14. What are the peculiarities of polymer concrete?

Chapter 17

HEAT-INSULATING AND ACOUSTIC MATERIALS

Materials with low thermal-conductivity, assigned for the thermal insulation of building constructions, industrial equipment and pipelines are called heat-insulation materials. Application of heat-insulation materials is one of the most important directions of reduction material capacity of construction and decreasing of fuel consumption.

Application of materials with low thermal-conductivity in enclosing structures, for example in large-panel dwelling-houses, enables to reduce the charges of steel and cement in 1.5-2 times as compared to structures without thermal insulation.

Application of the light-weight brick walls with effective heat insulation in place of the continuous bricking allows substantially shortening the need in bricks, cement and lime, reducing weight of structures and transport expenses.

Charges of thermal energy for houses heating are increased in inverse proportion to resistance of the wall heat transfer. This property depends on the density of materials used.

Three-layer wall structures are widespread in construction. These structures (panels, shells, etc.) consist of thin external layers, which are made from strength material, and heat-insulating middle layer. Such structures differ by relatively low weight and prefabricability that allows executing construction works in any time of the year. Asbestos cement, aluminium alloys, zinc-coated steel, waterproof plywood and other sheet materials can be used for facing of panels. There is a positive experience of application of three-layer reinforced concrete panels. Heat-insulating materials based on the local raw materials can be used for filling of framed external walls and internal partitions and also for ceiling heating.

The insulation of thermal equipment, technological apparatus and pipelines at construction of thermal power-stations results in reduction in 20-25 times of thermal charges.

Characteristic feature of heat-insulating materials is a type of initial materials. Depending on it there are inorganic and organic materials.

According to their form heat-insulating materials are divided into piece, roll and loose; according to character of structure - on hard, flexible and friable.

Density and heat-conductivity are basic properties which characterize efficiency of heat-insulating materials.

Also an important value has limiting temperature of application, strength, deformation, fire-proofness, biostability, vapour permeability, etc.

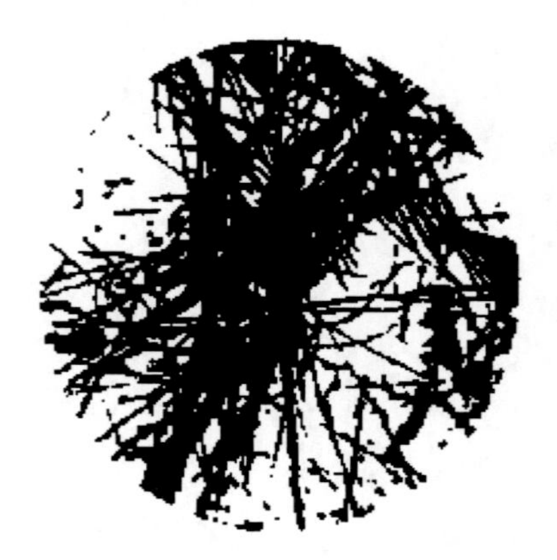

Figure 17.1. Microstructure of mineral wool.

Materials with the average density up to 600 kg/ m^3 can be used as heat-insulation. However, effective heat-insulation materials must have the average density no more than 400 kg/ m^3. Some of them (gas-filled plastics, mineral wool products based on synthetic binders, etc.) can be considerably lighter (no more than 50-100 kg/ m^3).

According to thermal conductivity heat-insulating materials can be divided into three classes: low heat-conducting – up to 0.06, middle – 0.06-0.115 and high – 0.115-0.175 W/(m·K). For high porous materials heat-conductivity is similar to the heat-conductivity of air – 0.025 W/(m·K).

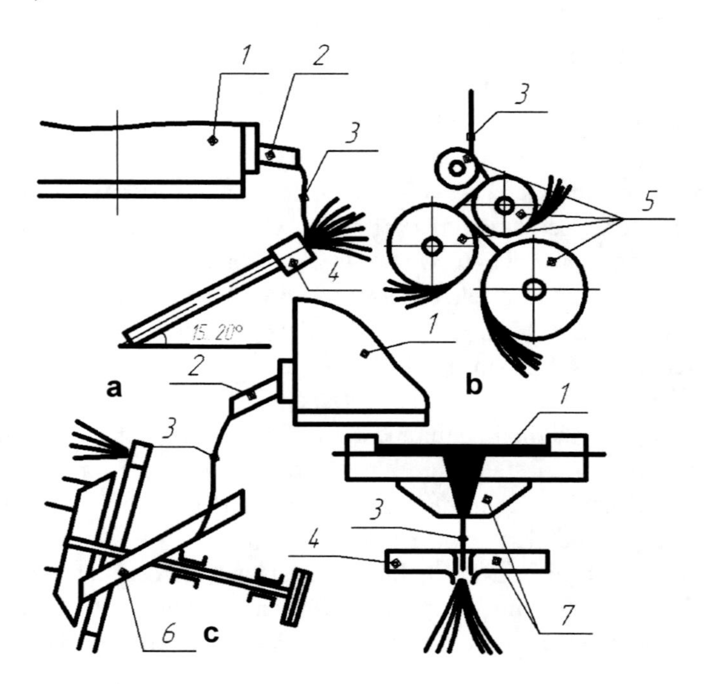

Figure 17.2. Methods of processing of silicate fusion in fibers:
a– steam-blow ; b-centrifugally rolling; c – vertical blow-spinneret;1- melt device; 2-tap hole; 3- fusion; 4-blowhead; 5,6-roll and blow centrifuges; 7- blow-spinneret device.

Reduction of pores sizes deteriorating heat transfer positive influences on heat-reflecting properties of materials. Also complication of chemical composition and formation of amorphous structure of materials improve heat-reflecting properties.

It follows to take into account that in operational conditions thermal conductivity of heat-insulating materials is in arcwise increase with growth of temperature.

The limiting temperature of application is an important operating property of heat-insulating materials. For example, limiting temperature of application of expanded perlite and vermiculite is 900 °C, mineral wool – 600°C, porous concrete – 400-700°C, and gas-filled plastics – 60-180 ° C.

Strength of heat-insulating materials must be sufficient for their warehousing, transporting, installation and application during required term. The compressive and bending strength for the most widespread heat-insulating materials is in the range of 0.1-1.5 MPa.

Range of application of heat-insulating materials depends on their fireproofness, biostability, chemical resistance, water absorbtion, gas and vapour impermeability and number of other properties.

Acoustic materials are used for providing of required auditory perception and reduction of sound level in apartments (sound-absorbing materials) or their sound-proofing (sound-insulating materials). Application of acoustic materials has positive influence on health of people and promotes increasing of the labour productivity.

The same as heat-insulating, acoustic materials are characterized by high porosity. Materials with the connected pores are characterized by good sound-absorbing properties. Their basic features are low values of dynamic module of elasticity and accordingly speed of sound distribution.

17.1. Inorganic Heat-insulating Materials

Inorganic (mineral) heat-insulating materials differ from organic by noncombustibility, relatively low hygroscopicity, rot resistance. They can be widely applied in thermal insulation of building structures, industrial equipment and pipelines. Mineral wool materials are most widespread among the heat-insulating materials. *Mineral wool* mainly consists of glass-like fibers with diameter 1-10 μm and length 2-20 cm (Figure 17.1), which are manufactured by processing of melts of blast-furnace slag and number of silicate rocks (diabases, basalts, marls, etc.). Silicate melts is converted into a mineral fibers as a result of influence of blowing of steam or gas (blowing method) or centrifugal force (centrifugal method) (Figure 17.2). Centrifugal - blowing method of fiberizating, which includes application of centrifugal force and blowing is the most widespread. Thermal conductivity of mineral wool at temperature 25°C changes depending on average density in the range of 0.042 – 0.046 W/(m·K), limiting temperature of application is 600°C. Mineral wool is granulated for prevention of compaction during transporting and storage. G*lass wool* is similar to properties of mineral wool. Glass batch are used for manufacture of glass wool.

Application of mineral and glass wool as ready products is the most rational (Figure 17.3). Synthetic polymers and bitumens are effective binders for such products. The basic types of products are non-rigid, semi-rigid and rigid board, cylinders and semi-cylinders.

Figure 17.3. Mineral wool products:
a– felt; b– semi-rigid board; c– heat insulating shells; d– longitudinal broaching mat; e– transversal broaching mat.

Mats without finishing and with finishing are made of bitumen paper, cardboard, asbestine fabric, glass linen and metal grid. There are mineral wool mats of transversal and longitudinal sewing. More effective is the longitudinal one, because it is based on the longitudinal placing of sewing material is possible to make the products with unlimited length. The mineral wool mats are applied as suspension heaters in light wall structures and also in ceilings. They can be successfully used for insulation of pipelines (Figure 17.4) and technological equipment. Application of sewed mineral wool mates allows improving conditions of work in the process of heater placing and also decreasing the dust formation.

The semi-rigid boards with average density 100 and 125 kg/m^3 are widely used. Also, high rigid mineral wool boards, which have strength 0.04-0.1 MPa, density 75-250 kg/m^3 and water absorption 15-60%, can be manufactured.

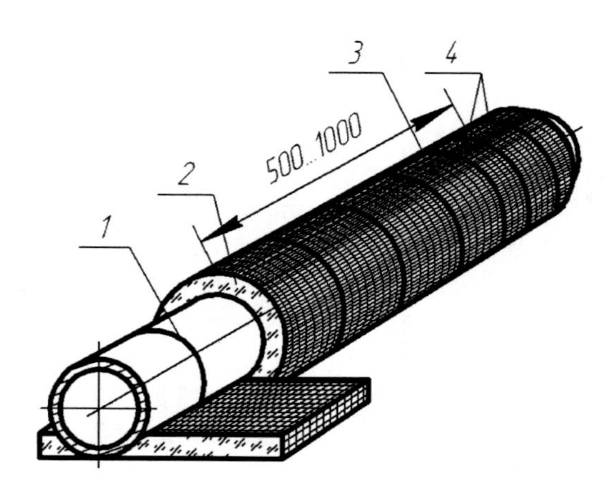

Figure 17.4. Insulation of pipelines with mineral wool mats:
1– hanger; 2-heat-insulation layer; 3- tyre; 4- lacing.

The basic type of the synthetic binder for mineral wool board is phenolic alcohols. It is possible to reduce the content of synthetic binder by additional introduction of bitumen in the mineral wool products.

Mineral wool products are widely used in thermal insulation of pipelines and industrial equipment. Along with mats and boards for this purpose cylinders and semi-cylinders on synthetic binder can be used. High temperature wool and products with the temperature of application up to 1250°C can be manufactured from kaolin and silica initial materials.

Basic physical and mechanical properties of products based on mineral and glass wool are shown in Table 17.1.

Glass-fiber materials occupy about 10 % of general amount of modern heat-insulating materials. Glass-fiber materials economically effective to apply only as light products with an average density 30-50 kg/m^3 and thermal-conductivity 0.037-0.04 W/(m·K).

In the last years the technology of *basalt fiber materials*, which combine high heat-insulating properties and temperature proofness (up to 750°C), longevity, incombustibility and non-toxicity, is developed. Industry produces different types of products based on basalt fibers - linens, cords, rolls, mats, board, etc. The optimum value of density for basalt fiber materials is equal 80-100 kg/m^3.

Asbestos can be used for manufacture of heat-insulating products, mastics and mortars in combination with Portland cement or other binders.

Lime-silica and some others heat-insulating products, which contain asbestos, are manufactured. They are used mainly for the insulation of equipment and pipelines which work at temperatures 400-600°C.

Table 17.1. Physical and mechanical properties of products based on mineral and glass wool

Types of products	Average density, kg/m^3	Thermal conductivity at 25 ±5°C, W/(m·K)	Tensile strength, MPa, not less than	*Compressibility (load 0.002* MPa), not more than, %	Temperature range of application,°^{0}C
Mineral wool materials					
Sewed mats	85-135	0.044	-	30-40	-18 +600
Boards on bitumen binder	75-250	0.046-0.064	0.075-0.008	5.5-45	-100 +60
Boards on synthetic binder	35-350	0.044-0.66	0.008-0.01	4-15	-100 +400
Cylinders and semi-cylinders on synthetic binder	75-225	0.048-0.052	0.015-0.025	-	-100 +400
Glass wool materials					
Boards on synthetic binder	40-200	0.05-0.057	-	10	-60 +180
Mats	25-50	0.045-0.047	-	40-50	Up to +500

Lime-silica products are made of fine mixture of quicklime and silica materials - diatomite, tripoli, marshalite and others in the conditions of autoclaving. Content of asbestos is in the range of 15-30%. The products of this group produce as boards and semi-cylinders with the minimum density 225 kg/m^3 and thermal-conductivity 0.112 W/(m·K) at 300°C.

For thermal insulation large interest is paid to the products based on the *expanded perlite* and *vermiculite*. Perlite belongs to the group of volcanic glass, and vermiculite - to hydromica. Both the rocks contain the certain amount of the bound water, which at temperature 800-1000°C is intensively removed as vapour. At firing of perlite the expanding coefficient (ratio between sizes of grains after expanding and initial grains l) is mainly equal 6-15; vermiculite 3-20. After firing, expanded perlite can be divided into sand with bulk density 75-250 kg/m^3 and crushed stone with density 300-500 kg/m^3. Thermal-conductivity of perlite sand is 0.041-0.07; crushed stone 0.075-0.09. Perlite sand can be used as fine aggregate in concrete and mortars for making heat-insulating products and also fire-proof plasters, which are used at temperature of 200-875°C.

Scaly structure is characteristic feature of expanded vermiculite. It is used as grains, which have coarseness up to 10 mm, bulk density 100-200 kg/m^3 and thermal-conductivity 0.064-0.076 W/(m·K). The possible temperature of application of expanded vermiculite is higher, than perlite one and achieves 1100°C.

Production of both non-fired and fired heat-insulation products is possible with application perlite and vermiculite fillers.

Non-fired products can be produced with the application of different binders: bitumen, polymers, liquid glass, gypsum, Portland cement. Their average density is in the range of 150-400 kg/m^3. Their range of application depends on the properties of light-weight aggregates and binders. In particular, *bitumen perlite* can be applied for heat-insulation of pipelines and different coatings; *polymer perlite*, which has comparatively higher strength, is possible to use as heaters of self-bearing and hangings light panels.

Bitumen perlite and polymer perlite concrete are the varieties of heat-insulating materials, which can be successfully used also for waterproofing arrangement. Heat-insulating, acoustic and decorative properties are combined in vermiculite products.

Perlite and vermiculite in combination with a ceramic bond or liquid glass allow obtaining the fired heat-insulating products with an average density 250-400 kg/m^3. Such products can be used for the thermal insulation of technological equipment at operational temperature up to 900-1100°C.

The *cellular materials* are manufactured by expansion of the different raw material masses in the process of forming or at thermal treatment. Gas- and foam concrete are the most widespread. Cellular concrete products of autoclave and non-autoclave hardening with a density not more than 400 kg/m^3 are applied for warming of building structures and thermal insulation of industrial equipment with the temperature of surface, which is insulated, no more 400 °C. They can be used in premises with relative humidity of air no more than 60% without the special protection against moistening.

Heat-insulating *porous ceramic materials* are manufactured from porous ceramic masses. Porisation of ceramics is executed due to the introduction of combustible additives (sawdusts, lignin, anthracite, etc.) into the batch or by mixing of ceramic mixtures with foam and gasifier. Application of diatomite, tripoli and some other sedimentary rocks containing silica are widespread for the manufacture of light-weight ceramics. Heat-insulating products from porous ceramics are made with a density of 350-500 kg/m^3 and compressive strength 1-2 MPa.

Foamglass is manufactured at temperature of 750-850°C by expansion of softened glass mass which contains gasifier. Foamglass is an effective inorganic heat-insulating material. High porosity (80-95%) allows to provide low density (100-700 kg/m^3) of foamglass. Also, favourable structure of this material and high content of closed pores provide relatively high strength, water resistance and low water absorption. Temperature stability of ordinary foamglass is 300-400°C and non alkaline foamglass is 800-1000°C. Foamglass is easily processed and can be of different colouring. This material is applied for the thermal insulation of refrigerators, heating systems and also as facing and acoustic material.

17.2. Organic Heat-insulating Materials

Gas-filled plastics are the most effective organic heat-insulation materials. Plastics, which consist of the system of isolated cells, are called *foam plastics;* system of connected pores – *cellular plastics*; system of repeating hollows - *honeycomb plastics* (Figure 17.5).

Foam plastics based on polysterene, polyvinylchloride, polyurethanes and phenol-formaldehyde resins are the most widespread in construction. Basic properties of porous plastics change depending on the type of polymers and method of manufacture (Table. 17.2). Ratio between strength of heat-insulating plastics and their density is considerably higher than for other heat- insulating materials. Foam plastics have lower water absorption and higher sound insulating ability than cellular plastics.

Weight of foam plastics is less in 4-5 times in comparison with such traditional materials as mineral wool boards, glass-fiber boards and foam-glass at improvement of heat-insulation properties in 1.5-2.2 times. *Foam urethanes* are one of the most effective foam plastics. They are used mainly as heaters in three- and double-layer panels. High heat-insulating properties of foam urethanes, their closed cellular structure, which provide minimal water absorption, absence of corrosive influence on metals and good adhesion to the most of materials enable application of this type of foam plastics for spraying on different (especially metal) structures.

Sprayed layer can be used simultaneously as heat-insulation and protection of metal against corrosion. Sprayed polyurethanes can be used for coating of engineering structures, mainly oil storage tanks. Their application is possible for the thermal insulation coatings of houses, refrigerators, pipes, etc. Basic obstacle for application of foam plastics is their high combustibility, including sprayed ones. Therefore a plastic thermal insulation must be reliably protected from the action of fire.

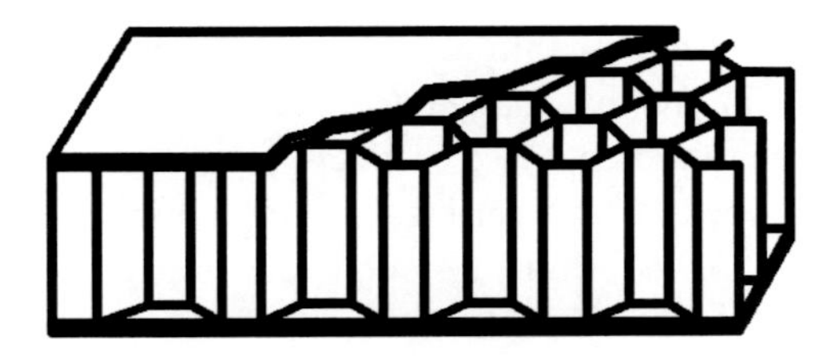

Figure 17.5. Honeycomb plastic.

Table 17.2. Basic properties of porous plastics

Type of plastic	Average density, kg/m^3	Ultimate strength, MPa		Limit temperature of application, °C	Thermal-conductivity, W/(m·K)	Water absorption for (30 days), % by volume
		Compressive strength	Bending strength			
Expanded polystyrene	30-200	0.15-3	0.4-7	60	0.11-0.2	1-15
Foamed polyvinyl chloride	50-270	0.23-2.5	0.4-4	60	0.15-0.19	3-10
Foamed polyurethane	30-200	0.15-3.5	1-5	150	0.12-0.21	5-18
Phenolic foam plastics	80-150	0.25-0.7	0.3-0.6	130	0.14-0.19	8-28
Urea-formaldehyde cellular plastic	10-25	0.02-0.04	-	110	0.11-0.15	75-85

In the last years foam plastics with decreased combustibility are developed. Their surface is coated with fireproof materials for this purpose.

Phenolic foam plastics are the least combustible. They are more economic than other, have a wide source of raw materials. However their strength is lower than foam urethane. Furthermore opened cellular structure of phenoplasts and, as a result, high water absorption requires more careful waterproofing of them.

For construction considerable interest is paid to by foam plastics on the basis of epoxy polymers - *foam epoxy*. This material is characterized by low thermal-conductivity and at the same time it has advantages of epoxy polymers: high adhesion to a number of materials, comparatively high mechanical properties, water and chemical resistance, etc.

Peat heat-insulating products can be manufactured as boards, panels, etc. Poorly decomposed peat which is fissioned at processing on separate fibres can be used as raw material for them. The products formed by pressing act are dried and resinous substances, which glue together fibers, are secreted from peat. Density of peat boards is 170-260 kg/m^3, bending strength – 0.3 MPa, thermal-conductivity – 0.052-0.075 W/(m·K). Water absorption achieves 180-190% during 24 hours. In the moistened state, especially at storage in stacks, peat products can have ability to self-ignition. Limit temperature of their application is 100°C. Peat heat-insulating products can be applied for the insulation of refrigerators, industrial equipment, etc. It is required to protect these products from moistening.

Sawdust concrete can be used as heat-insulation material due to it comparatively low density (not more than 550 kg/m^3). Sawdust concrete is a type of light-weight concrete. Ground wood or some phytogenous waste products can be used as the aggregate of this material. Compressive strength of sawdust concrete products, assigned for thermal insulation, is 0.5-1.5 MPa; water absorption 60-85%; thermal-conductivity 0.15-0.17 W/(m K) at humidity 15%. Sawdust concrete products, taking into account their high heat-insulating ability at the low values of strength and module of elasticity, are rationally applied for hangings and self-bearing external panels and large blocks at the proper protection from moistening.

Fibrolite is similar to properties of sawdust concrete, which includes wood wool with fibers length in the range of 200-500 mm as aggregate. Fibrolite can be manufactured as boards with density in the range of 300-500 kg/m^3 and bending strength from 0.4 to 1.2 MPa. Water absorption of fibrolite is 35-60% and thermal-conductivity is in the range of 0.09-0.15 W/m°C. Fibrolite differs by high acoustic absorption due to connected structure of pores; easy processability and high adhesion to plaster.

Wood wool for providing the normal hardening is processed (mineralized) in solutions of calcium chloride or other salts.

Wide application of local heat-insulation materials is the most rational in conditions of farm construction. Effective raw materials for the manufacture of local heat-insulation materials are mainly wastes of woodworking industry and agriculture.

17.3. Acoustic Materials

Materials which absorbing more than 40% of sound waves energy if frequency oscillations is in the range of 500-1000 Hz, i.e. coefficient of acoustic absorption is more than 0.4, are called *sound-absorbing materials*. These materials are widely applied in the modern construction practice. Mineral wool boards based on starch binder can be used as sound-absorbing material. Such boards (Figure 17.6) have density 350-450 kg/m^3, bending strength 1-1.8 MPa, coefficient of acoustic absorption 0.6-0.8. Also, carboxymethyl cellulose, polyvinyl acetate emulsion, phenol alcohols can be used as organic binders for sound-absorbing materials.

The acoustic boards based on mineral wool granules or mineral wool fibers are used for the acoustical ceiling construction public and office buildings with relative humidity of air not more than 70% at a temperature of 15-20°C. Surface of board can be painted by polyvinyl acetate emulsion.

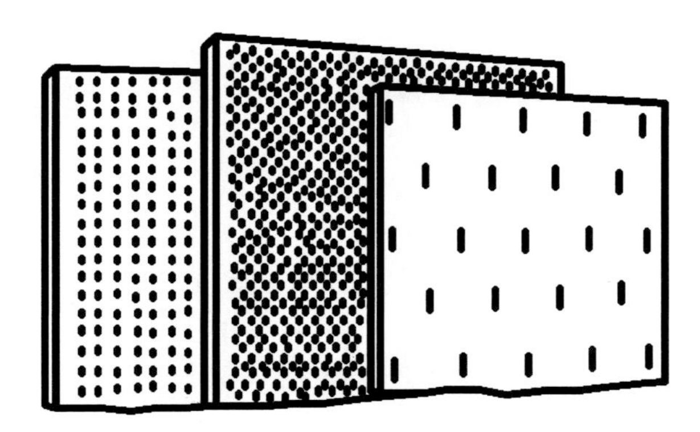

Figure 17.6. Sound-absorbing board.

Mineral wool boards with a gypsum facing and the gypsum perforated boards with the layer of fabric on which mineral wool is placed, are also applied as sound-absorbing materials. Perforated wood fiberboards, cement acoustic fiberboard with density of 400 – 500 kg/m^3; porous concrete boards with the system of connected pores; polyurethane cellular plastic and asbestos cement perforated boards can be also used as sound-absorbing materials. Ceramic boards and blocks based on brick chips with liquid glass are applied for sound proofing of the technological equipment at *operating temperature up* to 500°C. Also concrete based on porous aggregates (perlite, vermiculite, slag pumice, etc.) are applied as acoustic ones.

Different acoustic facings are applied. They can be used not only for the improvement of premises acoustic but also for the decorative finishing of interior surfaces. Acoustic ceilings from the profiled aluminium sheets and other sound-absorbing materials are widely used.

Sound-insulating materials are manufactured as mates, boards, gaskets, etc, with density not exceeded 300 kg/m^3, module elasticity – 15 MPa. For sound insulation porous rubber, elastic polymer materials, mineral wool, wood-fiber boards, asbestos and other materials are used.

Self-Assessment Questions

1. What is the importance of heat-insulating and acoustic materials in construction?
2. Tell about manufacture and properties of mineral wool and products on its basis.
3. Describe glass-fiber and basalt fiber heat-insulating materials.
4. Describe lime-silica heat-insulation materials and materials on the basis of perlite and vermiculite.
5. What are the peculiarities and properties of porous heat-insulating materials?
6. Tell about heat-insulating materials based on ground wood and wood wastes.
7. Tell about basic types of acoustic materials.

Chaptr 18

VARNISHES AND PAINTS, PAPER COATINGS

Natural or synthetic materials, which are applied in a viscous liquid state by thin layer at the structures and elements with the film formation for their protection against the harmful influence of the environment, also for decorative design and improving of the hygiene and sanitary conditions, are related to the varnishes and paints. They are divided into basic and auxiliary materials. Paints, varnishes and enamels are the basic materials. Primers and fillers (spacklings), used at the surface preparation for the painting, solvents, diluents are auxiliary materials. Paints are also divided by the next features: type of film-forming substances (oil, glyph, epoxy, lime, silicate, etc.); kind of liquid phase (aqueous and waterless); relation to the water action (water-resistant and water-nonresistant); main purpose (chemically resistant, thermal resistant, electric insulating, etc.) (Table 18.1, Figure 18.1).

Paints are used for the formation of the opaque coloured decorative and protective films, which hide the texture of material under paining; varnishes, - for creation of transparent colourless coating and finishing of painted surface and also for protecting from mechanical damages. Composition of paints consists of the binding or film-forming substances and colouring agents or pigments. They can contain fillers, due to which it is possible to reduce the content of pigments, and special components which improve technological and operating qualities of materials (hardeners - siccatives, diluents, etc.).

Varnishes are dispersions of natural or synthetic polymers, bitumens, drying oil, in volatile solutions. Plasticizers, siccatives and other admixtures can be added in composition of varnishes.

Enamel paints (enamels) are suspensions of pigments in polymer or oil varnishes; unlike paints, they contain less fillers and have more intensive glance.

Priming paints provide the required adhesion between the cleaned surface and coating. Usually the primer consists of the same materials as paints, but has less content of pigments.

Spacklings (putties) are applied for filling of voids and smoothing of surfaces before painting. They contain not only adhesive substance but also a large amount of filler (usually chalk).

The typical scheme of multilayer coating includes: primer - one layer; spackling - one or several layers; enamel (paint) – several layers; varnish - one layer. The sequence of the materials coating, usually, remains the same, but separate elements of coating can be excluded. For example, if the surface of the painted material is sufficiently flat, without the significant defects and the increased decorativeness of coating is not required; application of spackling and varnish is not required.

18.1. Basic Components of Varnishes and Paints

Film of varnishes and paints coating is formed by binder, which is included in the composition of both priming and facing layers. For obtaining of painting mixtures there are used mineral binders (lime, cement, liquid glass); glues, which are made of materials of animal origin (bones, casein, leather); vegetable glues (starches, dextrin, flour), synthetic glues (carboxymethylcellulose). Films formed by water-soluble glues except casein, can be again diluted at water treatment that is why mixtures on their basis are not applied for the external painting of premises at high humidity of environment.

Drying oils and synthetic insoluble polymers are the binders for waterless mixtures. They provide the forming of water-resistant coating as well as cement and liquid glass. *Drying oils* are the binders for oil paints. They are divided into three groups: natural, compacted (seminatural) and artificial (synthetic).

Natural drying oils are the products of heat treatment at 150-300°C of vegetable oils with the addition of 2-4% of siccatives - manganous or cobalt salts of fatty acids for hardening accelerating. Compacted drying oils are obtained by the dissolution of vegetable oils, with the following compacting by oxidizing or polymerization. Synthetic drying oils are produced by heat or chemical treatment of polymers often with admixtures of vegetable oils. Such variety of drying oil is glyptal one.

Table 18.1. Classification of painting mixtures depending on performance conditions of coating

Mixture	Basic assignment depending on performance conditions
Weather resistant	Coatings resistant to atmospheric influences under various climatic conditions, performed at open grounds
Limited weather resistant	Coatings performed under shelter and inside of the unheated and heated premises
Conservation	Coatings used for temporal protection of painted surface at the processes of production, transporting and storage of elements. Coatings can be used for protection against the action of fresh and sea water
Special	Coatings which have specific characteristics: resistant to the X-rays and other radiations, luminous, thermo-regulating, for painting of rubber, plastics, etc.
Drying oil and petrol - resistant	Coatings resistant to the action of oils, petrol, kerosene and other oil products
Chemically resistant	Coatings which are resistant to the action of acids, alkalines and other chemical reagents
Heat resistant	Coatings resistant to the action of high temperature
Electric insulating	Coatings at which voltage, electric arc and surface discharge act

All drying oils harden at the air conditions in the thin film during 12-24 h. Natural drying oils have the highest quality; they are used for the production of high quality oil paints for the external and internal painting of metal structures, roofs and joinery.

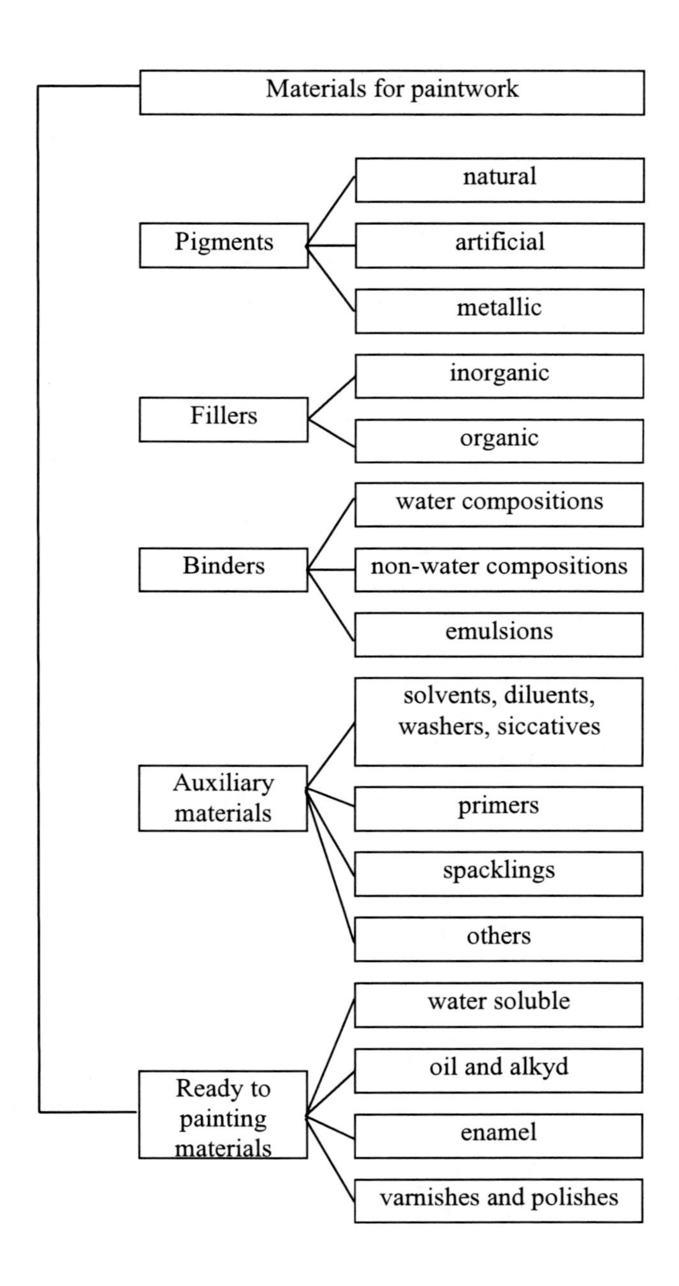

Figure 18.1. Classification of materials for production of painting mixtures.

Table 18.2. Application of mineral pigments

Pigments	Colour	Application
Chalk	White	Primers, fillers, water paints
Zinc white	White	Oil and silicate paints
Dry lithopone	White	Oil paints
Titanium white	White	Oil paints and enamels
Manganese peroxide	Black	Water and waterless painting mixtures
Carbon black	Black	Waterless painting mixtures
Graphite	Grey	Water and waterless painting mixtures
Iron minium	Red	Water and waterless painting mixtures
Dry ochre	Yellow	Water and waterless painting mixtures
Zink chrome	Yellow	Oil, enamel and glue paints
Chromium oxide	Green	Water and waterless painting mixtures
Ultramarine	Blue	Water and waterless painting mixtures
Azure iron	Blue	Water unalkaline painting mixtures

From synthetic polymers, which are used as binders for varnishes, paints and enamels, phenolformaldehyde, carbamide, perchlorovinyl, polyvinylacetate and other high molecular compounds are the most effective.

Insoluble binders are used in combination with organic solvents or as water emulsions. At first case there are obtained volatile resin mixtures, in other case - water-based mixtures. Organic solvents, as a rule, are toxic; their application requires special measures for safety and industrial sanitation. Water-based paints have no smell, harmless, they are easily applied at the surface, including moisture, and dry out rapidly. But they form less dense coatings, than volatile resin mixtures, less stable and frost-resistant; they can be subjected to microbiological destruction.

Basic properties which determine the quality of binders for varnishes and paints are viscidity, density, colour, transparency, time of hardening, adhesion, hardness, flexibility, ability to be polished.

Painting materials, besides binders contain *pigments* - finely dispersed materials, non-dissolved in binder substance and solvent and capable in the mixture with them to form the opaque coatings of diverse colors and tints. At the selection of pigments there are considered the color, light- and the weathering resistance, resistance to the binder action, to oil absorption, and also resistance to the influence to hydrogen sulfide and other chemical compounds.

Pigments are divided into white, black and with different tints of grey, red, yellow, green, blue and brown. In Table 18.2 mineral pigments widespread in construction are shown.

Pigments of red, blue and yellow colors are considered as the basic. Other colors can be obtained by mixing pigments between itself in different proportions (Figure 18.2). For example, green color is obtained at mixing of blue and yellow pigments, violet - red and blue, orange - red and yellow and so on. Ability of pigments to give at mixing the required colouring is called the *painting ability* or *intensity*.

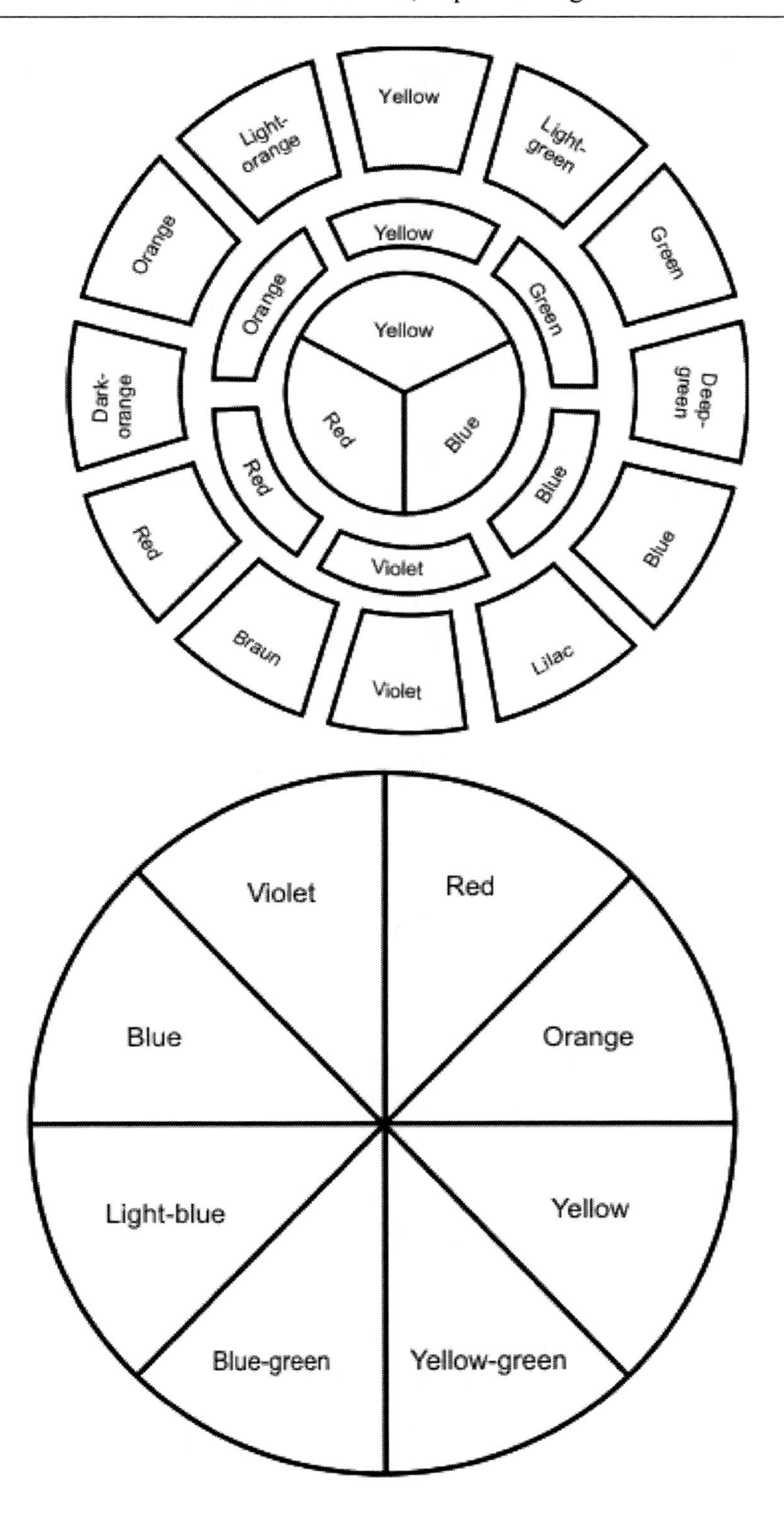

Figure 18.2. Colored circle for mixing of pigments.

Table 18.3. Basic properties of fillers of the paintwork materials

Filler	Index of refraction	Density, kg/m^3	Oil absorption, g/100 g	pH value of water extraction
Barium sulfate (heavy spar)	1.64-1.65	4460-4500	6-10	8-9
Calcium sulfate:				
Gypsum	1.53	2900-2990	20-25	-
Anhydrite	1.59	2950	20-25	-
Talc	1.58	2730-2880	20-50	9-10
Mica	1.59	2.740-2.880	20-50	9
Wollastonite	1.63	2780-2910	20-26	9-10
Asbestos	1.55	2800	-	-
Kaolin	1.6	2540-2600	13-20	5-8
Quartz	1.54-1.55	2200-2650	15-25	-
Diatomaceous silica (fossil meal)	1.4-1.5	1900-2300	100-220	7-10
Natural chalk	1.6	2710	10-14	9-10
Settled chalk	1.48-1.6	2650	30-50	9-11
Calcite	1.65	2700	15-22	9.7
Settled magnesite	1.5.-1.7	2900-3100	-	-
Dolomite	1.62-1.65	2850	15-19	10
Witherite	1.6	4300	14-16	-
Barium carbonate setteled	1.63	4300	17	-

Painting ability is determined by the whitenings, i.e., by mixing pigments with the consecutively growing dose of chalk for the aqueous, or white for the oil mixtures and by the comparative estimation of obtained paintings. The pigments of small intensity, for example ocher, become unobtrusive already at whitening 1:15, with the high intensity, for example azure keeps its tints at whitening to 1:2017 and more. Such intensive pigments for the dilution and the reduction of prices are mixed up with the fillers - chalk, kaolin, and other.

The second important characteristic of pigments is the *spreading capacity* or covering property - the dose of pigment, required for the complete coating previously laying on the surface layer of contrasting paint. The pigments of low intensity at the same time can have high spreading capacity. For example, ocher has large spreading capacity (60-90 g/cm^2) with the small painting ability, azure - vice versa. Spreading capacity is increased with the increasing of the difference of the refractive indices of light by pigment and binder.

For the improvement of a series of properties and the reduction of prices of the painting mixtures fillers are used. For the aqueous painting mixtures by the simultaneously coloring substances and as fillers the chalk and lime can be used. For the savings of white in the oil paints barites is applied. In chalk-glue mixtures for the giving of the best coating ability kaolin is added. Into oil, silicate, emulsive and some other mixtures for increasing the water and weather resistance, mechanical strength talc is introduced as filler. Asbestos admixture attaches to the paints enhanceable fire resistance and the strength (Table 18.3).

Table 18.4. Basic solvents of the paintwork materials

Solvent	Density, g/cm^3	Application
Acetone technical	0.789-0.791	For the nitro paints
White spirit	0.795	For oil and enamel paints, glyptal and asphalt varnishes
Purified benzene	0.87-0.88	For the quick-drying paints
Butyl acetate	0.879	For nitro paints and enamels
Butyl alcohol (Butanol)	0.81	The same
Kerosene	0.800-0.845	For the paints at the execution of auxiliary works
Coal xylene	0.860-0.866	For the glyptal and perchlorovinyl paints, bitumen varnishes
Turpentine	0.860-0.875	For oleoresinous varnish, enamel and bitumen lacquers
Coal solvent, oil solvent	0.865-0.885	For the glyptal, pentaphthalic and bitumen varnishes and enamels
Toluene coal	0.863-0.867	For the perchlorovinyl enamel paints

Solvents and *diluents* are applied for reduction of paintwork material's viscosity to the viscosity, required at the selected method of material spreading. Introduction of solvents retards the process of film formation; their excess can also result in the decline of the strength and density of coating, increasing of shrinkage and creep.

The volatile liquids with the boiling point at 50-200 °C are utilized as solvents (Table 18.4). According to the ability to dissolve this or other substance or only to dilute it until less viscidity, there are distinguished active solvents and substances which can be only diluents. The same solvent can be either active solvent or diluent for the different binders.

As diluents there are also applied drying oils, emulsions, adhesive solutions, which, unlike the solvents, contain film-forming materials and at the same time are applied for diluting of the pastes or dry inorganic paints.

Special admixtures can be added for adjustment of properties of the paintwork materials (except of the basic components - binders, pigments, fillers, solvents and diluents): siccatives, plasticizers, water-repellent agents, initiators, activators, hardeners, thixotropic agents, etc.

Siccatives accelerate drying of oil and enamel paints, varnishes and drying oils. As siccatives the solutions of lead- manganese salts of naphthenic acid or mixture of naphthenic acid with the drying or semi-drying oils in the gasoline or the turpentine are used.

Plasticizers give to the paint and varnish coatings the elasticity and enhanced resistance to the shock loadings. Hydrocarbons, their halogen derivatives, ethers, ketones, alcohols, amines and others belong to them. The most considerable at the production of paintwork materials are the following plasticizers: non-drying vegetable oils; various ethers - phthalates, sebacates, phosphates; resins of alkyd type, chlorinated paraffins and biphenyls.

It is possible to enhance the adhesion of coating film to the basement by the plasticizers; however the plasticization can decrease the hardness and enhance superfluously the plasticity of coating film. Plasticizers are tested on the strength, elasticity and other properties.

Plasticizer should be colourless, odourless, untoxic, and resistant to the action of ultraviolet rays.

18.2. Types of Paints and Varnishes

Among the water painting materials the most common are lime, silicate, cement paints. The separate group is formed by water-emulsion mixtures.

Lime paints are obtained at the basis of lime paste which is diluted to consistency of "milk" and alkali-proof pigments (ochre, chromium oxide (green rouge), iron minium, ultramarine blue and others). For the increasing of resistance potassium aluminum sulfate (alum) or cooking salt are added. The variety of lime paints are dry mixtures on the basis of the ground lime and calcium chloride with alkali-proof pigments.

Lime mixtures are used for painting of facades and internal brick, plaster and concrete surfaces.

Lime paints have comparatively low strength and longevity.

Glue paints are mixtures of pigment, filler - usually chalk, and finely divided glue. Before using the paints are dissolved in water to painting consistency. They are not water-resistant, intended for the internal finishing of plastered surfaces. On reason the wide application of synthetic film-forming substances at production of paints, glue mixture have lost their value.

Silicate paints are the suspensions of alkali-proof pigments and silica fillers in potassium liquid glass. The use of liquid sodium glass is the reason of appearance on the painted surface of white color salt efflorescence. Paints harden as a result of formation of low soluble silicates. It is recommended to use them in moist terms exploitation for painting of brick and plastered facades, and also for the internal finishing.

Cement paints are manufactured on the basis of Portland cement and alkali-proof pigments. For the improvement of properties there added up to 15% water-retaining admixture - dry hydrated lime, 2-4% accelerate of hardening - calcium chloride and 1% water-retaining agent - calcium stearate and others in their composition. Paints are weatherproof. They are used for the external and internal painting at stone, brick, concrete, plaster and other porous surfaces which are preliminary moistened. At dilution of cement paint in polyvinylacetate, perchlorovinyl and others water emulsions of polymers polymer-cement paints are obtained. They are characterized by good adhesion with different construction materials, raised weather-resistance, elasticity, ability to harden at low humidity.

Water-emulsion and *water-dispersion paints* are emulsions or suspensions of pigments and fillers in the film-forming aqueous dispersions of polymers with the additives of emulsifiers, stabilizers and other substances. Depending on the form of polymer water-emulsion paints there are polyvinyl acetate, rubber, polyacryl and others. They are used for the internal and external works. Paints for the internal works are not recommended to use in the premises with the increased humidity.

Paints on the basis of vinyl acetate polymers, copolymers of the styrene with the phenylbutazone, acrylate copolymer are the most widely used.

Table 18.5. Description of acryl paints "Ceresit"

Index	Paint	
	CT 42	CT 44
Composition	Aqueous dispersion of the acrylic resin with mineral fillers and coloring ingredients	
Density, g/m^3	1.4	
Resistance to the sedimentation, h	12	3
Temperature of basement at application of paint, °C	from +5 to +35	
Water absorption of the treated surface, $kg/(m^2 \cdot h^{1/2})$	no more than 0.5	no more than 0.5
Resistance to washing-off	not less than 5000 cycles	
Expense in l/m^2 at coating in one layer in two layers	 0.2-0.5 up to 0.4	 0.1-0.4 up to 0.3

Quantity of film-forming substance in the paints of this type is 40-55%. Facades of buildings, external wooden surfaces are painted mostly by polyacrylate emulsion paints, styrene- polyvinyl acetate and butadiene paints are used predominantly for interiors.

Water paints do not have a smell practically. Painting coatings have the high adhesion almost to all basements and good performance parameters. It is possible to paint humid surfaces by them; they are non-combustible, permit the water vapor and promote the comfort of dwellings. Physical properties of the water paints depend on the type of polymer, which is applied as binding agent. So, paints based on the polyvinyl acetate have the low water-resistance. They are applied mainly for coloring of ceilings and internal walls in the dry premises. Butadiene-styrene dispersions have the good water resistance; however they are characterized by limited light fastness.

Acryl paints are the most universal and compose the most considerable part of the water paints. They can be produced with a great number of different colors and save the color at intensive ultraviolet irradiation. Acryl paints allow creating elastic coatings; they are durable, and resistant to washing.

Basic specification of acryl paints "Ceresit "are resulted in Table 18.5.

Usually, water paints deteriorate their properties at freezing, therefore in cold time they must be kept in the heated premises.

The water-dispersed *organic-silicon (silicon) paints* impart the hydrophobic properties to the coatings that substantially increase their durability. Paints, based on the silicon resins combine the best properties of acryl and silicate paints.

Before the application, paint can be diluted by water till viscosity for required the spreading by paint sprayer or by roller and brush. The period of practical drying of the basic types of water-emulsion (water-dispersion) paints - is less than 1 hour.

Oil paints are the suspensions of pigments in drying oil. If required fillers, siccatives and other admixtures are added to their composition. Paints are produced doughy or ready to use. They are used for the different types of painting on a metal, wood, plaster, concrete, taking into account quality of drying oil and type of pigment. Thus, for instance, mixtures with the use oxidized drying oil are suitable for the painting of surfaces inside the premises, in contrast to the paints on the natural and gliptal drying oils, which are intended for the using both on the internal and external coatings.

At application of white lithopone, which darkles under the action of light, there is not possible to use oil paints for external painting works, and white lead which contain harmful matters are not recommended for painting of internal surfaces of housings apartments.

The whitened oil paints with the reflection coefficient by 50-70% are recommended to use in the official and working premises of the buildings of public and industrial destination, with the reflection coefficient by 30-40% - in the vestibules and on the staircases, the corridors, which do not have natural illumination, but it is less than 30% - in all auxiliary premises, if their light regime do not present the increased requirements.

It is required to paint only the dry surfaces with oil paints and coating them with the thinnest skim. Humidity of plaster should not exceed 8 %, and wooden surfaces - 12%. Painting on the incompletely dried previous layer is not allowed.

In the winter conditions, painting with the oil paints inside the apartments can be made at a temperature not below than 8 °C (temperature is measured in the distance at 0.5 m from a floor).

For the priming it is possible to use drying oil, into which 5-10% of pigments are introduced. At the treatment of wooden and concrete smooth surfaces into the drying oil dry pigments usually are added, at the treatment of the plastered and metallic surfaces - the paste paints are used. Ocher, Indian red or their mixture (for the metal is used only Indian red) are applied.

Enamel paints are the variety of painting mixtures for which solutions of synthetic polymers as the binders are used. Process of film-formation of enamel paints goes at evaporation of organic solvents.

In the construction perchlorvinyl, pentaphthalic, glyptal, epoxy and other enamel paints are widely used. Each of them has their rational range of uses. So pentaphthalic enamels are rationally used for coating of wood floors, perchlorvinyl for the painting of the facades of houses, glyptal - internal elements of houses on plasters, wood and metal.

Enamel paints are dissolved in petrol, solvent, turpentine, xylol, toluene and some other substances.

Enamels form a solid layer with intensive glance after drying. The types of enamels are conditioned by the polymeric binders applied.

Drying film in the enamel paints occurs as a result of the evaporation of volatile solvent and simultaneous hardening of polymeric bonding agent on the surface. Enamel paints are produced usually ready to use. They can include plasticizers, hardeners, accelerators of drying, stabilizers, antirots, surface-active substances (SAS) and other special admixtures. Enamel paints are used for the external and internal painting on the metal, the wood, the plastering. Light resistance, the chemical stability and resistance to weather effect are characteristic for them.

Enamels are coated at primed surface. They protect the surfaces of constructions from the action of the environment; give them decorative appearance and some special properties.

Perchlorovinyl enamels are resistant to the action of the most mineral acids and alkalis of different concentration at a temperature 60 °C, however not resistant to the action of oxidants, concentrated sulfuric and nitric acids, hydrocarbons. They dry out quickly and form the durable semiglance film. It is recommended to apply enamels at the temperature not below

8°C. Coatings from the perchlorovinyl enamels disintegrate at heating higher than 60-80°C with the release of hydrogen chloride vapor.

Epoxy enamels are characterized by the high adhesion to the metals and non-metals, hardness and resistance to the corrosive environments. Epoxy enamels are usually double-base; they consist of suspensions of coloring pigments in solution of epoxy resin and the hardener. The diethylenetriamine and polyethylenepolyamine can be the hardeners.

Epoxy paint-and-varnish materials can be cold and hot hardening. At coloration by the materials of cold hardening, the temperature of air should be provided not below 15 °C.

Epoxy enamels are toxic and inflammable materials, that is explained by properties of solvents and hardeners, which applied at their making. After drying out coatings do not influence harmful on the human organism.

Epoxy enamels are placed by the method of airstream atomization.

Glyptal enamels are the suspensions of the pigments and fillers in glyptal varnish with admixtures of the siccative and solvent. Enamels are diluted to the required viscosity by lacquer petroleum, turpentine, xylene, solvent or their mixture.

It is possible to apply enamels, consisting of the glyptal varnish and aluminum powder (6-12 %) for an anticorrosive protection. Their heat-resistance is up to 300°C.

It is not recommended to apply glyptal enamels at the surfaces, on which direct solar radiation, precipitations and biological factors affect.

Pentaphthalic enamels are made on the basis of the phthalic anhydride and pentaerythrite in mixture with pigments, fillers and solvents. They form more atmosphere resistant, compliant, waterproof, proof to the temperature oscillations and mechanical influences coatings compared to glyptal.

At adding aluminum powder to the pentaphthalic varnish heat-resistant coatings (up to 300 °C) are obtained. Enamels before the use are diluted to the working viscosity by the solvent, xylene, lacquer petroleum.

Organic-silicon enamels differ by resistance to the action of high and low temperatures and their cyclic difference, the high insulating properties, oil-, gasoline- and freeze-resistance.

Powder paints are the fine powders which including synthetic resins, hardening agents, pigments and other admixtures. They are polymerized at 140-200 ^{0}C and form uniform film. They have the high longevity and adhesion to the covered surfaces. Due to the absence of solvents, powder paints have less shrinkage and porosity. The method of coloring allows to form ideally flat and smooth surfaces which characterized by the high resistance to the different types of influences and good corrosion resistance.

Powder paints are classified: depending on the type of resins - epoxy, epoxy-polyester, polyester, acrylic, polyurethane; depending on the type of the formed surfaces - smooth, shagreen and other; depending on the degree of glance of surfaces - glossy, semigloss, semi-matt, matt.

Coloration by the powder paints does not require priming of surfaces. Before coloring of powder paint, preparation of surface is made by the methods of defatting and phosphatization. Paints are putted mainly on metallic and glass surfaces by the method of spraying. It is difficult to get the thin-coat finishes (10-20 μm) using the powder paints.

Varnishes which are used for construction works divided by the followings groups: oil-resin, **not oil** synthetic, on the basis of bitumens, nitro- and ethylcellulose, spirit.

Properties and application of varnishes is determined by properties of organic binders. For example, varnishes on the basis of urea-formaldehyde, polyester polymers are used for painting of parquet floors, for the finishing of plywood, millworks, wood particle hardboards. Pentaphthalic varnishes are applied for varnishing of external and internal wood or metal coatings; perchlorvinyl varnishes - for protecting of various building constructions from aggressive environments. Bitumen varnishes are used for the anticorrosive coat of metallic details.

Also, bitumen-resin varnishes (solutions of bitumens and vegetable oils in the organic solvents) can be used for anticorrosive coat forming. Heat-resistance of such varnishes is no more than 50°C, the alkaliproof is also small. Paints for anticorrosive protection of metal products and structures can be obtained at their mixing with aluminium powder.

Ethylene varnishes and paints are made on the basis of divinyl acetylene polymer. They have high anticorrosive and painting qualities, ability to harden at a temperature to -25°C. Ethylene varnishes are relatively quickly deteriorated at the action of the light. A proof and economic paint can be obtained by mixing of ethylene varnish (60%) and red iron oxide (40%)

Nitrovarnishes are widely used for varnishing of metals and wooden products.

18.3. Paper Coatings

Wallpapers are the roll finishing material with the printing pattern or embossing of different tones on single-color or multicolored background. They are applied for papering of walls and ceilings of dwellings and public buildings. Decorative coating of wallpapers can imitate the wood grain, fabric texture and surface of metal and other materials.

Wallpapers are made in the next types: paper, vinyl and textile on the paper basis and on basis of nonwoven composite materials.

Wallpapers are divided into 2 types: simplex-wallpapers (wallpapers, having one layer of paper) and duplex-wallpapers (wallpapers, consisting of two glued layers of paper).

Vinyl wallpapers are the paper basis with the vinyl coating of different density, which gives the strength and humidity resistance to the wallpapers. The vinyl wallpapers are used, as a rule, in apartments with enhanced hygienic requirements. It is possible to select 3 basic categories of the vinyl wallpapers, which differentiate a method of fabricating, by a density and surface appearance: foamed vinyl, flat vinyl, dense ("kitchen") vinyl.

Acrylic wallpapers are the analogue of the foamed vinyl wallpapers, but acrylic emulsion is putted on basis, in contrast to vinyl.

Textile wallpapers are paper linens, laminated with filaments of the natural or mixed fibers. Textile wallpapers have the enhanced heat-insulatinh and sound absorbing properties, light fastness. They belong to the group of nonflammable materials. Textile wallpapers on the basis of textile linens, which are glued on the foam-rubber, are produced also.

Velours wallpapers are paper linens, on which in the process of production a pattern is putted at first, and then velours naps. A soft velvet surface appears as a result. Textile and velours wallpapers are intended for pasting over of walls and ceilings of office, dwelling and administrative buildings.

Paper and vinyl wallpapers are made smooth and relief, embossed, structural and metallized. The last are laminated with a metallic layer on the basis of foil or metallized film.

The row of new types of wallpapers appeared at the last years. The special type of wallpapers is wallpapers for painting. They are produced on a paper and nonwoven composite materials bases, usually in the rolls of the large length and saturated with the special water-repelling composition. Wallpapers of this type allow reducing of requirements to the quality of wall preparation, possessing with an ability to vapour permeability. It is possible to repaint them with the dispersion paints to 5-15 times. Structural wallpapers are made from two layers of texture paper of white color, glued between them. The three-layered embossed wallpapers are also made with the additional layer of paper, giving the large inflexibility to the linen. The coarse fiber wallpapers consist of two layers of dense paper with the wood fibers between them (sawdust). The presence of sawdust provides a grainy texture of the coating. Wallpapers hold to 15 cycles of repainting. Glass fiber wallpapers have the glass fabric basis, which gives the high degree of fire-resistance, strength and elasticity to them.

Wallpapers are supplied in the rolls with edges or without them.

The stability of coloration to light, resistance to the abrasion, destroying effort in the wet state, whiteness, stability of embossing pattern, maintenance of harmful matters, isolated during the exploitation in to the an air, are standardized for the wallpapers. Wallpapers should not have mechanical damages of linen, wrinkles, folds, breaks of edge.

Symbolic notation of the wallpapers includes their type (paper, vinyl and textile wallpapers on the paper or no woven composite materials basis), useful width and length of linen in a roll, grade, resistance of coloring to the light, method of gluing and removal from base after gluing on (Figure 18.3).

Lincrust is a variety of wallcoatings. This is a roll material with raised patterns, consisting of the mixture of synthetic polymers and filling agents on the paper basis.

Polyvinyl chloride decorative finishing film is a roll finishing material, made by a roller-calender method from vinyl chloride polymer, plasticizers, pigments and different admixtures. The films are produced with a glue layer on a reverse and without a glue layer.

A film surface can be smooth or stamped and multicolored with a printed pattern.

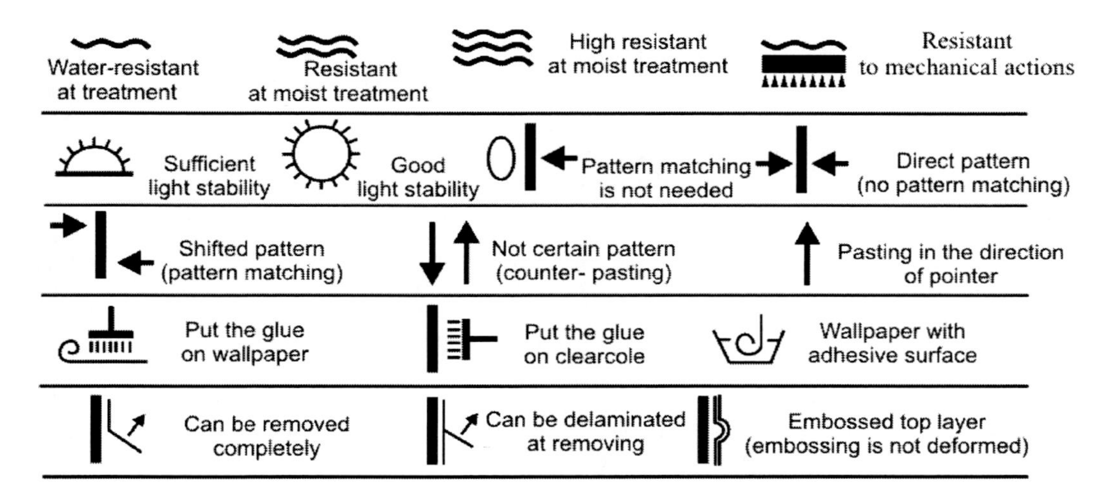

Figure 18.3. Symbolic notation of the wallpapers.

Self-Assessment Questions

1. What are the basic types of varnishes and paints?
2. What binding film-formation substances are applied for making paints and varnishes?
3. What are the examples of pigments? Tell about painting ability of pigments.
4. Tell about solvents and diluents, siccatives, plasticizers of paints and varnishes.
5. Give the descriptions of water painting mixtures.
6. Give the descriptions of oil painting mixtures.
7. Give the descriptions of basic types of enamel paints.
8. Tell about types and peculiarities of varnishes.
9. Tell about types and peculiarities of paper coatings.

Chapter 19

WOODEN MATERIALS

Free of a bark fiber fabric of tree trunk is meant by wood. The wood is widely applied in construction as saw-timber, plywood, joinery, glued structures, prefabricated wooden houses, elements from the wastes of woodworking and sawmilling. Wood is applied mainly as a roundwood and boards.

Wooden materials are distinguished from others by the row of positive features: comparatively high mechanical strength at a small average density, high processability, elasticity, low thermal-conductivity, considerable resistance to the alternate freezing and thawing and some to other corrosive actions.

The hygroscopic property, ability to decay, casting, swelling and cracking under cyclic moisture conditions, heterogeneity of physical-mechanical properties in various directions (anisotropy), flammability belong to the factors limiting application of materials from wood in construction. Inherent failings of wood are largely removed by its modification with polymeric substances, antiseptics, and fire-retarding additives.

The wood of coniferous trees is most widely used in construction. According to the scales of application in construction the softwood (coniferous wood) takes a place in the following row (after diminishing): pine-tree, fir-tree, larch, silver fir, cedar. Larch, which wood is valued due to the high density, strength and resistance against decay has the best physical-mechanical properties.

The wood of oak among the hard wood for responsible constructions on air and under water, for parquet, millwork is widely used.

The problem of the thrifty use of the wood resources become more important with every year, they are the sources of not only a construction wood but also of many various chemical and other products, and also as one of main natural factors.

With development of production of the precast concrete and other modern construction materials, the application of wood is limited by the only indeed the rational area, where its technical and economic advantages become sensible.

The problem of utilization of wastes of wood procurement and processing, production of various materials on their basis acquires all of greater value.

19.1. STRUCTURE OF WOOD

Composition and Structure

Wood is characterized by the stratified fibrous structure and consists of cells, which have a different form, size and assignment. Thus, 90-95% of wood of coniferous trees consists of *tracheides* - the stretched cells of wood in the line of a tree trunk with the length 2-5 mm and width 30-70 mm which conduct water and solutions from roots to the head in the time of tree life. The cells shell is formed mainly by the cellulose ($C_6H_{10}O_5$), which is a main component of bearing frame of tree. The polysaccharides- *lignin* and *hemicelluloses* are complex organic compounds. According to the composition they are close to the cellulose and also belong to the components of cell walls and intercellular substance

Usually wood includes 40-50% of celluloses, 20-30% of lignin and 15-30% of hemicelluloses, 1-3% are the concomitant components (resins, oils, tannin and other).

The atomic average chemistry of wood is practically identical for all of woods: 49.5% of carbon, 44.08% of oxygen, 0.12% of nitrogen and 6.3% of hydrogen. Mineral substances which give ash at combustion of wood are 0.2-1.7%. The salts of alkali-earth metals are mainly included into composition of the ash.

Wood is the main and the most capacious of mass part of trunk. Except of it, there is a core tube which has usually a diameter 2-5 mm approximately in a trunk center (Figure 19.1). It is the weakest part of trunk that is easily subjected to the decay.

Wood is outwardly crusty, which protects a tree from atmospheric and external mechanical actions. Bark includes two zones: external - *the cork* which carries out the protective function and internal - *bast* that actively takes part in the nutrients moving in a tree.

On the border between the bast and wood there is a skim of cells, fissionable and capable to grow which is called the *cambium*. The cambium predetermines the increase of wood and bark.

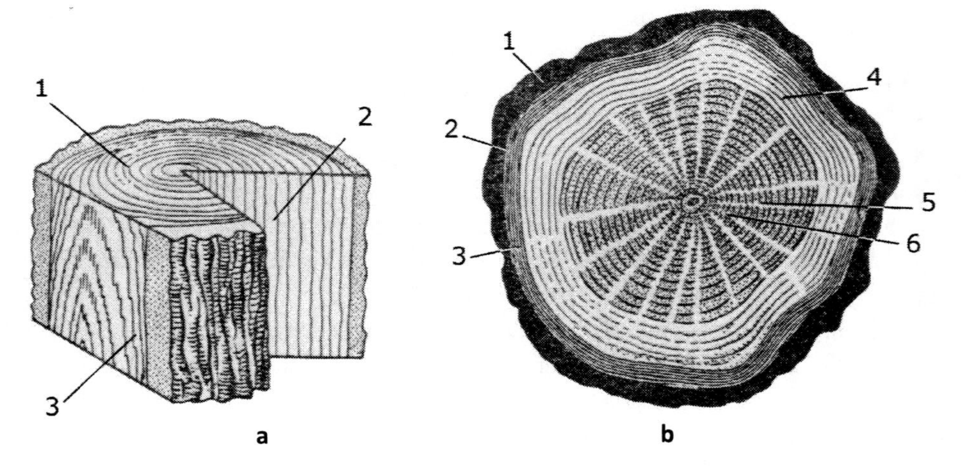

Figure 19.1. Structure of wood:
a- basic cuts of wood: 1- transversal; 2- radial; 3- tangential; b- structure of wood in a transversal section: 1- bark; 2- cambium; 3- bast; 4- sap-wood; 5- medullary sheath; 6-primary ray

Wood depending on the features of macrostructure is divided into three groups - heartwood, ripewood and sapwood. Wood of heartwood wood species (pine-tree, cedar, larch, oak, ashwood, poplar and other) has darker coloring of central part *of core* and lighter peripheral part - *sapwood.* The wood of all of trees consists only of sap-wood in the early age. A core appears, for example, in pine-trees in age of 30-35 years, in oak 8-12 years. It consists of dying off cells, impregnated with and stopped up by the stratums of resin, calcite, tannic and other substances. A core has an enhanced density and resistance against decay.

If the central part of wood has an identical color with peripheral and differs only with less humidity, it is called not a core, but *ripewood.* The ripewood as a core is denser part of the trunk.

Sapwood consists of younger cells and is intended for movement of moisture mineral permeates in it. With the age sapwood gradually passes to the heart or ripewood. At identical humidity many mechanical properties of the sap-wood correspond to the heartwood. It's resistance to decay is less, but it is easier saturated by the antirots. To the sapwood forest trees, which practically have identical wood according to the coloring and humidity both in a center and on periphery, a lot of wood species belong (cedar, alder, hornbeam and other).

Wood consists of separate annual layers which differ with a naked eye especially in coniferous trees. On the transversal cut of the trunk these layers have the appearance of concentric rings surrounding a core. Annual layers include two parts - summerwood and latewood. The summerwood appears in spring, it is lighter and softer than late, that appears only at the end of summer. This difference is especially strongly imaged in coniferous trees.

Composition of latewood largely determines the physical-mechanical properties of wood integrally. The resin ducts are concentrated in late softwood. Resin which fills them diminishes the water absorption of wood, increases resistance to decay. The presence *of rays* is characteristic for all woods-lines which radially diverge to the bark directly from a core or at some distance from it. They serve for conducting of water solutions of nutrients in horizontal direction in wood. Wood splits on primary rays easily and gives cracks at the shrinkage, because cells, which are included in these areas, are bonds between them are comparatively weak.

In the hardwood, *vessels*- tubular formations of cells with a diameter 0.1-0.4 mm and to 10 cm of length, directed in the line of a trunkare - are the weakest elements of structure, except of primary rays.

Defects of Wood

The defects of structure, violation of wholeness, damages and illness that means deficiencies which reduce quality of commercial timbers are taken to the defects of woods (Figure 19.2). In obedience to operating classification all the defects are divided in ten groups: knottiness, fungi paints and rots, chemical paints, insect damage, deformations, checking, defects of trunk form, defect of wood structure, wounds, undue laying in wood, mechanical damages and defects of treatment.

The basic defect that determines the type of wood *is knottiness*; knots are the basis of branches, located in wood of the trunk. Negative influence of knots consists in worsening of mechanical properties of wood as a result of violation of homogeneity and curving of fibers.

The knots hamper also the woodworking and in some cases are accompanied by internal rottenness. The type of knots (form, degree of growth, state of wood), their sizes and number, is specified in description of the knottiness.

Decay of wood appears in the fading of its color, diminishing of average density and strength.

Rots are caused by the development of the simplest vegetable organisms in the wood – *fungi*. The fungi which sit around the wood do not contain chlorophyll and can not synthesize the organic substances. Hereupon they are forced to feed the ready organic compounds and that is why they are settled on a tree.

Development of fungi in the wood takes place only at certain humidity (usually 25-70%) and temperature of air upon the average from 5 to 25°C. The decay does not take place in water, because access of oxygen, required for the vital functions of fungi is halted. Development of fungi is halted also at a temperature below 0°C and higher 40-45°C. The chemism of wood decay consists in its decomposition with the outburst of free carbon dioxide and water.

There are distinguished destructive and corrosive rottenness. The first one is generated by the fungi which destroy the cellulose of dead wood; the forest fungi, which parasitize on living wood and which use, mainly, a lignin, are the reasons of the second one.

A destructive rottenness is characterized by the prismatic splitting and darkening of wood, and the corrosive is accompanied with the stratification of wood by annual rings with its coloring in brown colors.

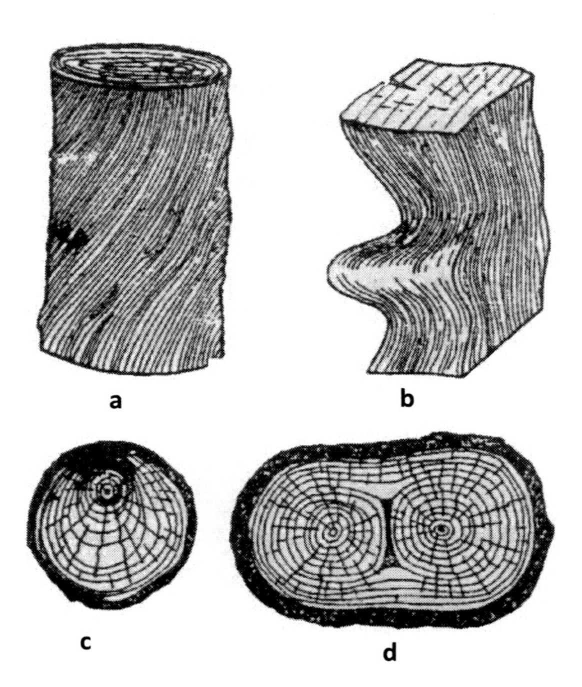

Figure 19.2. Basic defects of wood structure:
a- cross grain; b- twisting; c- compression wood; d- double core.

Along with the wood-destroying there is a group of wood fungi which give the various paints to wood and does not worsen almost its physical-mechanical properties. The variety of wood paints is mildew, which meets on the raw sap-wood of all the trees and caused by the carpophores of the mold fungi.

The insect damage of wood is called wormholes. The main masses of insect- vermin of wood are the different types of beetles.

There is a group of strong destroyers of wood from the class of shellfishes or crustaceous in the sea, they do not inhabite in the rivers and lakes.

The variety of worm-holes according to the degree of damage of wood is set (superficial to 3 mm, shallow to 5 mm, profound (rotten) is more than 5 mm) - and count up the number of openings.

Deformations and crackings - group of defects which are the consequence of change of form or violation of wood density. They arise up under the action of considerable internal tensions which appear in the process of tree growing, at the sharp change of temperatures, uneven deleting of moisture etc.

The number, character and sizes of cracks, and also their direction in relation to operating forces influence on the mechanical properties of wood. So, the most negative influence at a bend is shown by the crack of neutral area, which is in the plane, perpendicular to the force. Crack area of which coincides with directing effort has the least influence.

19.2. Physical and Mechanical Properties of Wood

Physical-Mechanical Properties

The wood humidity influences on the physical and mechanical properties, and also its availability. The humidity hesitates from 30 (oak) to 45% (fir-tree) for greenwood. The air-dry wood which for a long time has been laid on the air has the humidity 15-20%.

There are distinguished the hygroscopic (inherent) and free moisture in the wood. The hygroscopic moisture impregnates the cells and is retained by the physical-chemical bounds. The maximal amount of hygroscopic moisture, which can be imbibed by wood after its conditioning on the air, saturated with the aqueous vapor, is called *the saturation point* of cell walls or *limit of hygroscopic property*. Maximal humidity of cell walls of the greenwood or wetted by conditioning in water is called *the limit of saturation*. The humidity of wood which equals to the limits saturation and hygroscopic property at a temperature 15-20° C is practically identical and upon the average for all types of wood equals to 30%.

Unlike bound water the free one fills canals of vessels and intercellular space and is retained by the physical-mechanical links with the wood. The removal of free water requires less power consumption that is why its influence on wood properties is considerably less substantial. At first the free water mainly is removed, and then bound at the wood drying. The process of wood drying is halted for achievement of equilibrium humidity that means the humidity of ambient air. It is possible to find the magnitude of equilibrium humidity with the help of the special diagrams.

The *shrinkage* takes place at removal bound moisture from wood that means reduction of timber sizes. Opposite, at the wood humidification the cell walls are thickened, that causes *swelling*. The cross-grained moisture deformations are the most substantial. So, a complete linear shrinkage of wood in tangential direction is 6-10%, and longitudinally to fibers - only

0.1-0.63%. The value of shrinkage and swelling grows also with the increase of average density of wood.

The moisture deformations can be calculated by coefficients of shrinkage (K_{shr}) and swelling (K_{sw}) that characterize the proper deformations according to a decline or increase of content of bound moisture in wood at 1%.

Ratio between K_{shr} and K_{sw} is determined by the formula:

$$K_{shr} = \frac{100K_{sw}}{100 + 30K_{sw}}. \qquad (19.1)$$

The coefficients of volumetric shrinkage of some widespread woods and other physical-mechanical properties are resulted in Table19.1.

At drying, as a result of nonuniformity of distribution of humidity in cross-section of wood and anisotropy, the internal tensions appear in it. The development of these tensions can cause cracking and casting of timber.

For prevention of these defects the special value has the mode of wood drying. Drying is one of the most responsible and labour- intensive operations in technology of woodworking. The humidity should not exceed 8-10% for millwork, and for external structures – 15-18%.

At the calculation of drying processes, impregnation and other, the thermal properties of wood should be known. As a result of porous structure, wood conducts the warmth badly. Low thermal conductivity of wood, especially cross-grained, predetermines its wide application in the non-load-bearing structures of buildings that are heated. The timber, with thickness up to 15 cm, is equivalent by a thermal conductivity to the wall of brick with the thickness in 2.5 bricks.

Coefficient of linear expansion of wood along fibers is only (3-5) $10^{-6}\,{}^{0}C^{-1}$, that means in 3-10 times less than for metal, concrete and glass, due to what it is possible not to arrange expansion joints in wooden buildings. In transversal direction of fibers a change of linear sizes is in 7-10 times are more than in longitudinal.

Dry wood has very small conductivity, approximately the same, as well as good electric insulating materials. However with the increase of humidity, conductivity grows. At humidity which equals to the satiation limit, it is in ten of millions times larger than the conductivity of seasoned wood.

The density of wood is determined by the complex of substances which are the component parts of cell walls. As these substances have practically identical composition for all of wood, the real density of wood (density of wood substance) hesitates in narrow limits - from 1.49 to 1.56 g/cm^3 and equals to the average 1.53 g/cm^3.

The average density of wood depends on its humidity and porosity. The value of average density is specified for the standard 12%-th humidity ($\rho_{o(12)}$). In a range from zero to 30%-th humidity the next formula can be applied:

$$\rho_{o(12)} = \rho_{o(\omega)}[1 + 0{,}01(1 - K_{shr})(12 - \omega)], \qquad (19.2)$$

Table 19.1. Physical-mechanical properties of wood

Type of wood	Average density, kg/m^3	Coefficient of volumetric shrinkage	Ultimate strength lengthwise the fibers, MPa			
			Compression	Tension	Static bend	Radial cleavage
Softwood						
Larch	660	0.52	64	125	111	9.9
Pine	550	0.44	48	104	86	7.5
Spruce	445	0.43	45	103	79	6.9
Siberian fir	375	0.39	39	67	68	6.4
Siberian cedar	440	0.42	38	78	62	6.2
Hardwood						
Oak	690	0.43	57	123	108	10.2
Birch	630	0.54	55	168	110	9.3
Beech	670	0.47	55	123	108	11.6
Teil	495	0.49	45	121	88	8.6
Alder	520	0.43	44	101	80	8.1
Aspen	495	0.41	42	125	78	6.3
Poplar	440	0.42	39	88	62	6.1
Ash-tree	690	0.48	52	140	182	12.2

where K_{shr} - coefficient of volumetric shrinkage; ω - humidity.

At humidity of wood more than 30% it is possible to use a formula for the calculation of average density :

$$\rho_{o(12)} = \frac{A\rho_{0(\omega)}}{1+0{,}01\omega}, \tag{19.3}$$

where A - coefficient equals 1.222 for a birch, beech, larch, robinia and 1.203 for other wood.

According to the average density all types of wood are divided into three groups: light-weight (ρ_o<550 kg/m^3), middle-weight (ρ_o=550-750 kg/m^3) and heavy-weight (ρ_o> 750 kg/m^3).

Mechanical Properties

The indexes of mechanical properties of wood species, as its physical properties, depend on humidity; moreover only the bound water, which is in cell walls, influences. The increase of bound moisture content diminishes the indexes of all of mechanical properties sharply. The strength of wood can be calculated by a formula:

$$R_{12} = R_{\omega}[1+\alpha(\omega-12)], \tag{19.4}$$

where R_{12} and R_{ω} - ultimate strength of wood at 12% humidity and at humidity ω; α - coefficient of the strength decrease of wood at the growth of its humidity at 1% (for the ultimate compressive strength lengthwise fibers and static bend α =0.04; at tension lengthwise fibers α=0.01).

As a result of structural features, the mechanical properties of wood depend, also, on an angle between direction of operating effort and direction of fibers.

The most essential and characteristic mechanical property of wood is *compressive strength* lengthwise fibers. This property of wood is determining for piles, farms, columns and other structural timber.

In the most cases it is impossible to find out destruction at the action of compressive forces across the fibers; therefore it is limited by determination of proportional limit which is taken as conditional limit of the strength. Conditional compressive strength across the fibers upon the average for all wood is approximately in 10 times smaller than the compressive strength lengthwise the fibers.

The compressive strength across the fibers has a practical value in the places of angle joints or connections of timber details with metallic, for railway sleepers, etc.

The ultimate tensile strength lengthwise the fiber is in 2 and more times higher than at a compression (Table 19.1). For pine-tree and fir-tree, for example, it is equal on the average about 100 MPa. At tension across the grain, the ultimate strength is in 10-40 times less. Here the strength in a radial plane of all wood is higher than at tension in tangential area. It is caused that the break of weak pith rays passes in last case, while in a radial area it goes on early and dense late zone. The tensile strength especially strongly goes down at presence of knots and curly grain.

Wood rarely works on tension in constructions and elements. It is predefined with difficulty to prevent the destruction of details in the places of fixing. The indexes of the tensile strength of wood across the fibers are taken into account for prevention of its cracking at the intensive modes of drying.

Wood is widely used for structures which work on cross bending: in the floors, in bridge truss, trestles, platforms, stair, and others like that. The strength of wood at a static cross-bending is middle between the tensile strength and compressive along the fibers. On the average, it can be accepted equal approximately 90 MPa for different wood.

The strength of wood in some occasions is important at a shear and twisting for the calculation of structural timber. The most widespread type of tests on shear is splitting off along the fibers, resistance to which is approximately 0.15 of the compressive strength. The strength at twisting for the basic wood is almost in 1.5 times higher than resistance to splitting off.

The *wood hardness* is an important at treatment with machining tools and at deterioration actions. This property is determined on standards-cubes by indentation method. The most hardness (50-90 MPa) is inherent to an ash, beech, elm, larch.

The *creep* that results in noticeable deformations of constructions of the protracted loading is characteristic for the wood especially, during work in the moist conditions,

Wood working in dry premises, outdoors, and also in underground and underwater structures, in conditions which eliminate formation of fungi, is characterized with the high resistance. Mechanical properties of wood change considerably after staying in the river water during for a few hundreds of years. Salt water already through comparatively short time significantly worsens the properties of wood.

At the action of acids and alkalis, the mechanical properties of wood are worsened to the extent of their concentration increasing. *Corrosive resistance* of hard wood is lower than coniferous one.

Within the limits of one type of wood, the resistance depends on its density. Resistance is increased with the age of tree, at a movement from a sap-wood to the kernel and from the underbody of trunk to overhead. Protecting of wood from decay is carried out mainly by the chemical treatment with antisepsics, and from ignition - by the fire-retardants.

The directed change of the wood properties is arrived at its modification due to pressing after a previous steaming-out or heating, and also treatment of synthetic polymers. The modified wood has in few times larger strength, hardness, impact resistance, decreased hygroscopicity and water absorption.

Antiseptics are the toxic compounds which give resistance against the wood fungus, insects, etc.

It is possible to divide antisepsics into three groups depending on chemical and physical properties: oils and soluble in oils; soluble in organic solvents; water-soluble. Carboniferous and shale penetrating oils are mainly included in the first group of antiseptics; in the second - pentachlorophenol and copper naphthenate soluble in organic solvents. The basic representatives of the third group are sodium fluoride, hydrochloric zinc and other.

Substances which increase the fire-resistance of wood are called *fire-retarding agents.* The protective action of fire-retardant additives can be predefined by the extraction at heating of crystallization water as steam or other noncombustible gases which displace the air from the surface of wood and attenuate combustible gases (sulfuric and acid phosphorous ammonium, galloon). There are many fire-retarding additives (for example, borer, boric acid, sodium silicate, and hydrochloric zinc), that melt at heating and form protective dense tape, which covers the surface of wood and interferes to the oxygen access. Such fire-retarding additives as hydrate of potassium, some glues, facilitates at a high temperature creation of foam heat-insulating layer.

The mixtures of different fire-retarding additives are applied usually in practice. Wood is saturated with fireproof mixtures at the action of flame smolders, but does not burn. After the removing of fire a smoldering is halted. The various paints can also protect the wood from ignition.

19.3. Wooden Materials and Products

The materials and products which are obtained by tooling of tree trunk (timber and saw-timber, millwork and details), physical and chemical processing of fibred wood-pulp without the special introduction of binder substances (cane fiber boards, plastic lingo-carbohydrates and piezo-thermoplastics) or with addition of binders (glued wood, particle boards, sawdust concrete and other) are applied for the construction purposes.

Timbers

Round timbers are the segments of tree trunks with the chopped off knots and pollinated butt ends. Segments of trunks with diameter more than 12 cm consider *chumps*, from 8 to 11 cm - *rickers*, from 3 to 7 cm - *polewood.*

Round timbers are divided into sorts depending on the amount and type of defects. Rot is permitted in none of sorts. Logs which are utilized for making of bearing structures must have humidity not more than 25%.

Round commercial timber is stored at outside court in stacks in not more than 2 m high, which provides the normal natural drying. The butt ends are covered with the moisture protecting mixtures and lime solution to prevent the logs from cracking.

The saw-timbers (Figure 19.3) are obtained by the longitudinal sawing of logs. They are divided into slab boards, quarters, bars, boards and half logs.

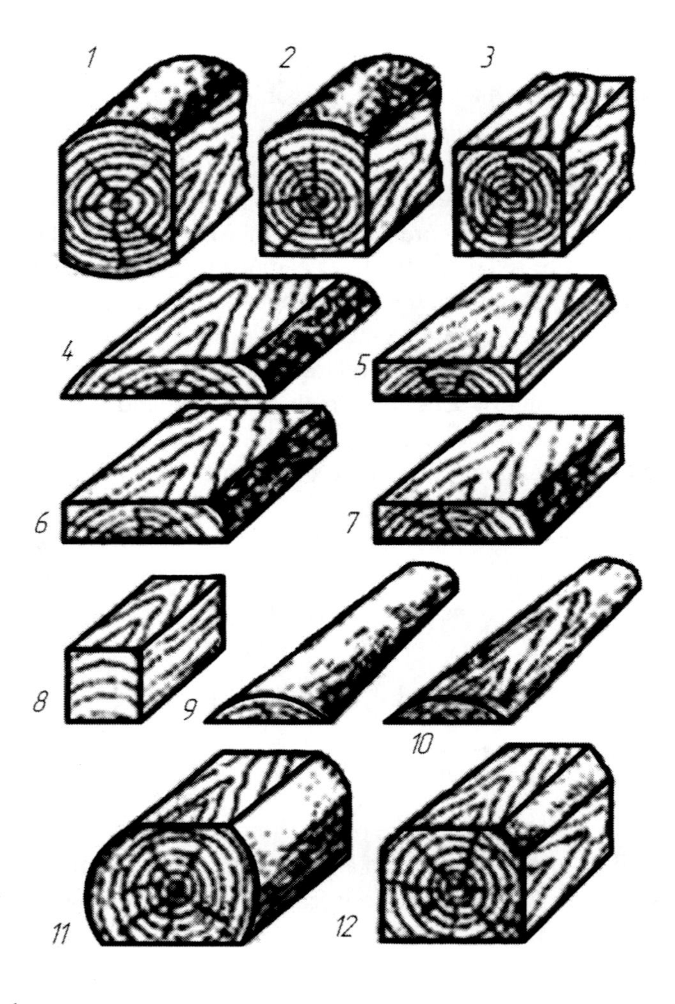

Figure 19.3. Saw-timber:
1- two-edged cant; 2- three-edged cant; 3- four-edged cant; 4- unedged board; 5,6- edged board; 8-scantling; 9, 10- half logs; 11, 12- railway sleepers.

Plates are obtained by tree sawing for the axis of trunk on two halves, *quarters* - in mutually perpendicular directions. Incomplete plates are the wastes at sawing of logs called half logs. *Bars* have a thickness and width more than 100 mm, scantlings - thickness to 100 mm and width no more than double thickness, *boards* - thickness to 100 mm, width - more than double thickness.

According to the thickness, the saw-timbers are divided into fine-bored (planks) to 32 mm and thick, more than 32 mm; according to the character of treatment – on edged, sawed from all four sides and unedged, sawed only from two sides.

Length of coniferous saw-timbers is in the range of 1-6.5 m, hardwood in the range of 0.5-6.5 m.

The sort of saw-timber is determined by the presence of wood defects, and also sawing exactness, cleanness of treatment and warping degree. The saw-timber with humidity less than 25% is saved in the closed premises or under a cover, and more than 25% - in stacks.

Boards and bars, cut on the set sizes with assumptions on tooling and shrinkage, which is used for making of details, are called work materials. According to the kind of work materials treatment there are sawn, glued, calibrated. There are utilized also millwork materials, which have a special cut form.

Along with the roundwood and sawn, the sliced, shelled, shredded timber is applied in construction. The veneer, plaster clapboard, roof slab, bundling chipping and other setting are made by an adzing. Decorative veneers are thin layers (0.6-1mm) of wood which differ with beautiful texture and color. It is made from wood of oak, ashwood, beech, African mahogany, larch and other. A texture paper and various decorative tapes are widely utilized in the place of natural planed veneer.

The peeled veneers which are obtained from the steamed out blocks in the type of continuous ribbon with a next scission on the sheets of necessary sizes is applied for making of plywood and laminates, and also facing elements from wood.

Parquet - is a material for arranging of floors in housings and public buildings, which made as block parquet, mosaic, as parquet boards. Block parquet consists of slats which are made from wood of oak, beech, ashwood, maple, pine-tree and other. Fabricated wood block flooring and shields consist of slats which are glued by a certain picture on basis. There are slots and cogs on the edges and butt ends of parquet board for connection between themselves. Veneer squares or the veneered facing dalle can be also glued in place of slats on parquet shields. Facial side of parquet board and shields usually has the transparent lacquered coverage. The artistic parquet blocks are applied for the floors of unique buildings. The mosaic parquet is made as carpets, which consist of separate slats, glued by a facial surface on a paper or other elastic material.

The basic types of elements and details of wood are wood moldings, millwork, window and door blocks, partitions and panels.

Wood moldings are boards and scanting for floors, baseboards, banisters, cases, boarding and other elements – are got by milling on machine-tools. They are characterized by a certain form of transverse section; for example, boards and scanting for floors have a slot on one edge, and cog on another.

Shields are made for the installation of partitions and ceilings, and also casing for concrete and reinforced-concrete structures,they consist of wooden framework to which the side plates are asigned to. The battenboards are got at gluing on the shields from both sides in one or two layers of veneer; they are used at wall, floor, shield furniture installation. The *window and door blocks* are also got from millwork for construction.

The factory-made houses are produced from bars, framed wooden elements and panels and also and other prefabricated articles (Figure 19.4). In the framed wooden houses the wooden framework is filled with the fibreboards and woodwool slabs. The water-resistant board materials are applied for revetment of external walls. In the factory-made houses for boarding are used edged boards, space between which is filled with heat-insulating materials.

The most effective are the panel and prefabricated houses in which the glued wood and new finishing materials are widely utilized.

Glued Wood

The glued wood belongs to the most effective construction materials. It can be stratified - from veneer (plywood, laminated-wood plastic), massive - from the lump raw waste lumber and woodworking (panels, blocks, squared beams, boards), combined (battenboard).

Construction plywood is the sheet material, obtained by gluing of three and more veneer slices with the mutually perpendicular placing of fibers. After the drying the veneer packages are glued together usually by the hot compaction on the hydraulic press.

There is distinguished plywood of raised waterproofness, glued on phenol-formaldehyde resin, medium waterproofness, - on the carbamide and albuminic casein glues and limited waterproofness - on protein glues.

The veneer sheets of plywood can have a thickness from 1.5 to 18 mm.

Along with ordinary one, the decorative plywood is made, faced with the film coating. The bakelized plywood which is made from a birch veneer has high structural strength, impregnated with phenol-formaldehyde resins, at the protracted conditioning. The plywood is also reinforced by the expanded lathes or encase with the metal skim to increase the strength, hardness and inflexibility.

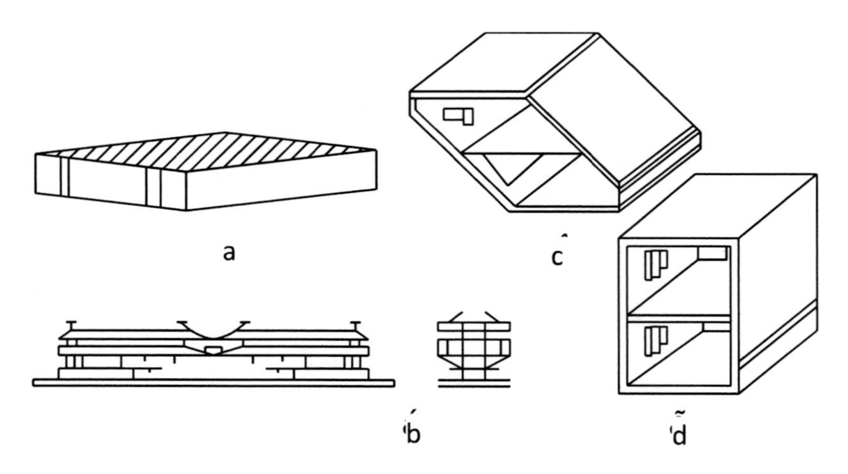

Figure 19.4. Elements of prefabricated wooden low-rise buildings: a,b- demountable of "block-package" type; c,d - folded; d - general view of block; a- block, folded in a package; b,c - transporting of block - packages and installation of package.

Plywood is widely used for revetment of walls, production of veneered doors, glue plywood sheets for external and internal walls and ceilings, panels, for built-in furniture and others like that.

The glued products from the lump timber waste classify at the type of glue which is used and character of treatment and characteristics. The strength of wood adhesion depends on porosity, correlation of summerwood and latewood in annual layers, its humidity, chemical composition, fiber angle. It is set that the strength of gluing is in linear relation to porosity, it grows also in measures of cellulose content increasing.

Advantages of the glued wood are low average density, waterproofness, possibility of products manufacturing from of small size material of irregular shape or large structural elements; influence of wood anisotropy is weak in the glued structures, they are characterized with the rot-resistance and by a low capacity for ignition, are not liable to the shrinkage and warping. Glued timber structures by the terms and labor expenditures of construction and also resistance to the action of corrosive air environment often successfully compete with steel and reinforced-concrete structures. Their application is effective at the erection of agricultural enterprises, exhibition and commercial halls, sporting complexes, buildings and constructions of prefabricated type.

As well as the reinforced concrete, the glued timber can be produced preliminary tensed, reinforced by the steel rods. In the reinforced structures as solid-webbed or hollow beams, the bearing ability almost in 2 times is higher, than in all-wood.

The nomenclature of products from the glued wood is wide. So, from the deal-ends, the panels are glued which have the cleavage strength along fibers in the glued juncture not less than 6 MPa. The static bending strength of toothed finger joint is not less than 35% of flaw-free wood strength. Such panels are applied for the installation of partitions, boarding of houses, floors boarding.

The glued shields used, mainly for floor boarding is manufactured from the wastes of lumber. For their making lump wastes of lumber are cut on the scanting, which are dried to the humidity at 10-12%, and then are glued together at pressing. The blocks are addiced from two sides on the thicknessing machine, cut on a perimeter to the set sizes with simultaneous making on the lateral edges of slot and cog.

The economy of industrial wood to average 20 % at using of the glued elements in building, and the cost of structures diminishes in 3-4 times comparatively with the cost of structures from whole wood.

Structures from the glued wood allow decreasing the mass of buildings in 2.5-3 times, to reduce the efforts on 25-30%, and terms of construction in 1.5-2 times.

Beams, frames, arches, arched girders, and spatial structures belong to the basic types of the glued industrial timber.

19.4. Materials Based on the Ground Wood

The ground wood appears both directly at sawmill and woodworking (sawdust, cutting waste) and after the special processing of raw waste lumber and non high quality wood (chips, shaving, fibred mass). It is possible to divide the construction materials from the ground wood into two groups:

- without application of the special binders or with their small addition;
- with the use of organic or mineral binders.

The wood particles are bound as a result of approaching and fiber splicing, their cohesiveness and action of physical and chemical bonds which arise up at piezo-thermal treatment in the materials of the first group.

Cane fiber boards are materials which are formed from wooden fibred mass with next thermal treatment. The boards can be also obtained from the fibers of bast plants or from the other fibred raw material which has sufficient durability and flexibility.

Depending on assignment the hard and soft boards are made.

Hard boards are intended for the facing of internal surfaces of buildings, the solid doors and other elements of house-building, soft - for heat-insulation of non-load-bearing structures and sound-proofing of partitions.

The cane fibers are tooling added, they can be endued with the texture of wood, skin and others like that.

Finishing boards are faced with the synthetic polymers with the gasket of texture paper. They are produced also with a matte surface or painted water-emulsion polyvinyl acetate paints. Boards, painted by enamels, have a glossy surface, enhanced water resistance.

The lignin carbohydrate and piezothermoplastics are the varieties of *wood plastics*, obtained from sawdust or raw waste lumber of agricultural production with the use of hot compaction. At the working of wood plastics there is a partial hydrolysis of polysaccharides of wood and formation of organic acids which are the catalysts of destruction of lignin carbohydrate complex under the action of temperature. Chemically active products (lignin and carbohydrates) interact between themselves at pressing. The stronger and dense material than initial wood appears in a result.

The substantial lacks of production of lignin carbohydrate plastics is a necessity of powerful press equipment and protracted cycle of pressing.

Piezothermoplastics can be made from sawdust by the two methods: without previous treatment of raw material and with its hydrothermal treatment.

According to the first method of piezothermoplastic production, the technology is similar to the method of obtaining of lignin carbohydrate plastics.

According to the second method, the standard sawdust less than 4 mm are processed in the autoclaves with vapor at a temperature 170-180°C and pressure 0.8-1 MPa. The hydrolyzed compacted mass is partly dried out and at certain humidity consistently exposed to the cold and hot pressing. Specific pressure of the cold and hot compaction is 15 MPa; temperature is 160°C.

Piezothermoplastics are divided into isolation, semisolid, hard and extra-hard.

At an average density 700-1,100 kg/m^3 piezothermoplastics, made from birch sawdust, have the static bending strength 8-11 MPa. At an enhanced average density - to 1,350-1,330 the kg/m^3 the ultimate strength arrives at 25-40 MPa.

The physical-mechanical properties of piezothermoplastics allow applying them for floors, doors, and also as finishing material.

The resin-bonded chipboards are the most widespread among the materials with organic binder. This material is obtained by the hot compaction of the ground up wood, mixed with synthetic polymers (Figure 19.5). Its advantages are homogeneity of physical-mechanical properties, possibility of high mechanization and automation of the manufacturing.

Industry produces flat and extrusion boards. At the first, the particles are placed parallel, in the second - athwart to the plane of board. It is arrived at the extrusion method of pressing.

Range of application of resin-bonded chipboards is various. As structurally finishing materials they are applied at arranging of floors, ceilings, walls, partitions, doors, built-in furniture and others like that. Easy boards are applied for heat-insulation.

The main representatives of group of materials on wood fillers and mineral binders are cement wood (sawdust concrete), woodwool and xylolite.

C*ement wood* is a light-weight concrete on the phytogenous aggregates, preliminary treated with the solution of mineralizer. It is used mainly as panels and blocks for the construction of walls and partitions, floor and roof slabs of buildings, heat-insulation and sound insulating slabs. Cement wood structures are exploited at relative humidity of premises not more than 60%, it is required to use the vapor sealing layer at greater humidity.

The systematic action of corrosive environments and temperatures on the cement wood more than 50°C and below zero of 40°C is shut out.

External surface of structure from to the cement wood which contacts with the precipitations, regardless of the humidity conditions of exploitation must have a texture (finishing) layer.

Depending on an average density in the dried up state a cement wood concrete is divided into heat-insulation (with density up to 500 kg/m^3) and constructional (500-850 kg/m^3).

The best type of aggregate for the cement wood is a wood mass (hogged chips) with attitude of the most size of particles toward the least 5-10, in thickness particles 3-5 mm and by the most length to 25 mm. Such form of particles allows approaching moisture deformations along and across the grains and reducing the negative influence on the structure formation and strength of this type of concrete.

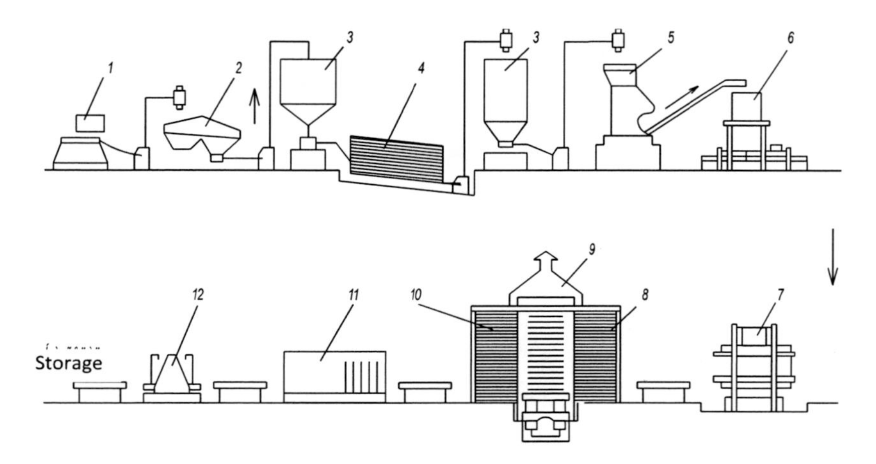

Figure 19.5. Scheme of production of resin-bonded chipboards:
1- chipping machine; 2- vibrating screen; 3- bins; 4- drier; 5- mixer; 6- cloth-spreading machine; 7- "cold" press; 8- feeding apparatus; 9- "hot" press; 10- unloader; 11- chamber for board conditioning; 12- trimming machine.

Hogged chips from coniferous and deciduous wood are wet with the water or in solutions of mineral salts to decline the amount of harmful extract substances. Last, neutralizing the action of harmful substances in wood, simultaneously forces the cement hardening.

Technology of cement wood products includes the preparation of basic materials, obtaining of mixture and it's placing in forms, hardening and drying, finishing and storage. Along with the non-reinforced products from the cement wood the products, reinforced by the steel reinforcement are manufactured.

At relative air humidity higher than 60% reinforcement is placed in a protective layer from a cement concrete which provides reliable bypassing of steel. It is recommended also to protect reinforcement by the special coatings.

A cement wood has the best heating engineering descriptions, than claydite concrete, that allows erecting the walls of the less thickness. The replacement of traditional materials with the cement wood allows reducing mass of buildings in 1.3-1.5 times. At the equivalent thickness of wall according to the conditions of heat-transfer, the mass of one square meter of cladding structure from the cement wood is in 7-8 times below, than from a brick and in 2-3 times below, than from claydite concrete.

Woodwool consists of the mineral binder and wood shaving as aggregate and simultaneously reinforcing component.

Woodwool with an average density 400 kg/m^3 is applied for a thermal isolation. A heat-insulation woodwool has thermal conductivity 0.09-0.12W/(m·K).

At the average density not more 400 kg/m^3 and more the woodwool as constructive-heat insulating material simultaneously can be utilized for the erecting of walls and ceilings.

Woodwool differs with the high acoustic absorption, the predefined connecting character of pores, and also good capacity for treatment, nailing, tripping with a bonding plaster and concrete. Negative properties of woodwool are considerable air permeability, big water absorption, low water resistance and ability to the fungi defeating in the moisture state.

Xylolite is an artificial construction material that is obtained as a result of hardening of mixture, which consists of magnesia cement, sawdust and solution of chloride or magnesium sulphate. This material is mainly used for arrangement of monolithic or precast floor coverages. Advantages of xylolite floors are their low coefficient of heat-absorption, hygienics, sufficient hardness, high abrasion resistance and possibility to the various coloring.

The recommended xylolite mixture consists of caustic magnesite and sawdust in ratio by a volume 1:1.5-1:1.4. The content of caustic magnesite at 100 m^2 of coverage with the thickness at 10 mm is 410-620 kg and crystalline chloride of magnesium - 260-400 kg.

Self-Assessment Questions

1. What value for construction has the wooden materials?
2. Tell about composition and structure of wood.
3. What are the defects of wood?
4. Tell about physical properties of wood?
5. Tell about mechanical properties of wood?
6. What the methods are applied to protection the wood from rot and ignition?
7. Tell about the timber, applied in construction.
8. Tell about the materials from the glued wood.
9. Tell about the materials on the basis of the grown up wood.

Chapter 20

MATERIALS BASED ON WASTE PRODUCTS

Increasing of public production volume is a peculiarity of the scientific and technical progress. The industrial production grows from year to year in the whole world, and the amount of the waste products increases proportionally to its growth, enlarging approximately in 2 times for 8-10 years.

Continuously increasing of wastes volume, which forms at the mineral wealth and fuels mining, their processing and application, is one of the sources of all the greater contamination and cluttering the environment. The growing mass of wastes from year to year is one of the main factors of the environment quality declining and destruction of natural landscapes.

The enormous amounts of industrial wastes are accumulated in dumps. The enormous areas of the landed grounds are alienated for the storage of wastes. The hundreds of thousands of ground hectares, suitable for agricultural production are occupied with the dumps of industrial enterprises.

Transportation and storage of wastes consumes the considerable resources from the basic production. The resources, occupying 8-10 % of the mined coal cost, produced energy and steam, are expended for the organization and operation of dump enterprises of coal and energy industries.

In accordance with the operating norms all the industrial wastes depending on the content of harmful chemical substances in them are divided into four classes by hazard:

Class	Characteristic of substance (wastes)
First	extraordinarily hazardous
Second	highly hazardous
Third	moderately hazardous
Fourth	low-hazardous

The production of construction materials is the most capacious among the industries - consumers of industrial wastes, which are the by-products of different manufactures.

Taking into account that expenses on material resources in the estimated cost of the most construction materials production is the greater part, obviously, it is possible to predicate that application of waste products - industrial by-products which can be considered as raw materials is one of ways of efficiency increasing of construction materials production.

20.1. APPLICATION OF WASTE PRODUCTS OF METALLURGY INDUSTRY

The bulk of the waste products of metallurgy industry generates as slags.

Slags are the by-products of high temperature processes of the metal smelting, firing of the solid fuel, and also some chemical manufactures. Metallurgical slags appear at interaction of fuel, barren rocks, containing in the ore and fluxing agents. They are subdivided into the slags of the ferrous and nonferrous metallurgy.

Depending on the character of process and type of the furnaces among the slags of the ferrous metallurgy blast, steel-smelting (open-hearth, converter, Bessemer and Thomas, cupola, electrosmelting) and the productions of ferro-alloys are distinguished.

The yield of blast-furnace slags is the largest: it is 0.6-0.7 tones per 1 tonne of cast iron. The output of slags is considerably less at the steel smelting – 0.2-0.3 tones per 1 tone at a open-hearth method; 0.1-0.2 tones – at the Bessemer and Thomas one and 0.1-0.04 tones at smelting in a cupola and electric furnace.

The slags are obtained at smelting copper, lead, nickel, tin and other in the nonferrous metallurgy. The output of slags per 1 tone of the nonferrous metal is 15-25 tones.

It is determined by the chemical analysis of metallurgical slags that four oxides predominate in their composition: CaO, MgO, Al_2O_3 and SiO_2. In addition, there are always present ferric and manganese oxides and also sulphureous compounds in the small amount in metallurgical slags.

At the estimation of slags as raw material for construction materials the important parameter of their chemical composition is the percentage of basic and acid oxides - lime factor (F_L). At $F_L>1$ slags are attributed to the basic, at $F_L<1$ - to acid slags.

Oxides, contained in slags, form various minerals. The possibility of existence in slags up to 40 double and triple compounds among which silicates, aluminum silicates, aluminate and ferrites occupy the lead positions is established by the analysis of state diagrams of the corresponding oxide systems.

Belite $2CaO \cdot SiO_2(C_2S)$, rankinite $3CaO \cdot 2SiO_2(C_3S_2)$ and solid solution of helenite $2CaO \cdot Al_2O_3 \cdot SiO_2(C_2AS)$ and okermanite $2CaO \cdot MgO \cdot 2SiO_2(C_2MS_2)$ - melilites crystallize in slags at the slow cooling. At raised content of Al_2O_3 and SiO_2, the anorthite $CaO \cdot Al_2O_3 \cdot 2SiO_2$ (C_2AS_2) and pseudowollastonite α-$CaO \cdot SiO_2(\alpha$-$CS)$ are also present.

Glassy phase contains almost in all the metallurgical slags in one or another quantity along with the products of crystallization. In molded slowly cooled basic slags the amount of glass is insignificant, and in granulated blast-furnace slags it arrives 98%. The glass, being thermodynamic unstable phase, in a great extent determines chemical activity of slags. It is found out that the slag glasses react with water considerably more intensive, than crystals of minerals.

From all of types of metallurgical slags, *blast-furnace slags* are the most widely used in the production of construction materials, that is caused by their leading position in general balance of metallurgical slags, and also by the affinity of their composition to composition of cement, by ability to acquire hydraulic properties at quenching, etc. The bulk of blast-furnace slags is obtained at smelting of the rerolling and foundry-irons.

Estimation of hydraulic properties of the blast-furnace granulated slag is determined by the coefficient of quality K, defined by formulas:

- at MgO content up to 10%

$$K = \frac{CaO + Al_2O_3 + MgO}{SiO_2 + TiO_2}; \tag{20.1}$$

- at MgO content more than 10%

$$K = \frac{CaO + Al_2O_3 + 10}{SiO_2 + TiO_2 + (MgO - 10)}. \tag{20.2}$$

Ability of slags at the tempering to set and harden at the certain temperature-humidity conditions depends on their chemical and phase composition. At the ordinary temperature and without activating additives the ground slags do not possess the ability to harden that is explained by the absence or low content phases of enough active under these conditions. β-dicalcium silicate is practically the unique crystalline component of slags, able to harden but slowly at normal temperature. A series of other minerals, acquires the hydraulic properties only at the conditions of raised temperature and pressure of water vapour at the presence of activating agents. The slag glasses react with water considerably more intensive than the crystals of minerals. High internal chemical energy of glass provides its raised solubility, resulting in formation of oversaturated solutions, their crystallization and, as a sequence of the last one, hardening and formation of artificial stone.

The mechanism of the slag glasses hydration consists in penetration in glass of the negatively charged hydrophilic ions, damaging the electrostatic equilibrium of the system and resulting in destruction of a slag. The films formation on the glass particles surface of the hydrated silica prevents the hydration at the ordinary conditions without activating admixtures. Introduction of alkaline compounds and sulfates into water solution, containing the ions of Ca^{2+}, $(OH)^-$ and $(SO_4)^{2-}$ promotes destruction of these films and baring of new surfaces of slag grains. During the *alkaline activation* the hydrated silica bounds into the calcium hydrosilicates and aluminosilicates, at the sulfate one - the calcium sulfate directly reacts with alumina, calcium hydroxide and water with formation of hydrosulfoaluminates. The effect of alkaline and sulfate activation increases with the increasing slags basicity. At the sulfate excitation activity of slags grows also as far as their aluminates content grows. Lime, alkali, soda and other salts of alkaline metals and weak acids, Portland cement can be used as alkali activator. Dehydrated or semi-hydrated gypsum, anhydrite, sulfate of sodium are sulfate activators.

The increasing in slag hydraulicity causes slags micronizing and increasing reacting surface of their grains because of it. Thus chemical activation and water-heat treatment in autoclaves act especially strongly at the development of binder properties of slags.

Clinker-free slag binders are the products of fine slag grinding, containing the activating admixtures of their hardening. The activators are thoroughly mixed up with slag either by their combined grinding (sulfate-slag, lime-slag binder), or mixing with water solutions (slag-alkaline binder).

Sulfate-slag cements are hydraulic binding agents obtained by the intergrinding of blast-furnace granulated slags and sulfate hardening activator - gypsum or anhydrite with small additive of alkaline activator - lime, Portland cement or burnt dolomite. The gypsum-slag cement, containing 75-85% of slag, 10-15 dehydrated gypsum or anhydrite, up to 2% of calcium oxide or 5% of Portland cement clinker is the most wide-spread among the binders of sulfate-slag group. The high activation is provided at the use of anhydrite, burnt at a temperature about 700°C, and high-aluminous basic slags. As basicity of slags diminishes, it is rational to increase the lime concentration (from 0.2 g/l of CaO for basic slags to 0.4-0.5 g/l for acid slags).

Other variety of sulfate-slag cements is slag clinker-free cement, consisting of 85-90 % of slag, 5-8% of anhydrite and 5-8% of the burnt dolomite. The degree of dolomite burning depends on the slags basicity. For basic slags, burning is conducted at a temperature at 800-900°C until partial decomposition of $CaCO_3$, and for acid ones - at 1000-1100°C to complete dissociation of $CaCO_3$.

Strength of sulfate-slag cement substantially depends on the grinding fineness. The high specific surface area (4000-5000 cm^2/g) of binder is achieved by wet grinding. The strength of sulfate-slag cement is similar to the strength of Portland cement at high enough grinding fineness and rational composition. However the defect of sulfate-slag cements is rapid decline of strength at storage, binding of raised amount of water during hydration is characteristic for this binder. The last one causes considerable change of optimum water-cement ratio (W/C) towards larger values (to 0.5-0.65) in the concretes. Reduced plasticity of sulfate-slag cements causes significant decline of concrete strength on their basis as far as increasing of the aggregate content. Optimum temperature of hardening for these cements is 20÷40°C - the strength reduces substantially at the lower and higher temperatures.

As well as other slag binders, sulfate-slag cement has small heat of hydration until 7 days that allows applying it at the erection of massive hydraulic constructions. Its high resistance to the influence of soft and sulfate waters promotes it also.

Lime-slag cements are hydraulic binding agents, obtained by intergrinding of blast-furnace granulated slag and lime (Figure 20.1).

They were the first slag binders, used in construction. They are applied for the preparation of building mortars and concretes with low strength (2-5 MPa). The admixture of gypsum stone (5%) can be added to adjust the terms of setting and improve some other properties of these binders at its production. It is possible to get the lime-slag cements of higher quality applying basic slags with raised content of alumina and quicklime. The lime content is 10-30%.

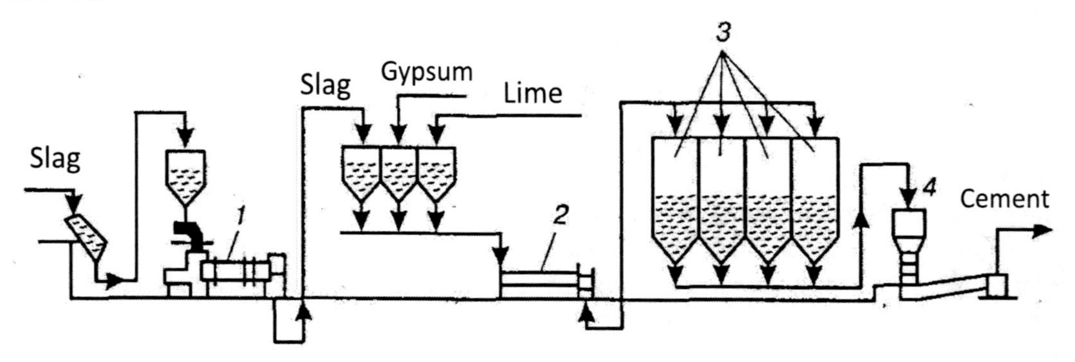

Figure 20.1. Scheme of the lime-slag cement manufacture:
1 – drying drum; 2- mill; 3- cement silos; 4- packing machine.

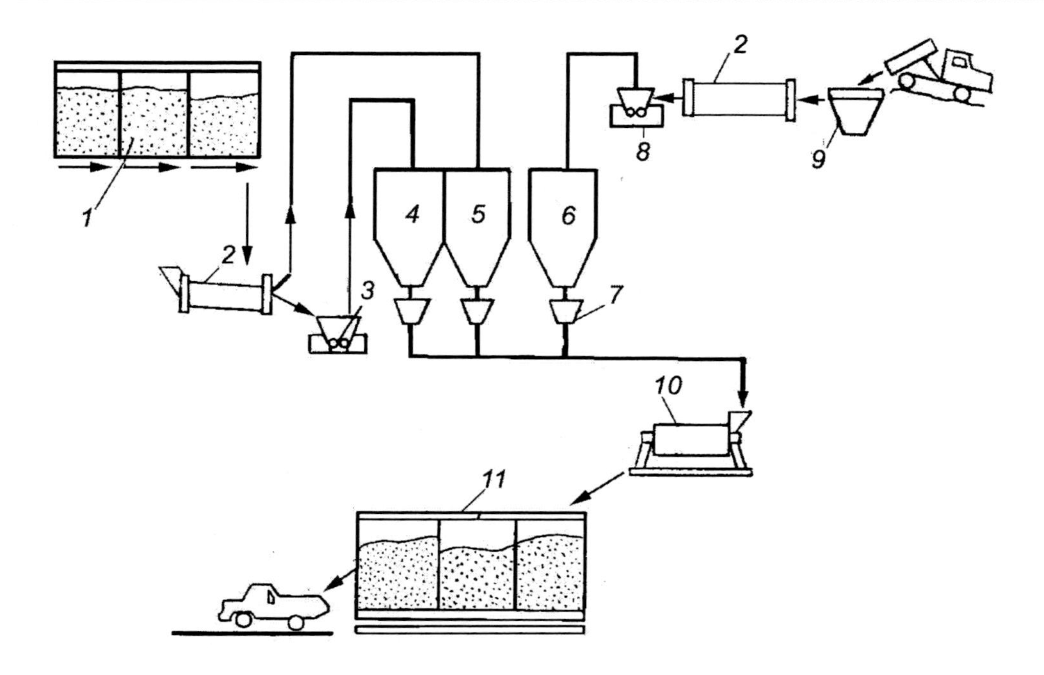

Figure 20.2. Scheme of the slag-alkaline binder manufacture:
- storage; 2 – drying drum; 3- rolls; 4, 5 – measuring hoppers for slag and active mineral admixture; 6 — measuring bin for alkaline component; 7 - batchers; 8 –rolls for grinding of alkaline component; 9 - storage of alkaline component; 10 – grinder; 11 – storage of binder.

Strength of the lime-slag cements is lower than sulfate-slag cements strength. Their ultimate compressive strength is 5-20 MPa. Initial setting time should come not earlier than in 25 min, and the final setting time is not later than 24 hours after the beginning of mixing. At the decline of temperature, especially after 10°C, growth of strength sharply slows and, vice versa, the increase of temperature at sufficient environmental humidity promotes intensive hardening. Air hardening is possible only after the comparatively long hardening (15-30 days) in moist conditions. Low frost resistance, high resistance in aggressive waters and small exothermicity are characteristic for the lime-slag cement

Blast-furnace granulated slag in slag clinker-free binders can be substituted by the slags of steel-smelting production and non-ferrous metallurgy. Strength of binding materials based on metallurgical slags at normal conditions is low, but it sharply increases at autoclaving and achieves at strength of Portland cement of middle and high strength. It is a result of high reactive ability of slag minerals under conditions of steam environment, high-pressure and temperature of steam 170-200°C.

Slag binders for autoclave concrete are the products of fine grinding of open-hearth, cupola and some other low-active at the normal hardening slags with additions of hardening accelerators like cement or lime (10-20 %) and gypsum (3-5 %). Their strength is increased at heat-moist curing in autoclaves under the pressure of 0.8-1.5 MPa at a temperature 174.5-200 °C. The compressive strength of steam-cured standards from plastic mortars of composition 1 : 3 achieves 20-30 MPa and more. They are obtained, mainly, the same as lime-slag and sulfate-slag cements. The metallic inclusions are separated from slags by magnetic separators before the crushing and grinding. The binding agents are ground to the sieve residue – No. 008 no more than 10-15 %.

Slag-alkaline binders are hydraulic binding agents, obtained by grinding of granulated slags composed of alkaline components or mixing of the grind slags with solutions of alkaline metals compounds (sodium or potassium), giving the alkali reaction (Figure 20.2).

The granulated slags - blast-furnace, electrothermal phosphate, slags of non-ferrous metallurgy, are applied to obtain slag-alkaline binders. Required condition of slag activity is a presence of glassy phase, able to react with alkalines. The grinding fineness should correspond to the specific surface area not less than 3000 cm^2/g.

Caustic and calcinated soda, potash, liquid glass and other are applied as alkaline components. Industrial byproducts are also used: alkaline fusion cake (soda production); soda- alkali fusion cake (production of caprolactam); soda-potash mixture (production of alumina); cement dust, etc. Optimum content of Na_2O in binders is 2-5% by slag mass.

All the alkaline compounds or their mixtures, giving an alkaline reaction in water can be used for slags with the lime factor (F_L) $F_L>1$, for slags with $F_L<1$ only caustic soda and alkaline silicates with the module 0.5-2.0, nonsilicate salts of weak acids and their mixture can be used only under the conditions of steam curing. The destruction and hydrolysis dissolution of slag glass, formation of alkaline hydroaluminosilicates and creation of environment, promoting formation and high stability of low-basic calcium hydrosilicates are intensified by the presence of alkalines. Small solubility of new formations, stability of structure in time are the decisive conditions of durability of the slag-alkaline stone.

Initial setting time of these binders is not earlier than 30 min, and the final is not later than 12 hours from the start of mixing.

The compressive strength at the age of 28 days of the slag-alkaline binders is 30-60 MPa and more. The admixture of cement clinker (2-6%) can be added into the binder for the acceleration of strengthening and diminishing of deformability. Compressive strength limit of high-early-strength slag-alkaline binder in the age of 3 days is not less than a 50% of 28 day strength.

The slag-alkaline binders are sensitive to the action of steam treatment. At the temperature of steam curing 80-90°C the cycle of treatment can be shortened up to 6-7 hours, active part of the regime lasts 3-4 hours. It is possible to reduce considerably maximum temperature of steam curing.

Concrete based on the slag-alkaline binders has lower porosity that provides their high watertightness, frost resistance, relatively low parameters of shrinkage and creep. In spite of the intensive growth of strength in the early terms of hardening, the heat generation of these binders is not high (in 1.5-2.5 times less than Portland cement one).

The slag-alkaline binders have high corrosive resistance. Alkaline components can be used as antifreeze admixtures, therefore binders and concrete intensively harden at subzero temperatures.

A series of special slag-alkaline binders is developed: high-strength, fast-hardening, non-shrinkable, corrosion resistant, heat-resistant and oil-well.

Specific capital investments at the production of these binders are in 2-3 times lower than those for Portland cement production.

At the production of 1 ton of slag-alkaline binder the expense of equivalent fuel is 110-160 kg and expense of electric power approximately is 80 kWh.

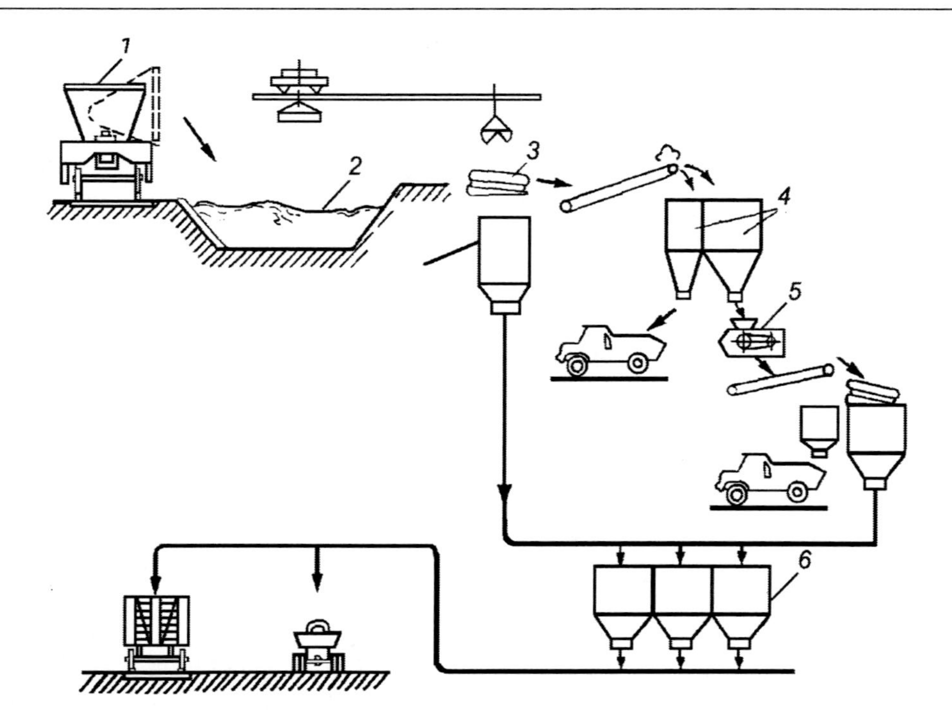

Figure 20.3. Scheme of manufacture of the cast crushed stone by trench method:
1 - slag ladle; 2 - trench for slag fusions; 3 - screen; 4 - bins; 5 - crusher; 6 – bins for finished products (slag crushed stone).

Metallurgical slags are considerable reserve for providing of construction industry with concrete aggregates. *Slag aggregates* depending on the value of bulk density can be heavy-weight ($\rho_0 > 1000$ kg/m^3) and light-weight ($\rho_0 \leq 1000$ kg/m^3), and by fineness of grains - fine (<5 mm) and coarse (> 5 mm).

Slag crushed stone is obtained by crushing of metallurgical dump slags or by special treatment of slag fusions (cast slag crushed stone). The blast-furnace slags, steel-smelting, and also copper-smelting, nickel and other slags of the non-ferrous metallurgy are mainly applied for the crushed stone production.

Slag crushed stone which made from slag fusions is the effective type of heavy-weight concrete aggregates. Physical-mechanical properties of such slag crushed stone can be not less than dense natural crushed stone. At the production of this material a fusion of the slag from the slag carriage buckets interflows by the layers 250-500 mm thick at the special casting grounds or in trapezoid trenches (Figure 20.3).

At the air curing during 2-3 h the temperature of fusion in a layer reduces up to 800°C and slag crystallizes. Then it is cooled by the water that results in development of numerous cracks. Slag massives at the linear grounds or in trenches are developed by excavating machines with the subsequent crushing and screening.

The physical-mechanical properties of the slag crushed stone are given below:

The slag crushed stone is characterized by the high freeze - and heat resistance, and also by wear resistance.

Average density of pieces, kg/m^3	2200-2800
Absolute density, kg/m^3	2900-3000
Compressive strength, MPa	60-100
Water absorption, % by mass	1-5
Bulk density of crushed stone, kg/m^3	1200-1500

Granulated slag is an effective fine aggregate for ordinary and fine-grained concretes and can be used as coarsing additive for improvement of natural fine sands. The porous varieties of granulated slag are applied as aggregates of the light-weight concrete.

Slag pumice is one of the most effective types of artificial porous aggregates. It is obtained by porization of slag fusions as a result of their rapid cooling by water, air, steam, and also influence of mineral gasifiers.

At present hydroscreen method is the most effective one (Figure 20.4) based on the sharp cooling of slag fusion in the system of the consistently set hydroducts, consisting of chamfers and hydromonitor nozzles through which water is supplied. The screen is set between the hydroducts.

Slag pumice is used as the aggregate for the light-weight concretes with a wide range by average density and mechanical parameters. It is used as a porous aggregate for constructive – heat-insulating light-weight concrete with density 1300-1600 kg/m^3 and strength 5-7.5 MPa and construction concrete with a density 1500-1800 kg/m^3 and strength 10-20 MPa, applied for the production of different structures (Figure 20.5).

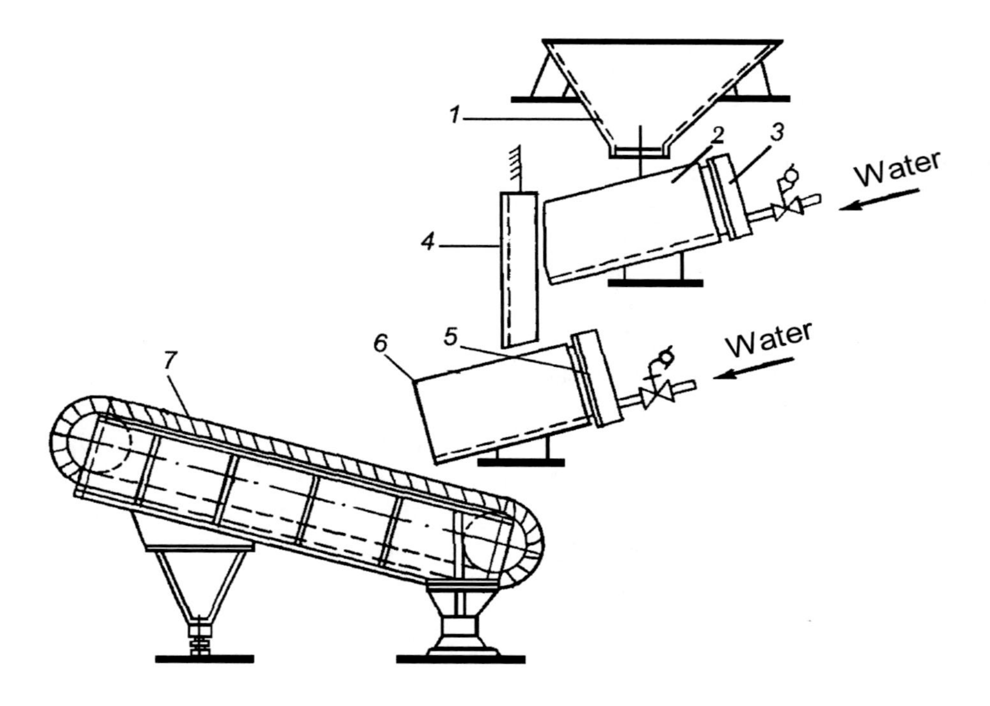

Figure 20.4. Scheme of the hydroscreen device:
1 - funnel; 2 - first tray; 3 - first hydromonitor nozzle; 4 - screen; 5 - second hydromonitor nozzle; 6 - second tray; 7 - re-loader.

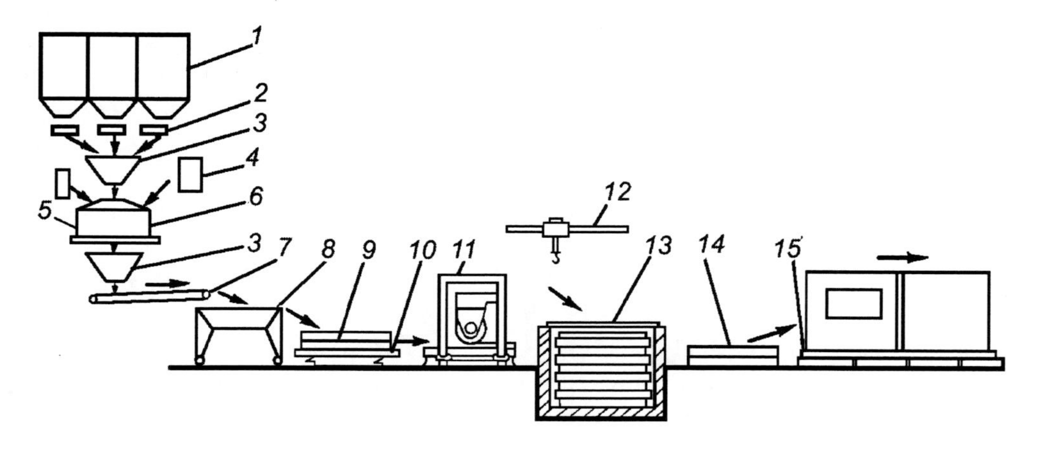

Figure 20.5. Scheme of manufacture of wall panels from porous slag-pumice concrete:
1 - bins for initial materials; 2 - weight batchers; 3 – receiving cone; 4 – water tank; 5 - mixer for preparation of water aluminium suspension; 6 - concrete mixer; 7 - band conveyor; 8 - concrete placer; 9 - formwork; 10 - vibroplatform; 11 - smoothing device; 12 - bridge crane; 13 – steam-curing chamber; 14- stripping of formwork; 15 - conveyer of panels finishing.

Concrete can be manufactured with a different average density depending on the type of slag aggregates: extra heavy-weight (ρ_0>2500 kg/m^3) based at some slags of steel-smelting and non-ferrous metallurgy; heavy-weight (ρ_0=1800-2500 kg/m^3) at the cast and dump slag crushed stone, sand and granulated slag; light-weight (ρ_0<1800 kg/m^3) at slag pumice (coarse aggregate) and granulated slag (fine aggregate). Along with coarse-grained concrete fine-grained slag concrete can be used.

Depending on the structure there are distinguished ordinary dense, no-fines and foam slag concretes.

It is possible to obtain the heavy-weight concretes of high compressive strength, applying ordinary or slag binder in combination with the slag aggregates. Thus the strength for steam-cured concrete achieves 10-30 MPa, and for autoclaved one 30-60 MPa. Replacement of coarse aggregate made of dense rocks in the heavy-weight concretes for slag crushed stone, obtained by crushing of dense metallurgical slags, does not reduce practically, and sometimes increases concrete strength due to their more developed and active surface. Concrete based on slag crushed stone has higher tensile and bending strength, than granite based.

General properties, inherent to the light-weight concretes, are characteristic for light-weight slag concretes, such as achievement of maximal strength at water consumption, providing minimum concrete mixture yield, and also at the use of fractionated porous aggregates; strength growth with the increasing of binder content to a certain limit. The large deformability and comparatively lower adhesion to the reinforcement, than in Portland cement based concrete is the peculiarities of light-weight slag concretes based on the clinker-free slag binder.

Strength of the cellular slag materials based concrete varies depending on the average density. So, heat-insulating ash-slag cellular concrete with ρ_o=400-500 kg/m^3 has compressive strength 0.6-2.0 MPa, and structural heat-insulating concrete (ρ_o = 600-1200) – 3.0-12.5 MPa. Maximal strength of the cellular concretes is achieved at the proportion of slag binder and silica component within the limits of 1:0.5-1:1.2 depending on the peculiarities of raw materials. The grinding fineness also influences on the strength of slag materials. So, at the increasing of specific surface area of the slag binder from 3500 to 6500 cm^2/g its strength increases at 50-60%. The indexes of durability and other properties are considerably improved at reduction of water-cement ratio up to 0.25-0.35 that is achieved at vibro-treatment at preparation of cellular mixture and at the stage of casting.

Walls made of cellular slag concrete panels (Figure 20.6) are in 1.3-2 times lighter than those made of claydite-concrete at the lower cost of the first. Specific capital investments in the production of structures made of autoclaved slag concrete at 30-40 % lower, than in the production of similar structures from other types of concrete.

The slag materials are widely used in the production of heat-resistant concrete as binder, aggregates and fine ground admixtures. Binders based of metallurgical slags exceed Portland cement by their heat resistance that is explained by comparatively low content of hydrate of lime in the slag cement stone. It is possible to get heat-resistant concretes, suitable for exploitation to 1200 °C, applying the blast-furnace cement.

The concrete, based on the slag-alkaline binder are attributed to the slag-alkaline concretes. The approximate composition of slag-alkaline heavy-weight concretes, % is following: ground granulated slag - 15-30; alkaline component – 0.5-1.5; aggregates - 70-85. At hardening alkalines react not only with slag but also with aggregates, first of all, with clay and dust particles. At hardening of such concrete, nonsoluble alkaline hydroaluminosilicates - analogues of natural zeolites, contributing to compaction and increasing of the material strength are formed. In this connection requirements to the aggregates for slag-alkaline concretes reduce considerably. Besides the traditional aggregates (crushed stone, gravel, sand) a lot of dispersible natural materials and byproducts of different branches of industry can be used for this purpose.

Such natural materials as local soils and low-strength rocks (fine sands, sandy loams, gravel-sandy and clay-gravel mixtures) which are impermissible for the production of Portland cement concretes due to the high dispersion and pollution can be used. Content of clay particles can achieve 5 %, and dust-like - 20%. It is not acceptable to apply the aggregates, containing grains of gypsum and anhydrite.

Compressive strength of heavy-weight slag alkaline concrete is in the range of 20-100 MPa. The tensile strength is 1/10-1/15, and flexural strength - 1/6-1/10 of compressive strength. The strength of steam-cured elements achieves 100 % and more of 28 day strength. Autoclaving activates strength growth; in this connection duration of heat- moist treatment can be considerably shortened as compared with cement concrete elements. Recommended duration of curing of the slag-alkaline concrete elements, at the thermal treatment is 2-3 hours.

Softening coefficient of slag-alkaline concrete is 0.9-1.0, and it sometimes exceeds 1.0.

Modulus of elasticity of this concrete is the same as of Portland cement one, maximum compressibility is 1-2 mm/m, maximum tensility – 0.15-0.3 mm/m. Wearing resistantance of slag-alkaline concretes is 0.2-1.2 g/cm^2 that correspond to the indexes of wearing of the rocks like granites and dense sandstones.

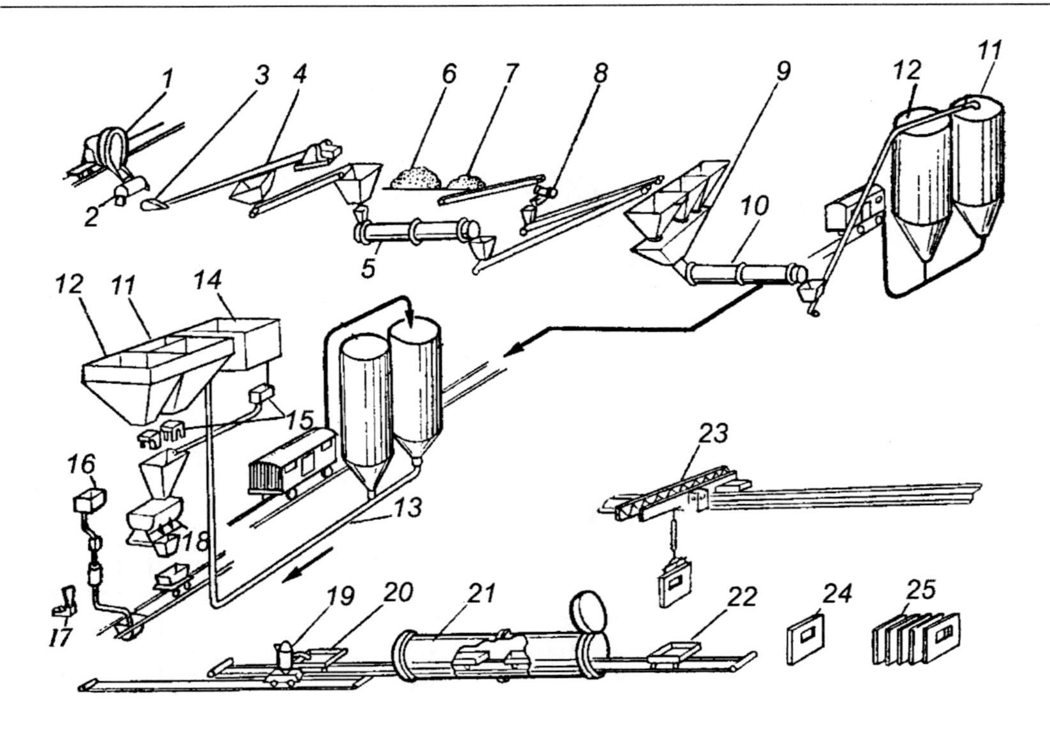

Figure 20.6. Scheme of manufacture of walling panels from the autoclaved gas-slag concrete:
1 - slag ladle; 2 - granulation pool; 3 - scraper; 4- bin with grate; 5- drying drum; 6 - lime; 7 - gypsum; 8 - crusher for lime and gypsum; 9- disk feeders; 10 - mill; 11 – lime-slag bin; 12 - ash bin; 13 - air tube conveyer; 14 - water tank; 15 - batchers; 16 – bin with surface-active substances; 17 - scales for aluminium powder; 18 - mortar-mixer; 19 - mixer-placer machine; 20- formwork carriage; 21- autoclave; 22 - stripping of formwork; 23- bridge crane; 24 - area of elements finishing; 25 - storage of finished products.

The structure of the slag-alkaline stone is characterized by the presence of the finest closed pores round by shape that is a result of raised surface tension of alkaline solution until hardening. Such structure of the hardening binder predetermines high watertightness and frost resistance of concrete.

The sufficient density of slag-alkaline concretes and permanent alkaline environment provide high preservation of steel reinforcement. The stable pH-value of the environment (pH>12) and good adhesion of concrete with the reinforcement allow to produce reinforced structures, including prestressed structures.

Increased corrosion resistance is characteristic for the elements made of slag-alkaline concrete, because there is no high-basic calcium hydroaluminates in its composition, causing the sulfate corrosion of cements in the products of their hardening, and also free lime is absent, leaching of which results in destruction of cement stone in soft water. Because of that, according to the resistance in the environment with low hydrocarbonate hardness, mineralized sulfate and magnesia waters, the slag-alkaline concrete exceeds the concrete not only at ordinary Portland cement but also one based on sulphate-resistant cement. In addition, they are resistant to the action of petrol and other petrochemicals, concentrated ammonia, solutions of sugar and weak solutions of organic acids; differ also by their high biological resistance.

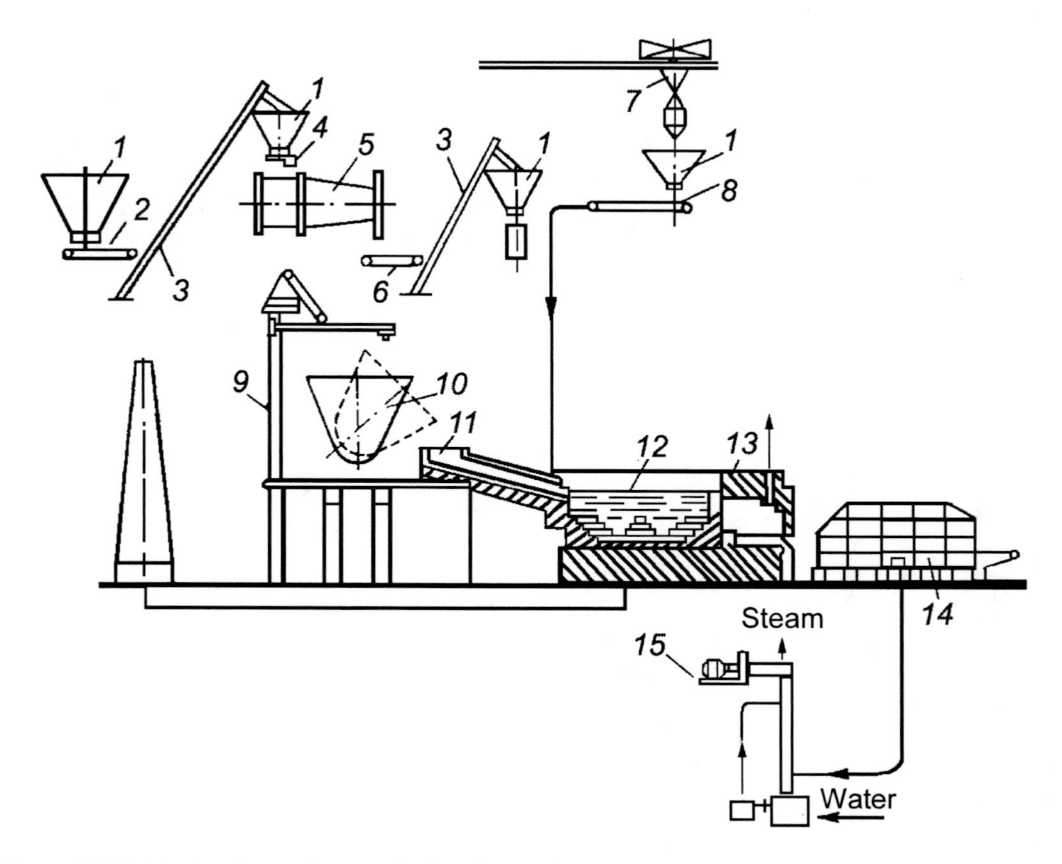

Figure 20.7. Scheme of manufacture of mineral wool from molten slags:
1- bin for sand; 2- feeder; 3 - screw conveyor; 4- disk feeder; 5 - drying drum; 6 - conveyer; 7 - telpher; 8 - dosing screw; 9 - pile-driver for punching of slag crust; 10- slag ledge; 11 - tray for slag unloading into furnace -slag receiver; 12 - furnace -slag receiver 13 - furnace -feeder; 14 - fibre-extracting chamber; 15 - smoke exhauster.

Experience of application of the slag-alkaline concrete for the winter concreting, have indicated that slag-alkaline mixtures do not freeze at temperatures to - 10-15°C.

Materials from slag fusions can be used for manufacture of slag wool, cast materials; glass and slag glass-ceramics.

The technological process of slag wool manufacture (Figure 20.7), as well as other varieties of mineral wool, consists of two basic stages: obtaining of fusion and its processing in fiber.

The various heat-insulating elements and materials are made of slag wool with help of organic and inorganic binders or without them with the use of polymers, bitumens, emulsions and pastes as binders. The basic types of elements are soft, semi-rigid and rigid slabs, cylinders, semicylinders. The physical-mechanical properties of elements from slag wool are shown in Table 20.1. The bulk of elements are used for the thermal insulation of non-load-bearing structures, pipelines and sound-proofing.

Table 20.1. Physical-mechanical properties of products made of mineral wool

Types of products	Average density, kg/m^3	Thermal conductivity at 25±5^{0}C, W/m^{0}C	Tensile strength, MPa, not less than	Compressibility under loading 0.002 MPa, no more than%,	Maximum temperature of application, ^{0}C
Sewed mats	85-135	0.044	-	30-40	-180-+600
Vertically-stratified mats	50-125	0.047-0.057	-	2-5	-120-+300
Slabs based on bituminous binder	75-250	0.046-0.064	0.075-0.008	4.5-5.5	-100-+60
Slabs based on synthetic binder	35-350	0.044-0.066	0.008-0.01	4-15	-100-+400
Cylinders and semicylinders based on synthetic binder	75-225	0.048-0.052	0.015-0.025	-	-100-+400

The slag cast elements: paving for roads and floors of industrial buildings, tubings, kerb-stone, anticorrosive tiles, pipes (Figure 20.8) and other are made of the molten metallurgical slags. Cast elements from slag fusion are more profitable economically, than stone casting, close to it by the mechanical properties. The average density of the cast elements from slag achieves to 3000 kg/m^3 and compressive strength – to 500 MPa.

The slag casting exceeds the reinforced concrete and steel, according by wearing resistance, heat resistance and a series of other properties. The cast elements made of slag are more effective, than steel ones in different linings, for example hoppers and streamers for the transportation of abrasive materials (ores, agglomerate, crushed stone, sand, etc.). Their service time is in 5-6 times higher than the service time of steel lining. No less than 2-3 tones of metal are saved at every tone of the slabs casted from a slag.

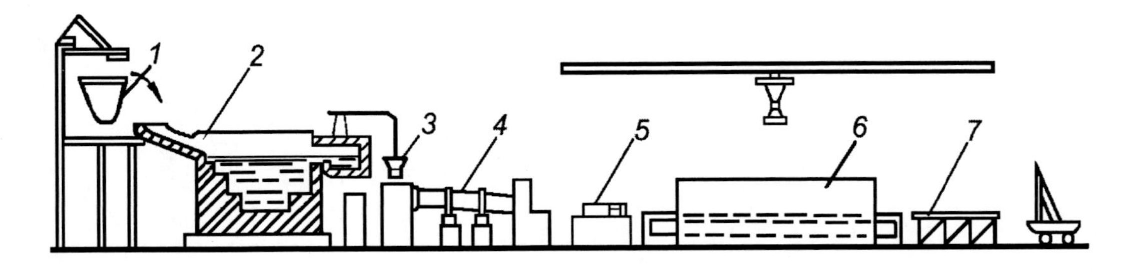

Figure 20.8. Scheme of manufacture of the cast pipes by centrifugal method:
1 - slag ledge; 2 – bath furnace; 3 - batcher; 4 - rotary machine; 5 - hydraulic pusher; 6 - annealing furnace; 7- control stand.

Metallurgical slags are applied as basic raw material at the obtaining slag glasses, and also as admixtures, intensifying the processes of glass manufacturing.

The *slag glass ceramics* — is a variety of glass- crystalline materials, obtained with the directed crystallization of glasses.

The production of slag glass-ceramics consists in obtaining of slag glasses, shaping the elements from them and subsequent their crystallization. Charge mixture for the obtaining glasses consists of slag, sand, alkali–containing and other admixtures. The use of melted metallurgical slags is effective; as if it saves up to 30-40 % of all the heat, spent for obtaining of glass.

The slag glass-ceramics differ from the most of construction materials by higher physical-mechanical properties (Table 20.2). So, their strength in several times exceeds the strength of initial glass and is similar to the strength of iron and steel. At the same time the slag glass-ceramics are in 3 times lighter than they are. The heat-resistance of the slag glass-ceramics reaches 150-200°C.

The parameters of chemical resistance and abrasive resistance for these materials are especially high. The slag glass-ceramics can be exposed to the different methods of mechanical treatment: polishing, cutting, drilling by diamond or carborundum tools. These materials can be strengthened by tempering at 50- 100 %.

Along with slags the large-tonnage waste products of a series of metallurgical industries are water suspensions of dispersible particles – slurries. The nepheline slurries are made in a large amount. They are wastes of alumina production, which consist mainly of dicalcium silicate (50-90%) and according to the content of oxides CaO, SiO_2, Al_2O_3, Fe_2O_3, occupy an intermediate place between Portland cement, blast-furnace slag and alumina cement.

Table 20.2. Comparative description of slag glass-ceramics and other materials

Property	Slag glass-ceramics	Stone casting	Granite	Steel	Alumi-num
Average density, kg/m^3	2600-2750	3000	2600..2800	7800	2700
Ultimate strength, MPa:					
compressive	500-650	230..300	100-250	420-550	250
bending	80-130	30-50	-	400-1600	220
tensile	50-70	15-25	-	300-1400	100-150
Module of elasticity, 10^3 MPa	90-100	93-110	40-60	210-220	65-78
Water absorption, %	0	0.05-0.22	0.2-0.5	0	0
Temperature initial softening, ^{0}C	900	1.050	-	1200-1500	658
Acid resistance, %	99.1-99.3	97-99	-	-	-
Alkali resistance, %	85-96.3	90-91	-	-	-
Wearing resistance, g/cm^2	0.01	0.04-0.08	0.21	-	-
Impact strength, kJ/m^2	3-4	2-3	-	500-1000	800

The presence of minerals, possessing a hydraulicity in the slurries (C_2S, C_2F and other), and their hydrates predetermines possibility of obtaining of binder matters from them at drying, grinding and introduction of activators.

Nepheline cement is the product of intergrinding nepheline slurry (80-85 %), lime or other activators, for example Portland cement (15-20 %) and gypsum (4-7 %). Initial setting time of nepheline cement should be not earlier than 45 min, and the final setting time - not later than 6 h after its mixing. Ultimate strength of this cement type is 10-25 MPa in 28 days.

Nepheline cement is an effective binder for masonry and plasters, and also for concrete of normal and especially autoclave hardening.

In the cement industry the industrial experience of application of nepheline slurry is accumulated as a basic raw component of Portland cement clinker. Complex production of alumina, soda products and cement based on the nepheline raw material can be organized at the factories. Approximately 10 tones of cement are obtained from 1 ton of alumina at the complex processing of nephelines.

20.2. Application of Waste Products of Fuel - Energy Industry

Products, obtained as waste products at mining, cleaning and firing of solid fuel, are the wastes of fuel - energy industry. This group of waste products is divided by source of forming, the type of fuel, number of plasticity of mineral part of wastes, content of combustible part, grain distribution, chemical- mineralogical composition, degree of fusibility, interval of softening and degree of swelling.

The Waste Products of Coal Production and Cleaning

The basic types of solid fuel are black and brown coals. Mining and overburden rocks, wastes of coal cleaning can be used for construction materials manufacturing.

Mining non-utilized rocks are often presented by argillites, siltstone, sandstones and limestones.

Metamorphosed argillites, siltstone and sandstones have high density and, as a rule, soak in water very hard. They can be attributed to lowplastic or unplastic clay raw material.

The waste products of coal cleaning, characterized by the low variations of composition and properties represent a huge interest for application in the production of construction materials. Content of coal in them, not extracted in the process of cleaning, can reach to 20 %. The wastes of coal cleaning are presented usually as pieces by coarseness 8-80 mm.

In accordance with the model charts of technological process of coal cleaning from mines after grinding it is exposed to the hydroclassifying by coarseness, then it is cleaned by gravitation method. The concentrate, industrial product and rock (wastes of the gravity cleaning of coals) are separated. At dehydration of concentrate there is separated the sludge with the size of grains less than 1 mm, which is delivered at flotation. The concentrate and wastes of flotation are obtained after the flotation cleaning.

The waste products of coal production and cleaning can be used mainly in the production of wall ceramic materials and porous aggregates. They are close to traditional clay raw materials by chemical composition. The sulphur, contained in sulfate and sulfide compounds, as a harmful admixture, is present in them.

At Al_2O_3 content in mineral part of waste products more than 15% and content of carbon less than 15%, wastes can be applied as raw material for the obtaining *ceramic masonry elements* with strength 7.5-30 MPa without the clay admixture. At Al_2O_3 and carbon content more than 15 % in raw material mixture, the clay is added. At content of Al_2O_3 in wastes less than 15 % and carbon more than 15% they are useless as basic raw material and can be used as leaning and fuel-containing admixture in clay batch.

The wastes of production and concentration of coal are applied as leaning and burning fuel-carrying admixtures in the production of ceramic masonry elements based on kaolin and hydromica clays, sandy clays and clayey shales. The lump wastes are ground before the introduction to the ceramic sludge. The grinding of wastes is made in hammer crushers, ball or other mills. For the sludge with the size of particles less than 1 mm a foregrinding is not required, it is dried a little to humidity 5-6 %. The admixture of wastes is 10-30 % at the obtaining of brick by the solf-mud method.

Introduction of carboniferous rocks to a certain limit can increase binding ability of ceramic batch and especially the resistance to compression stress. At comparatively high content of these rocks in a batch (up to 20-30 %) binding strength of clay raw material reduces sharply. The facilitation of moisture migration conditions promotes dryings properties of raw material. Introduction of optimum amount of fuel-carrying admixture results more uniform burning and improves the strength parameters of products (up to 30-40%), saves a fuel (up to 30 %). It is also eliminated the necessity in introduction of coal to the batch and promotes the productivity of furnaces.

The possibility of hollow brick and ceramic stone manufacture based on wastes of coal cleaning both solf-mud and dry molding is determined.

The replacement of clay rocks, obtained in stone-pits, by the processed wastes of coal cleaning leads to declining of expense of technological fuel approximately at 80 % and to the prime cost of elements at 19-28 % at the production of masonry products.

Carboniferous rocks are effective raw material for the production of *porous aggregates*.

The analysis of technological methods of production of artificial porous aggregates from the fuel-containing wastes of production and cleaning of coals shows that a method of agglomeration is the most effective one. Besides the simplicity of technology it enables to use effectively the fuel, contained in wastes. Essence of agglomeration process consists in burning of fuel in conglomerate batch in a horizontal layer, and as a result, the air, entering an area of burning, is heated and intensifies the process of sludge's fuel burning and hot gases, going out from the area of burning, dry a little and heat the next layer of the sludge. After burning of fuel, the area of burning moves to the below lying layer of sludge.

The products of burning, barren rocks, concomitant the deposits of coals are the burnt rocks. Their varieties are - clay and clay-sandy rocks, burnt in the Earth interiors at underground fires in coal layers.

The possibilities of application of the burnt rocks in the production of construction materials are very various. They find a wide use in road construction mainly at the basements arrangement. At the satisfactory physical-mechanical properties the burnt rocks are used not only for the lower but also for top layers of basements, and also for under layer of coating. The burnt rocks are most effectively used after the treatment with organic binder.

The burnt rocks, as well as other burnt clay materials, have the activity in relation to lime and can be used as hydraulic admixtures in binder of lime - pozzolan type, Portland cement, portland-pozzolan cement and steam-curing materials. High adsorption activity and adhesion with organic binders allow applying them in asphalt and polymer compositions.

Ash-and-slag Waste Products from the Thermal Power Stations

Fly ash as dust-like waste and lump slag, and also ash-and-slag mixture are formed at the combustion of solid types of fuel at thermal power-stations. They are the products of high temperature (1200-1700 °C) treatment of mineral part of fuel.

The fly ash is the fine-dispersed material, caught by the electrostatic cleaners, consisting mainly of particles 5-100 μm by size. Its chemical-mineralogical composition corresponds to the composition of mineral part of the burned fuel. For example, at combustion of black coal the ash is the burnt clay substance with includings the dispersible particles of quartz sand, at combustion of shales - marls with the admixtures of gypsum and sand. At burning of mineral part of fuel clay substance is dehydrated and the low-basic aluminates and calcium silicates appear.

The basic component of fly ash is a glassy silica-alumina phase, making up 40-65 % of all mass and having the appearance of particles of spherical form up to 100 μm in diameter. The α-quartz and mullite can be present among the crystalline phases in ashes, and at raised content of Fe_2O_3, also hematite is.

The ash and fuel slags are used as raw material components of *Portland cement clinker* and *active mineral additives* at the production of Portland cement, and also composite ash and slag cements. In composition of the raw material mixture at the production of clinker, ash and slags replace the clay and partly limestone components, in some cases this replacement improves the chemical- mineralogical composition of clinker and conditions of its firing.

The basic part fuel ashes can be used as active mineral additive in the production of cement. In this case they should contain no more than, %: SiO_2 -40, SO_3 -3, losses at ignition -10. Fly ash can be added, as well as other hydraulic mineral admixtures, in an amount no more than, %: into Portland cement -20, into pozzolanic cement - 55.

Introduction of ash into the cement in an amount of 20% reduces a little its strength in the initial terms of hardening, on 28 days the decline of strength is minimal, and at the protracted terms of hardening the strength of cements with an ash becomes often higher, than without ash. The increasing of the ash content (more than 20 %) usually results in the substantial decline of strength characteristics of cement - especially in the early terms of hardening.

The additives of ash to the Portland cement at the conditions of hydrothermal treatment of mortars and concretes make a positive influence. The increase of ash dispersion promotes the intensity ash Portland cement hardening.

A substantial effect is observed at combined introduction to cement of fly ash and admixtures - superplasticizers. Thus there is obtained binder with low water consumption, characterized at high dispersion (specific surface area 4000-5000 sm^2/g) low normal water consumption (16-20%) and strength up to 100 MPa. The concrete based on low water consumption binder are characterized by the intensive strength setting already through a few hours, and until 1 day their strength can achieve 60 MPa.

Fly ashes and slags acquire an ability to harden (Table 20.3) when mixed with admixtures of activators (lime, gypsum, alkalines). Intergrinding of ashes and slags with activate admixtures promotes strength of binder substances and increases the strength of products.

Only *lime- ash binder* (at the use of fly ash) and *lime-slag binder* (at the use of granulated slags) harden at normal conditions among the binders based on slag and ash. It is desirable to execute steaming of mortars and concrete based on ash-slag binder at 90-95^{0}C.

Satisfactory results of strength are achieved under the pressure of steam 0.8 MPa at autoclaving. The higher values of pressure of water vapor result in the sharp increase of materials strength. The heat treatment with steam of ash-slag materials can be substituted by heating with infra-red rays and electrical curing.

Rational range of application of non-clicker slag-ash binders is concretes hardening at a steaming and in the conditions of autoclaving. The masonry elements, blocks for foundation, structures for the different elements of buildings and constructions are made of such concrete. They can be applied also at erection of underground and submerged structures, exposed to the influence of fresh and sulfate waters. It is possible to use the slag-ash binders, hardened at normal temperatures in mortars for masonry and plasters, and also in low-strength concretes. This type of binder is not recommended to apply in environment at the lowered temperature, in structures, exposed to drying out and moistening, frequent freezing and thawing.

Slag and ash binders, as long-term experience showed, can successfully replace the lime-silica and cement-lime binders in the production of cellular concrete elements. Pulverized fuel slags and fly ashes allow also replacing the fine-grained quartz sand in composition of the cellular concrete elements.

Table 20.3. Optimum compositions of slag and ash binder

Types of slags and ash	Approximate content by mass, %			Strength of binders, MPa	
	Slag, Ash	Quicklime	Dehydrated gypsum	Steaming-or electrical curing 95-98^{0}C	Autoclaving at a temperature 175^{0}C
Granulated slags					
Medium calcium (CaO from 45 to 20%)l	65-55	30-40	5	20-30	30-40
Low calcium (CaO from 20 to 10%)	65-55	30-40	5	20-30	30-40
Hyperacid (CaO<10%)	65-55	30-40	5	20-30	30-40
Fly ash					
High calcium (CaO>20%)	55	40	5	15-20	25-35
Hyperacid (CaO<10%)	55	40	5	10-15	20-25

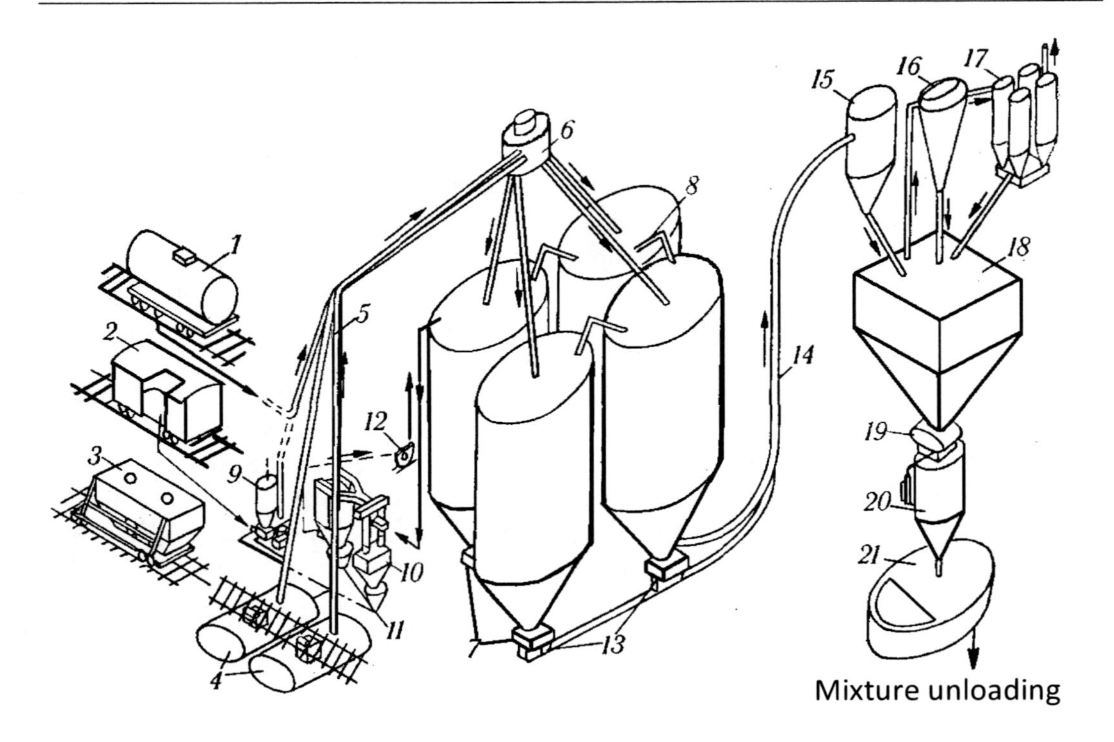

Figure 20.9. Scheme of introduction of fly ash in concrete mix:
1 - railway tank; 2 – open wagon; 3 - hopper type wagon; 4 – intake device; 5, 14 - pipelines; 6 - distributive device of bulk materials; 7 - pneumatic unloader; 8 - silos; 9 - discharger; 10, 11 - cyclones with dust-collecting bags; 12 - aspirator; 13 - jet pump; 15 - storage bin; 16 - cyclone; 17 - block of multicyclones; 18 – feed bin; 19 - device for portioned supply of fly ash; 20 - batcher; 21 - concrete mixer.

Efficiency of introduction of dry fly ashes as active mineral additives and microfillers at concrete and mortars production is determined.

Concrete mixtures with ashes have a greater cohesion, better pumpability, less bleeding and segregation. Here, concrete has higher strength, density, watertightness and resistance to some types of corrosion, less thermal conductivity.

The acid ashes not possessing with the binder properties are the most effective as active additives in concretes; their pozzolanic activity appears at reaction with the cement binder. It is possible to shorten substantially the expense of cement depending on this parameter in relation to particular cement, water consumption and workability of concrete mixture, conditions and duration of hardening. The most stable economy of cement is given by dry fly ash. Optimum content of ash, kg/m^3, for concretes is: steamed - about 150; normal hardening - 100.

Considerable practical experience of application of fly ash in concretes is accumulated in hydraulic engineering. The efficiency of replacement of 25-30 % portland cement by fly ash for the concretes of internal areas of massive hydraulic structures and 15-20 % for concrete in underwater parts of structures is well-proved at present.

Fly ash is widely used in the production of *precast reinforced-concrete structures*. Dry ash is used (Figure 20.9) in the concrete with compressive strength 10-40 MPa in an amount of 20-30 % by cement mass. However at excessive content of ash, the inflation of surface of the steamed elements is possible.

Ash is applied as a component of *building mortars*, in which the properties of mineral admixture, plasticizer and microfiller are combined. Ash improves the plasticity and water-retaining ability of mortar mixtures, properties of hardened mortars.

Ash-slag raw material can be used at production of aggregates of both heavy-weight and light-weight concrete. Slags from incineration of anthracite, black and brown coals, peat and slates; ash, crushed stone and sand from the fuel slags, agloporite on the basis of ash of thermal power stations (Figure 20.10), ash fired and non-fired gravel, clay-ash haydite can be used as *porous aggregates* for light-weight concretes.

Ashes and slags of thermal power stations are also effective raw material for manufacture of *lime-sand bricks* (Figure 20.11), *ash ceramics, mineral wool, glass*. Application of the fuel ashes and slags in the production of these materials is provided with the complex of their properties: chemical interaction with lime, dispersion, sinterability, calorific value, ability to give the silicate melts. Depending on the purpose of ash-and-slag raw material and applied technologies a leading value is acquired by one or another from the indicated properties.

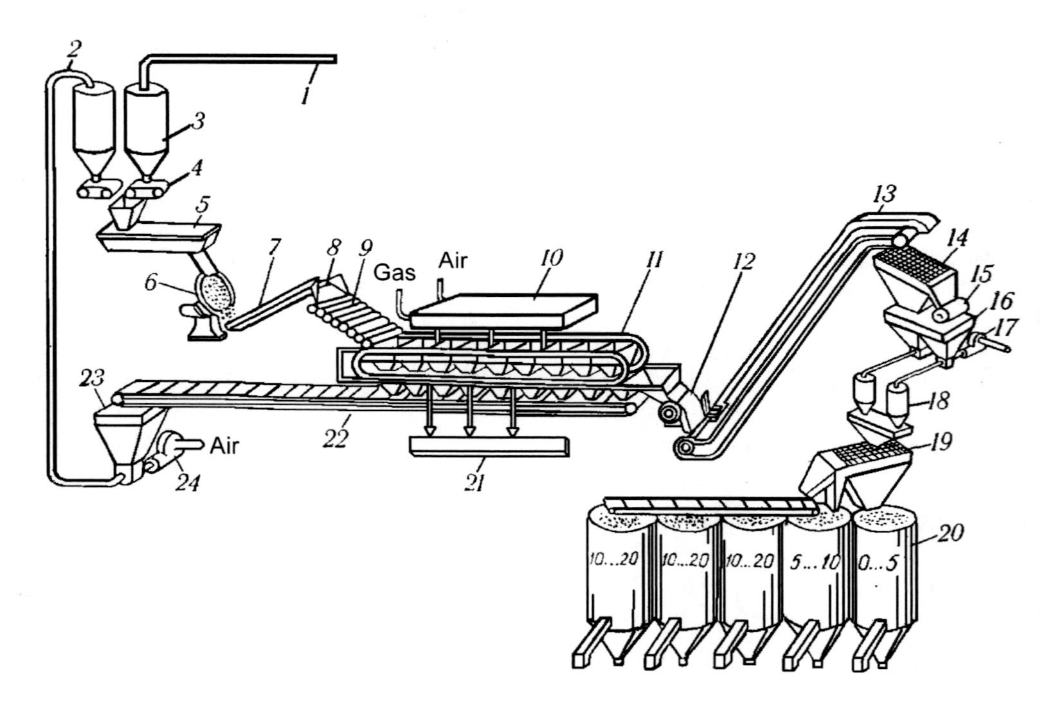

Figure 20.10. Scheme of manufacture of agloporite gravel from the ash of thermal power-station: 1 - pneumatic transport of ash; 2 - pneumatic transport (return); 3 – bin of ash; 4 - batcher; 5 - screw mixer; 6 - granulator; 7 - belt conveyor; 8 - tray; 9 - roller placer; 10 - furnace; 11- band sintering machine; 12 - rotary crusher; 13 - apron conveyor; 14 - unbalanced-throw screen; 15 –crusher; 16 – receiving bin;17 - dust centrifugal blower; 18 – bag collector; 19 - unbalanced-throw screen; 20 - bin of finished product; 21 - collector for cooled gases; 22 - belt conveyor for collection of spillage; 23 – receiving bin of spillage; 24 - high-pressure blower.

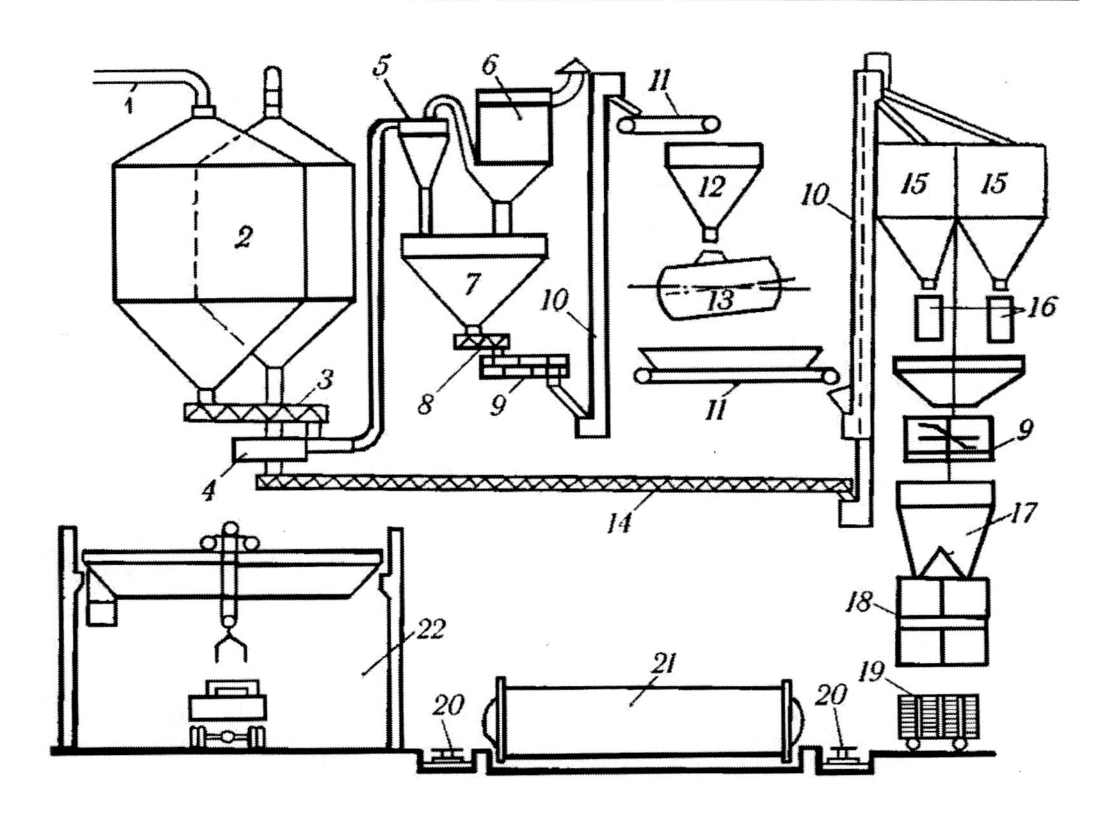

Figure 20.11. Scheme of lime-sand brick manufacture from high-calcium ashes:
1 - pneumatic conveyor; 2 - silo storage; 3- feed-screw; 4 - pneumatic pump; 5 - cyclone; 6 - bag collector; 7 - feed bin; 8 - feed-screw; 9- mixer; 10- elevator; 11- conveyer; 12 - batcher; 13 - reactor; 14 - feed-screw; 15 - bins of ash and cement; 16 - batcher; 17 – press bin; 18 - press; 19 - scalding cart; 20 - transmission cart; 21 - autoclave; 22 - storage of the finished product.

The *highway engineering* is one of the basic users of fuel ashes and slags, where ash and ash-and-slag mortars can be used for the arrangement of sublayers and bottom layers of basements, partial replacement of binder during the soil stabilization with the cement and lime, as mineral powder is in asphalt concretes and mortars, as admixtures in road cement concretes.

The nonconsolidated and consolidated ash and slag mixes are applied in the highway engineering.

Nonconsolidated ash and slag mixes can be used mainly as material for the arrangement of sublayers and bottom layers of basements of roads of regional and local value.

It is possible to promote the efficiency of application of waste ash-and-slag mixes in the highway engineering by their consolidation with lime, cement, lime-slag binder or ground granulated blast-furnace slag (Table 20.4), that increases the deformation module, allowing decreasing the thickness of basements and reduce the materials consumption and also expenses on their transportation.

Table 20.4. Properties of ash and slag mixes, consolidated with lime

Property	Strength class		
	1	2	3
Minimal calculation modules, MPa:			
Deformation	150	100	75
Elasticity	500	300	200
Minimal compressive strength of the water-saturated specimens, MPa, after days:			
28	2	1	0.5
90	4	2	1
Minimum coefficient of frost resistance after 90-day's hardening	0.75	0.7	0.5
Minimal coefficient of compacting	0.98	0.98	0.98

Ashes, obtained at firing of coals and pyroshales, are widely used as powder-filling material of *roofing and waterproofing mastics*.

20.3. Application of Waste Products of Chemical and other Industrial Branches

The wastes of chemical-technological industries can be classified according to: content of representative chemical component and technological assignment in the production of construction materials.

According to the content of characteristic chemical component there are distinguished: phosphorus- and fluorid-containing slags, gypsum and lime containing products, ferrous, siliceous and other materials.

The chemical products according to the technological assignment in the production of construction materials are divided in such groups:

1. The raw materials (for the production of cement, gypsum, lime, etc.).
2. Intensifiers of technological processes (fusing agents, reducer of hardness, granular forming agents, etc.).
3. Admixtures-modifiers of material properties (alloying substances, plasticizers, hardening accelerators, etc.).

Classification according to the technological assignment is very conditional, because the same chemical product, being the waste product of production, depending on a particular range of its application, can be attributed to the different groups. For example, phosphogypsum for the production of gypsum binder or sulphuric acid belongs to the first group of chemical waste products, and entered at burning of cement clinker - to the second one, at the cement grinding for adjusting of setting time - to the third one.

The slags of electrothermal production of phosphorus, gypsum-containing, lime and ferrous waste products, polymer products and others are the most valuable raw materials for the production of construction materials from the production wastes of chemical industry.

Phosphoric slags are the by-product of production of phosphorus by the thermal method in electric furnaces. At the temperature 1300-1500 °C calcium phosphate reacts with the coke carbon and silica, resulting in the phosphorus and slag fusion formation. Slag pours off from furnaces in the flaming state and granulates with wet-processing. There are 10- 12 tones of slag on 1 t of phosphorus.

The chemistry of phosphoric slags is close to composition of blast-furnace. Total content of calcium oxide and silica in them achieves at 95 % at their ratio 0.9- 1.1.

The peculiarities of phosphoric slags are the contents of P_2O_5 and CaF_2 and reduced amount of Al_2O_3 in them. The differences between the content of P_2O_5 and CaF_2 and coefficients of basicity, determining physical and chemical properties of slag fusions and peculiarities of granulation, substantially influence on phase composition, structure and properties of slags.

Most of the phosphoric slags are used in the cement industry. Phosphoric slag meets the requirements, specified to active mineral admixtures of artificial origin. Comparatively low content of Al_2O_3 causes the lower hydraulicity of phosphoric slags in comparison with blast-furnace slags.

It is possible to obtain slag pumice, cotton wool and cast elements from the phosphoric-slag fusions. *Phosphoric slag wool* is characterized with the long thin fibres and average density 80-200 kg/m^3. It has bulk density 600-800 kg/m^3 and glassy fine-pored structure.

From the slags of electrothermal production of phosphorus the *slag glass-ceramics* with the strength up to 400 MPa are obtained, possessing the raised resistance in corrosive environments and at high temperatures.

Possibility of application of phosphoric slags as a basic component of ceramic masses is set, for example at a facing tile manufacturing. The slag, being a fluxing agent, promotes formation of the required amount of liquid phase and improves sintering of ceramics.

The gypsum wastes can be formed at the production of the followings products: mineral acids (phosphogypsum, boron gypsum, fluorogypsum), organic acids (citric gypsum, etc.), the chemical treatment of wood (hydrolytic gypsum), treatment of water solutions of some salts and acids (silica gypsum, titanium gypsum, etc.), cleaning of industrial gases, containing SO_2 (sulfogypsum); production of salts from a lacustrine leach (leach gypsum).

The phosphogypsum is the most gypsum-contained waste product of chemical industry. So, 3.6-6.2 tones of phosphogypsum appear per 1 t of extraction phosphoric acid in account of dry substance.

Phosphogypsum contains from 80 to 98% of the gypsum and can be attributed to gypsum raw material. High dispersion of phosphogypsum ($S_{sp.s.}$= 3500-3800 sm^2/g) allows to exclude the crushing and coarse grinding from a technological process. At the same time, high humidity of phosphogypsum (up to 40%) complicates its transportation and preparation and results in the considerable fuel consumption for drying. Presence of water-soluble admixtures in phosphogypsum, in particular phosphorus- and fluorine-containing ones complicates processing of production wastes in comparison with processing of natural gypsum stone, causes the necessity of washing, neutralization and other technological operations and stipulates higher thermal consumptions accordingly. At ordinary technology gypsum binders based on phosphogypsum are low-strength that is explained with the high water consumption of phosphogypsum, stipulated by the large porosity of the formed semihydrate. If the water consumption of ordinary gypsum plaster is 50-70 %, for obtaining the paste of normal consistency from phosphogypsum binder without additional treatment, the required amount of water is 100-120 %. It is possible to reduce a little the negative influence on building properties of phosphogypsum of the admixtures contained in it, by re-crushing of phosphogypsum and shaping of elements with the method of vibrocompaction.

The basic methods of phosphogypsum preparation in production of *gypsum binders* can be divided into 4 groups:

1. Washing of phosphogypsum by water;
2. Washing in combination with neutralization and sedimentation of admixtures in water suspension;
3. Method of thermal decomposition of admixtures;
4. Introduction of neutralizing, mineralizing and crystallization regulative admixtures before burning and after it.

The methods of the first and the second groups are related to formation of large amount of polluted water (2-5 m^3 per 1 t of phosphogypsum), large expenses on their removal and cleaning. Most of methods of thermal disintegration of admixtures (the third group) are based on burning of the phosphogypsum until formation of soluble anhydrite with its further hydration and reburning to the semihydrate. They do not have the wide use yet, the same as the as well as the methods of the 4th group. The rare admixtures are required to realize the last one and they do not provide permanent properties of binder.

Knauf enterprise offers three variants of manufacture of binder from phosphogypsum depending on the area of its further use. The principle scheme of these three variants is presented in the Figure 20.12.

The *dry mixtures*, where the filling materials, plasticizers, retarding admixtures and, if it is required, other components are contained in their composition, are effective range of application of binder from phosphogypsum. Application of dry gypsum mixtures instead of the cement and lime ones allows to increase the labour productivity: at the floors arrangement - in 2-3 times, at wall plastering - in 1.3-1.5, tamping of oil and gas wells - in 1.5-2.5 times.

The phosphogypsum can be applied as *mineralizator* at burning of Portland cement clinker and as admixture for adjusting of cement setting in place of natural gypsum in the cement industry. Admixture of 3-4 % phosphogypsum in the batch allows increasing the saturation coefficient of clinker from 0.89-0.9 to 0.94-0.96 without the decline of the productivity of kilns, to promote resistance of lining in the area of sintering because of uniform formation of steady daubing and promotes the obtaining of the easy grinding clinker. The mechanism of mineralizing action of phosphogypsum is stipulated by the catalytic influence of SO_3 at temperatures below 1400°C, causing the decline of fusion viscosity, increase of its amount and formation of intermediate compounds, bonding CaO. The certain positive influence is rendered by the admixtures of phosphoric anhydride and fluorine.

The convenience of phosphogypsum is determined for replacement of gypsum at the milling of cement clinker. High content of sulphuric anhydride and presence of admixtures of water-soluble compounds of phosphorus and fluorine causes higher effect of retardation of phosphogypsum on setting time, than a gypsum stone. It allows decreasing of retarding admixture dosage in comparison with ordinary for gypsum.

Admixture of phosphogypsum does not influence at the cement strength, there can be an insignificant decline of the strength only in the early terms of hardening. Wide application of the phosphogypsum as the admixture at the production of cement is possible only at its pre-drying and granulation. Humidity of granulated phosphogypsum should not exceed 10-12 %.

The phosphogypsum can be used as basic raw material component in the production of cement that provides the effective process of simultaneous obtaining of cement clinker and sulphuric acid (Figure 20.13), essence of which consists in thermo-chemical decomposition of calcium sulfate in reducing medium. The reactions go on scheme:

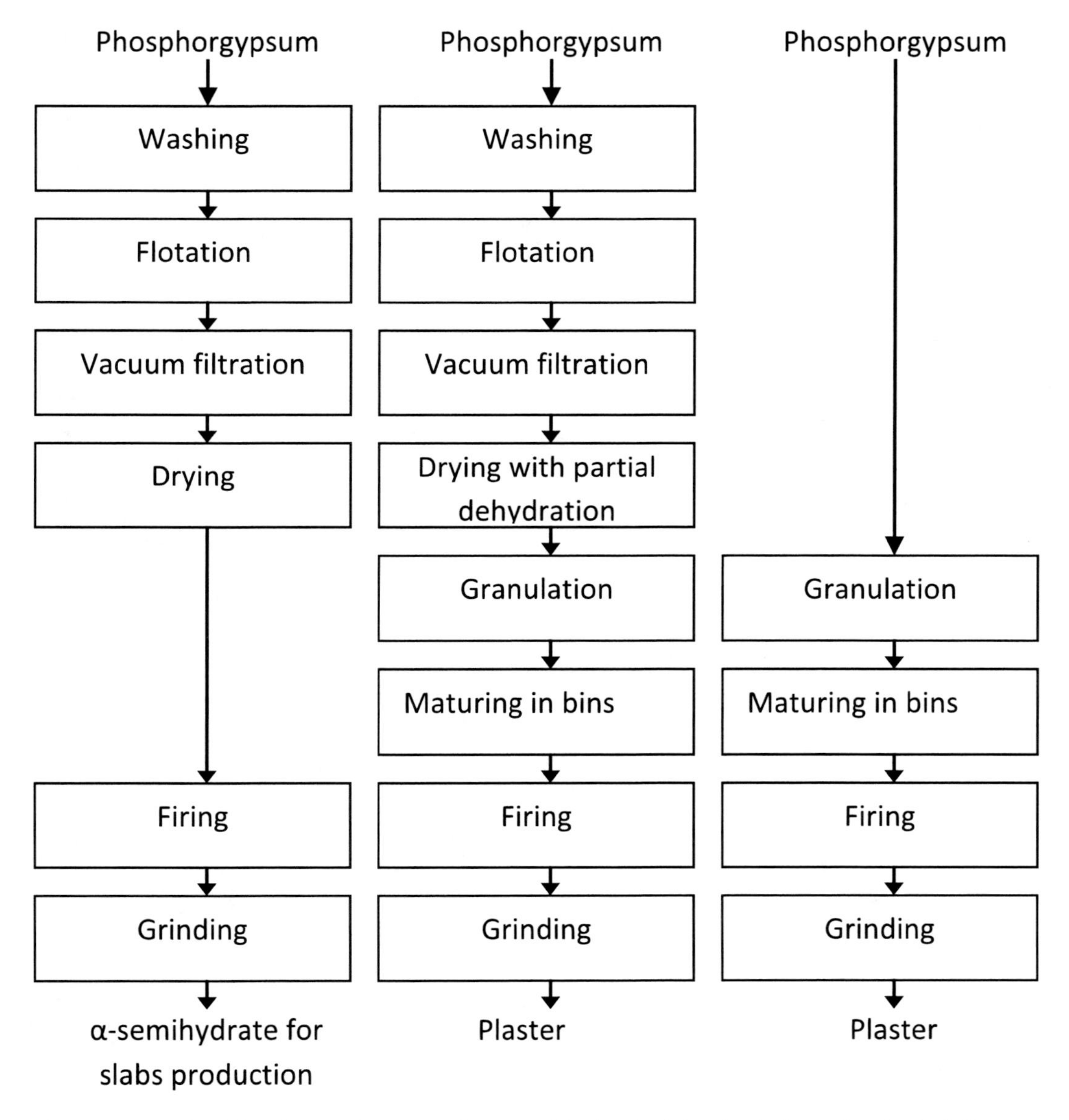

Figure 20.12. Principle scheme of process of gypsum binder obtaining by the Knauf method of (three variants).

$CaSO_4+2C=CaS+2CO_2\uparrow$;

$3CaSO_4+CaS=4CaO + 4SO_2\uparrow$.

Sulphureous gas is caught and transferred in sulphuric acid. The calcium oxide reacts with SiO_2, Al_2O_3 and Fe_2O_3, forming the clinker minerals.

Along with production of binders and products on their basis, other ways of phosphogypsum utilization are known. Experiments have shown that admixture up to 5 % of phosphogypsum in batch at the production of brick intensifies the process of drying and promotes the high quality of the elements.

Along with phosphogypsum, the other gypsum by-products (boron gypsum fluorogypsum, sulfogypsum (Figure 20.14), etc.) can be also successfully used for the construction material manufacturing, and foremost binder matters.

It is possible to use the lime, ferrous and other waste products of chemical industry along with the gypsum-containing wastes.

Carbide lime can be used for the obtaining of lime-silica binder and steam-cured materials on their basis.

The feldspar sands, burnt blast rocks, overburdens of iron-ore deposits, dumping blast-furnace slags and ore cleaning wastes can be used as silica components. On soda factories, pulp-and-paper mills and nitrogenous-fertilizer enterprises a large amount of raw materials, containing calcium carbonate is accumulated as waste products. One of industrial directions of these resources application is obtaining of belite-lime binder and lime-sand brick on its basis. Presence the wastes of chloride and calcium sulfate in their compositions considerably promotes the reactivity of raw meal, allowing operate the binder burning at 950-1000°C.

The pyrite drosses - waste products, formed at incineration of sulphuric pyrites (pyrite) on enterprises of the sulphuric acid production, are widely used among the ferrous waste products.

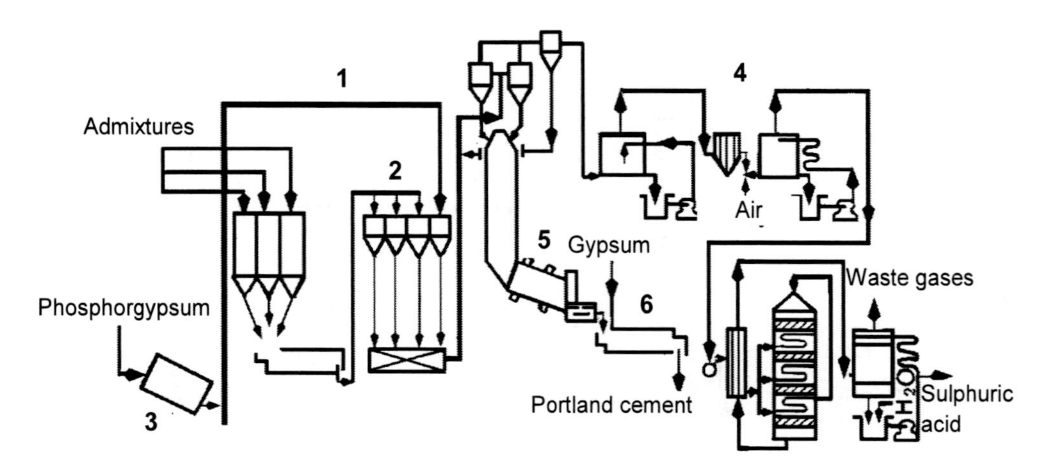

Figure 20.13. Scheme of complex manufacture of sulphuric acid and Portland cement based on the phosphogypsum
Divisions: 1 —grinding; 2 — preparation of raw powder; 3 — drying of phosphogypsum; 4 — sulphuric acid; 5 — furnace; 6 —cement milling.

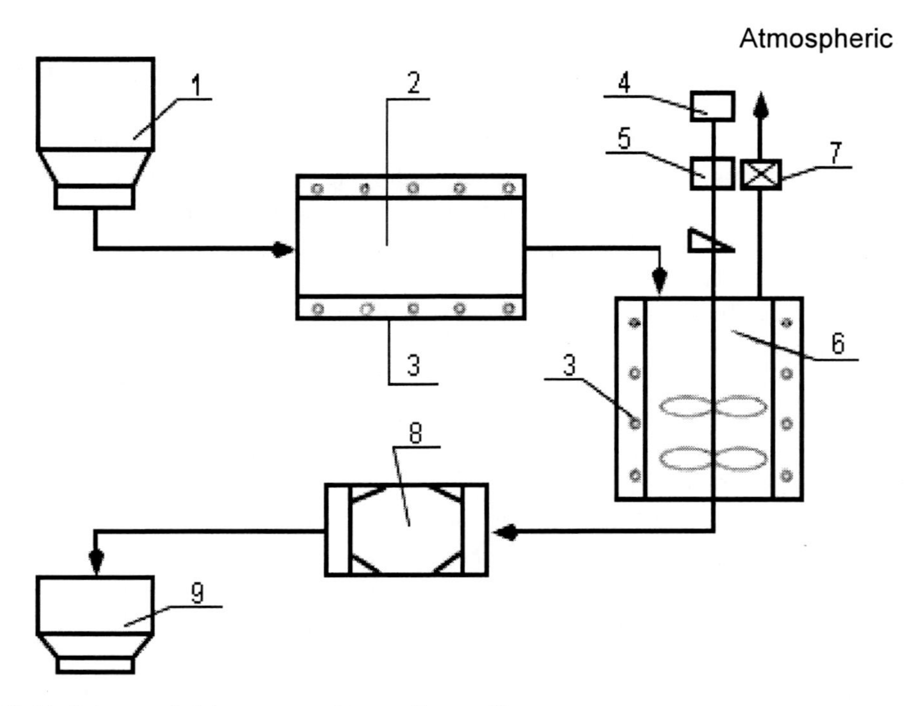

Figure 20.14. Scheme of alabaster manufacture from sulfogypsum:
1 - bin of sulfogypsum; 2 - kiln drier; 3 - electric heating units; 4 - electric motor; 5 - reductor; 6 – gypsum boiling furnace; 7- blower; 8 - ball mill; 9- bin of finished product.

The pyrite drosses are used as correcting iron-containing admixture in cement production. Because of change technology of sulphuric acid obtaining, the output of pyrite drosses is presently diminished and they are replaced by other iron-containing waste products in the cement production.

It is possible to utilize the products, with large sulphur content, for making of *sulphuric cement*, as impregnating compounds, at making of products from rubber, ebonite, plastics as filler of asphalt mixes. The obtaining of sulphuric cement is possible by melting of sulfur-containing rock with addition of fine-dispersed acid resistant filling agent and plasticizer. Compositions with sulphuric cement are heated to 145-155^{0}C before the use and quickly used. Concretes on the basis of sulphuric cements and heavy-weight aggregate have an average density 2300-2400 kg/m^3 and compressive strength 30-35 MPa. It is especially effectively to apply them in construction of chemical enterprises, highway engineering and hydroengineering, where a high-early-strength concrete with a high inoxidizability is required.

The enterprises of construction materials, located near-by hydrolytic factories, can utilize *lignine* - one of the most capacious wastes of hydrolytic industry.

Now there are the followings basic directions of application of hydrolytic lignine: as fuel-burning admixture in the production of ceramic materials; cutting replacer in building elements; raw material for the obtaining of phenol- lignine polymers; plasticizer and intensifier of grinding.

A series of waste products of chemical industry as solutions and sludges can be used as admixtures at the production of concretes and mortars, road-construction and other materials.

The materials on the basis of large-tonnage waste products of metallurgical, fuel and energy and chemical industry, were considered above. Along with them, the production wastes of wood processing, metal mining industry, industry of construction materials itself, municipal economy, can be used with success for the production of various construction materials. So, the simultaneously extractive rocks, dry and wet cleaning wastes at mining and processing of iron-ores, fluxes and refractiries, nonferrous metals; siftings at the production of crushed stone from igneous, metamorphic and sediment rocks are suitable for the production of construction materials.

At the certain chemical-mineralogical composition the *crushed waste rock* can be used for the obtaining of different binders. The carbonate and marl wastes are the raw materials for the production of air-hardening and hydraulic lime, Roman cement, composite binders. The silica-alumina materials in composition with carbonates can be added in the complement of raw meal for the obtaining of Portland cement clinker. In some cases they have certain advantages before traditional raw materials.

The *cleaning wastes of mining and other industries* can be widely used in the production of autoclave construction materials. Researches, conducted in the last years, have shown that it is expedient to utilize not only quartz sands but also clay sands, loams, aeolian soil, some types of clays, feldspar and clayey sands, a series of other rocks for the production of autoclaved materials.

The various types of wastes form in the production of different artificial construction materials in the process of the technological processing of raw material, and also as spoilage and others. Majority of these wastes at impossibility of their returning in a basic production can be used for the obtaining of other construction materials.

The hard domestic waste, waste caused by destruction of old buildings, constructions and road coating, construction waste, threadbare tires, waste paper, rag, broken glass, belong to the *wastes of municipal economy*.

Growth of volume of solid domestic waste in different countries changes from 3 to 10% per year. Solid domestic waste contain in average to 40% of literary garbage, 3-5% of ferrous metals, 25-40% food waste, 1-2% plastics, 4-6% textile, up to 4% of glass. It is possible to make about 750 kg of paper from one tone of literary garbage. It is also widely used for the carton production. The use of one tone of literary garbage in the production of paper and carton allows to save up to 4 m^3 of wood.

Utilized rubber elements such as conveyer bands, hoses and worn-out car tires are the large-tonnage wastes. It is possible to select for the repeated use about 700-750 kg of rubber, 130-150 kg of chemical fibers and 30-40 kg of steel from 1 tone of the waste at the complex application of rubber- caoutchouc materials and metal, contained in worn-out tires.

A rubber chips and micronized rubber powders can be applied as ingredients of rubber compounds. Thus get the rubbers, with the row of technical properties exceeded the materials, not containing the reclaims.

The technologies of roll and tiled materials based on the rubber-cord waste are developed. The worked-out rubber is applied also in the production of waterproof construction materials, materials for slabs, glues, mastics and sealants.

It is also effectively to use the plastic waste products to the reprocessing. The waste products are preliminary collated and purified from extrinsic particulates, and then exposed to grinding, agglomeration and granulation. Different building elements can be obtained from the chips. It is expedient to introduce the secondary raw materials into the polymer compositions in an amount of 40-50% of primary along with plasticizers, filling agents and stabilizers.

Packing and bottle polymer containers can be processed in finishing tiles and other elements.

It is possible to manufacture press-compositions with the required properties from the secondary polyethylene and polystyrene raw materials in mixture with the sand. The high strength indexes of such materials allow in combination with high waterproofness, for example, in Japan to utilize slabs from them for coating of sea-bed with the purpose of creation of the stations on fish breeding.

High waterproofness of the most of polymer waste products allows to utilize widely them in different materials, applied for sealing of joints between the panels of buildings, and also for coverage of parts of buildings, working under the water or in the conditions of increased humidity.

The large amount of so-called *concrete scrap* appears as a result of dismantling of buildings and structures, and also the accumulating of subquality products on the enterprises of the precast reinforced concrete. Processing of concrete scrap is directed mainly on the obtaining of the secondary aggregates and reinforcing steel.

Self-Assessment Questions

1. Tell about problem of industrial wastes utilization.
2. What are the peculiarities of blast-furnace slags?
3. What are the types of slag activation?
4. Tell about sulfate-slag and lime-slag cements.
5. Tell about technology and properties of slag-alkaline binders and concretes.
6. Tell about peculiarities of aggregates based on slag.
7. What are the types of concrete based on slag? Tell about their peculiarities and properties.
8. Tell about materials from slag fusions.
9. Tell about wastes of coal cleaning and their fields of application.
10. Tell about application of ash and slag mixes.
11. What are the peculiarities and fields of application of phosphogypsum?
12. Tell about ways of application of waste products of chemical industry.
13. What are the ways of utilization of wastes of municipal economy?

REFERENCES

[1] Materials in Construction: Principles, Practice and Performance by G.D. Taylor. 608 p.

[2] Construction Materials by J.M. Illston. Taylor & Francis; 3 edition (30 Aug 2001) 584 p.

[3] Concrete science : treatise on current research / V.S. Ramachandran, R.F. Feldman, J.J. Beaudoin. London : Heyden, 1981. 427 p.

[4] Friedrich W. Locher: Cement : Principles of production and use, Düsseldorf, Germany: Verlag Bau + Technik GmbH, 2006, 535 c.

[5] Neville, A.M. (1996). Properties of concrete. Fourth and final edition standards. Pearson, Prentice Hall. 456 p.

[6] Fundamentals of Concrete Science. Dvorkin L.I., Dvorkin O.L. Stroy-beton, St-Petersburg, 690p., 2006. (in Russian).

[7] Building Materials based on Industrial Wastes. Dvorkin L.I., Dvorkin O.L. Fenix, Rostov-on-Don, 363 p, 2007. (in Russian).

[8] PRACTICAL CONCRETE SCIENCE in questions and answers. Dvorkin L.I., Dvorkin O.L. and others. Stroy-beton, St-Petersburg, 322 p., 2008, (in Russian).

[9] Ceramic Materials: Science and Engineering. C. B. Carter; M. G.. Norton. Springer; 1 edition (April 20, 2007). 716 p.

[10] Materials Science of Polymers for Engineers, Tim A. Osswaldhttp://www.amazon.com/Materials-Science-Polymers-Engineers-Osswald/dp/1569903484/ref=sr_1_1?ie=UTF8&s=books&qid=1276876834&sr=1-1 - #, Georg Menges, Hanser Gardner Publications; 2 edition (June 2003), 622 p.

[11] Introduction to Glass Science and Technology (Rcs Paperbacks Series), J E Shelbyhttp://www.amazon.com/Introduction-Glass-Science-Technology-Paperbacks/dp/0854046399/ref=sr_1_6?ie=UTF8&s=books&qid=1276877187&sr=1-6 - #, Royal Society of Chemistry; 2nd edition (March 29, 2005), 291 p.

[12] The Asphalt Handbook (Manual), Asphalt Institute, 2007, 788 p.

[13] Black & Decker The Complete Guide to Masonry & Stonework, Creative Publishing international, 2010, 320 p.

[14] Thermal Insulation Building Guide, Edin F. Strother, William C. Turner, Krieger Publishing Company, 1990, 500p.

INDEX

B

C

F

G

H

I

J

K

L

O

P

Q

R

T

U